Geometrical Methods for the Theory of Linear Systems

NATO ADVANCED STUDY INSTITUTES SERIES

Proceedings of the Advanced Study Institute Programme, which aims at the dissemination of advanced knowledge and the formation of contacts among scientists from different countries

The series is published by an international board of publishers in conjunction with NATO Scientific Affairs Division

A	Life Sciences	Plenum Publishing Corporation
B	Physics	London and New York
C	Mathematical and Physical Sciences	D. Reidel Publishing Company Dordrecht, Boston and London
D	Behavioural and Social Sciences	Sijthoff & Noordhoff International Publishers
E	Applied Sciences	Alphen aan den Rijn and Germantown U.S.A.

Series C – Mathematical and Physical Sciences

Volume 62 – Geometrical Methods for the Theory of Linear Systems

Geometrical Methods for the Theory of Linear Systems

Proceedings of a NATO Advanced Study Institute and AMS Summer Seminar in Applied Mathematics held at Harvard University, Cambridge, Mass., June 18-29, 1979

edited by

CHRISTOPHER I. BYRNES
Harvard University, Cambridge, Mass., U.S.A.

and

CLYDE F. MARTIN
Case Western Reserve University, Dept. of Mathematics, Cleveland, Ohio, U.S.A.

D. Reidel Publishing Company

Dordrecht : Holland / Boston : U.S.A. / London : England

Published in cooperation with NATO Scientific Affairs Division

Library of Congress Cataloging in Publication Data
Nato Advanced Study Institute, Harvard University, 1979.
Geometrical methods for the theory of linear systems.

(NATO advanced study institutes series : Series C, Mathematical and physical sciences ; v. 62)
1. System analysis–Congresses. 2. Geometry–Congresses.
I. Byrnes, Christopher I., 1949- II. Martin, Clyde. III. AMS Summer Seminar in Applied Mathematics, Harvard University, 1979.
IV. North Atlantic Treaty Organization. V. American Mathematical Society.
VI. Title. VII. Series.
QA402.N33 1979 003'.01'512 80-20465
ISBN-13: 978-94-009-9084-5 e-ISBN-13: 978-94-009-9082-1
DOI: 10.1007/978-94-009-9082-1

Published by D. Reidel Publishing Company
P.O. Box 17, 3300 AA Dordrecht, Holland

Sold and distributed in the U.S.A. and Canada
by Kluwer Boston Inc.,
190 Old Derby Street, Hingham, MA 02043, U.S.A.

In all countries, sold and distributed
by Kluwer Academic Publishers Group,
P.O. Box 322, 3300 AH Dordrecht, Holland

D. Reidel Publishing Company is a member of the Kluwer Group

TABLE OF CONTENTS

DEDICATION

This volume is dedicated to Roger Brockett, Robert Hermann, and Rudolph Kalman in acknowledgement of their pioneering contributions in the use of geometric and topological methods in control theory.

PREFACE

The lectures contained in this book were presented at Harvard University in June 1979. The workshop at which they were presented was the third such on algebro-geometric methods. The first was held in 1973 in London and the emphasis was largely on geometric methods. The second was held at Ames Research Center-NASA in 1976. There again the emphasis was on geometric methods, but algebraic geometry was becoming a dominant theme. In the two years after the Ames meeting there was tremendous growth in the applications of algebraic geometry to systems theory and it was becoming clear that much of the algebraic systems theory was very closely related to the geometric systems theory. On this basis we felt that this was the right time to devote a workshop to the applications of algebra and algebraic geometry to linear systems theory. The lectures contained in this volume represent all but one of the tutorial lectures presented at the workshop. The lecture of Professor Murray Wonham is not contained in this volume and we refer the interested to the archival literature.

This workshop was jointly sponsored by a grant from Ames Research Center-NASA and a grant from the Advanced Study Institute Program of NATO. We greatly appreciate the financial support rendered by these two organizations. The American Mathematical Society hosted this meeting as part of their Summer Seminars in Applied Mathematics and will publish the companion volume of contributed papers.

Many people were involved in the preparations for the meeting and in the preparation of this volume. We would like to specifically thank the conference secretaries, Joyce Martin and Maureen Ryan, and a special thanks to Joyce who typed this volume, often from almost illegible copy. As the editors of this volume, we would like in particular to thank Roger Brockett who served with us as codirector and provided the leadership that is necessary for such a conference to succeed.

Christopher Byrnes
Harvard University

Clyde Martin
Case Western Reserve University

C. I. Byrnes and C. F. Martin, Geometrical Methods for the Theory of Linear Systems, ix.

INTRODUCTION TO GEOMETRICAL METHODS FOR THE THEORY OF LINEAR SYSTEMS

C. Byrnes, M. Hazewinkel, C. Martin, and Y. Rouchaleau

In this joint totally tutorial chapter we try to discuss those definitions and results from the areas of mathematics which have already proved to be important for a number of problems in linear system theory.

Depending on his knowledge, mathematical expertise and interests, the reader can skip all or certain parts of this chapter 0. Apart from the joint section, the basic function of this chapter is to provide the reader of this volume with enough readily available background material so that he can understand those parts of the following chapters which build on this--for a mathematical system theorist perhaps not totally standard--basic material. The joint section is different in nature; it attempts to explain some of the ideas and problems which were (and are) prominent in classical algebraic geometry and to make clear that many of the problems now confronting us in linear system theory are similar in nature if not in detail. Thus we hope to transmit some intuition why one can indeed *expect* that the tools and philosophy of algebraic geometry will be fruitful in dealing with the formidable array of problems of contemporary mathematical system theory. This section can, of course, be skipped without endangering one's chances of understanding the remainder of this chapter and the following chapters.

The contents of this introductory chapter are:

C. I. Byrnes and C. F. Martin, Geometrical Methods for the Theory of Linear Systems, 1-84.

1. SOME PROBLEMS OF CLASSICAL ALGEBRAIC GEOMETRY

The purpose of this section is to give insight into certain of the problems and achievements of 19th century algebraic geometry, in a historical perspective. It is our hope that this perspective, which for several reasons is limited, will go some of the distance towards explaining some natural interrelations between algebraic geometry and analysis, as well as a natural connection between algebraic geometry and linear system theory.

1.1 Plane algebraic curves

To begin, perhaps the most primitive objects of algebraic geometry are varieties, e.g., plane curves in $\mathbb{C}^2$ (say the variety defined by the equation $y = x^2$), and the most primitive relations are those of incidence, e.g., the intersection of varieties. To fix the ideas, let us consider the problem of describing all plane curves in $\mathbb{C}^2$ and the problem of describing their intersections. Since any two distinct irreducible (i.e., the polynomial $f(x,y)$, whose locus is the curve, is irreducible) curves intersect in finitely many points, the first problem of describing such an intersection is to compute the number of such points in terms of the two curves.

Now, whenever one speaks of a scheme for the description or classification of objects, such as plane curves, one has in mind a certain notion of equivalence. And, quite often, this involves the notions of transformation. For example, if $SL(2,\mathbb{C})$ is the group of 2×2 matrices with determinant 1, then $g \in SL(2,\mathbb{C})$ acts on $\mathbb{C}^2$ by linear change of variables and it has been known since the introduction of Cartesian coordinates that a linear change of coordinates leaves the degree of a curve invariant. That is, if $f(x,y)$ is homogeneous, then

$$f^g(x,y) = f\left(g^{-1}\begin{pmatrix} x \\ y \end{pmatrix}\right) \tag{1.1.1}$$

has the same degree as f. So, for homogeneous f, we may begin the classification scheme by fixing the degree. Now any f which is homogeneous of degree 1 is a linear functional, and these are well understood. If f is homogeneous of degree 2, then one can check that the discriminant

$$\Delta(f) = b^2 - 4ac,$$

where

$$f(x,y) = ax^2 + bxy + cy^2 ,$$

is invariant under $SL(2,\mathbb{C})$; i.e.,

$$\Delta(f) = \Delta(f^g), \qquad \text{for all } g \in SL(2,\mathbb{C}) . \tag{1.1.2}$$

This explains, in part, why the discriminant is so important in analytic geometry, but there really is a lot more to the story. First of all, (1.1.2) asserts that the discriminant of f, $\Delta(f)$, is the same regardless of the choice of coordinates used to express f (provided we allow only volume preserving, orientation preserving changes of coordinates). But this is also true for Δ^2, $\Delta^2 + 3$, etc. In 1801, Gauss [2,4] proved an important result: *any polynomial in* a, b, c *which is invariant under* $SL(2,\mathbb{C})$ *is a polynomial in* Δ. That is, let V denote the 3-dimensional space of quadratic forms in 2 variables, let R denote the ring of polynomials on V (i.e., polynomials in a,b,c) and let $R^{SL(2,\mathbb{C})}$ denote the subring of invariant polynomials, i.e., the polynomials satisfying (1.1.2).

1.1.3 Theorem (Gauss). $R^{SL(2,\mathbb{C})} = \mathbb{C}[\Delta]$ *and, if* $\Delta(f_1) = \Delta(f_2) \neq 0$ *then* $f_1^g = f_2$ *for some* $g \in SL(2,\mathbb{C})$.

Thus, Gauss classifies homogeneous $f(x,y)$ of degree 2 by the table:

Quadratic Form	Complete Invariant
f s.t. $\Delta(f) \neq 0$	$\Delta(f)$
f s.t. $\Delta(f) = 0$	rank of $f = \begin{pmatrix} a & b/2 \\ b/2 & c \end{pmatrix}$

(1.1.4)

Clearly, the same kind of question is equally important for homogeneous forms of degree r, in $n \geq 2$ variables. In 1845, Cayley posed the general problem, in the same notation as above [2]:

1.1.5 Cayley's Problem: *Describe the algebra* $R^{SL(n,\mathbb{C})}$ *as explicitly as possible; e.g., is* $R^{SL(n,\mathbb{C})}$ *finitely generated by some invariants* $\Delta_1,\ldots,\Delta_\ell$?

Now, the case $n = 3$ is particularly relevant for our discussion of plane curves. For, one may always "homogenize" a polynomial, and this process allows one to express the number of points of intersection of 2 plane curves in a beautiful formula, due to Bézout. Returning to our example,

$$X = \{(x,y) : y = x^2\} ,$$

to homogenize $f(x,y) = y-x^2$ is to substitute x/z, y/z for x,y and then to clear denominators with the result being the homogeneous polynomial $\tilde{f}(x,y,z) = yz - x^2$ satisfying

$$\tilde{f}(x,y,1) = f(x,y) . \tag{1.1.6}$$

Geometrically, since $\tilde{f}(x,y,z)$ is homogeneous the locus of $\tilde{f}$ contains the line connecting any nonzero solution with the origin. Indeed, the intersection of $\tilde{f}(x,y,z) = 0$ with the plan $z = 1$

is given by the zeroes of f, as in (1.1.6), and the locus of $\tilde{f}$ contains all lines through this curve. However, there is more, the line $(0,y,0)$ also lies in the locus of $\tilde{f}$.

Next, if one considers the projective plane

$$\mathbb{P}^2 = \{\text{lines thru } 0 \text{ in } \mathbb{C}^3\}$$

then, by homogeneity, the locus of $\tilde{f}$ is a collection of points in $\mathbb{P}^2$--one for each of the points in $f(x,y) = 0$ and one more, the line $(0,y,0)$, which may be regarded as the point at ∞. To make this more precise, we give "homogeneous coordinates" to a point $P \in \mathbb{P}^2$; i.e., regarding P as a line in $\mathbb{C}^3$, choose some non-zero $(x,y,z) \in P$ noting that any other choice (x',y',z') is a non-zero multiple of (x,y,z). The equivalence class $[x,y,x]$ is called "homogeneous coordinates" for P and to check membership of $P_0 = [x_0,y_0,x_0]$ in the locus of a homogeneous $f(x,y,z)$ it is enough to evaluate $f(x_0,y_0,z_0)$.

As an example of these ideas in control theory, consider the transfer function

$$T(s) = \begin{pmatrix} \frac{1}{s} \\ \frac{1}{s^2} \end{pmatrix} \tag{1.1.7}$$

and the coprime factorization

$$\begin{pmatrix} N(s) \\ --- \\ D(s) \end{pmatrix} = \begin{pmatrix} s \\ 1 \\ --- \\ s^2 \end{pmatrix} . \tag{1.1.8}$$

Now, for an arbitrary $s \in \mathbb{C}$, (1.1.8) is a point in $\mathbb{C}^3 - \{0\}$ although $T(s)$ does not determine this point canonically. Rather, $T(s)$ determines the line through

$$\begin{pmatrix} N(s) \\ D(s) \end{pmatrix}$$

as depicted below:

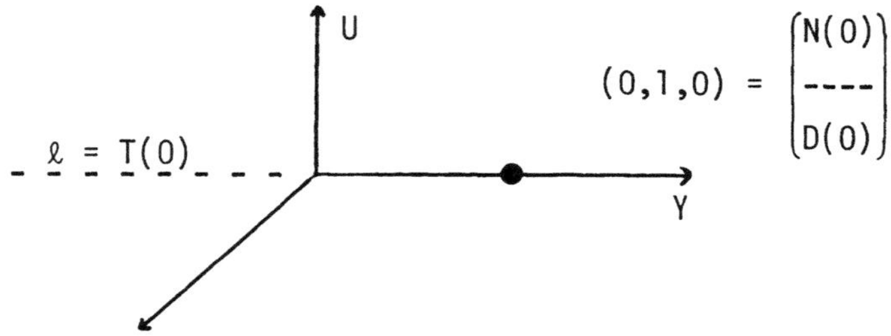

(1.1.9)

Since $T(\infty) = U$, T extends to a map of the extended complex plane

$$T : \mathbb{C} \cup \{\infty\} = \mathbb{P}^1 \to \mathbb{P}^2 \quad . \tag{1.1.10}$$

And one easily checks, using the homogeneous coordinates in (1.1.8), that $T(\mathbb{C} \cup \{\infty\})$ is the curve defined in our example, viz., the locus of $\tilde{f}(x,y,z) = yz - x^2$. Moreover, if $\mathbb{P}^1 \subset \mathbb{P}^2$ is the space of lines in Y, (1.1.9), then $\mathbb{P}^1 \cap T(\mathbb{C} \cup \{\infty\})$ is easily computed, under T^{-1} it is the set $\mathrm{sing}(T)$ of poles of T:

$$T^{-1}(T(\mathbb{C} \cup \{\infty\}) \cap \mathbb{P}^1) = \mathrm{sing}(T) \tag{1.1.11}$$

and thus consists of one point of multiplicity 2, (see Professor Martin's lectures for the geometry of a general transfer function).

1.1.12. Theorem (Bézout [9]). *If* $X_1, X_2 \subset \mathbb{P}^2$ *are irreducible curves of degree* d_1, d_2, *then, counting multiplicities,*

$$\#(X_1 \cap X_2) = d_1 \cdot d_2 \quad .$$

We shall prove this in the case where X_2 is a line $\mathbb{P}^1$. By a change of coordinates in $\mathbb{C}^3$, X_2 corresponds to the set of lines in the plane $z = 0$. And, by a change of notation, if X_1 is the locus of $f(x,y,z)$, homogeneous of degree d, then Euler's relation is

$$f(x_1,x_2,x_3) = d \cdot \sum_{i=1}^{3} \frac{\partial f}{\partial x_i} \cdot x_i \quad . \tag{1.1.13}$$

Intersecting $f(x_1,x_2,x_3) = 0$ with $x_3 = 0$ gives the equation, of degree d,

$$d \cdot \sum_{i=1}^{2} \frac{\partial f}{\partial x_i} \cdot x_i = 0$$

defining, counting multiple roots, d lines in the (x_1,x_2) plane.

1.1.14. Remark. $\mathbb{P}^2$ provides a model for a (non-Euclidean) geometry, where the points are just the points of $\mathbb{P}^2$ and the lines are just the loci of linear functionals on $\mathbb{C}^3$, i.e., planes in $\mathbb{C}^3$. Thus, for example, 2 distinct points P_1,P_2 regarded as lines in $\mathbb{C}^3$ determine a line $\mathbb{P}^1$ in $\mathbb{P}^2$, viz. the plane in $\mathbb{C}^3$ spanned by P_1 and P_2. Moreover, any 2 lines in $\mathbb{P}^2$ intersect in a point.

1.2 Riemann Surfaces and Fields of Meromorphic Functions

Thus, by homogenizing curves in $\mathbb{C}^2$, we take a lot of the mystery out of the points at ∞. Indeed, one can give a beautiful expression for the intersection number of 2 curves. Plane projective curves also arise in potential theory and in the calculus.

In two papers published in 1869 [1], H. A. Schwartz considered the problem of finding, for purposes of solving Dirichlet problems, conformal maps of bounded regions to the unit disk or, equivalently, to the upper half plane. For example, Schwartz considers the problem of finding a conformal map of the unit square onto the upper half plane, $\mathcal{H}$, where f maps 3 corners

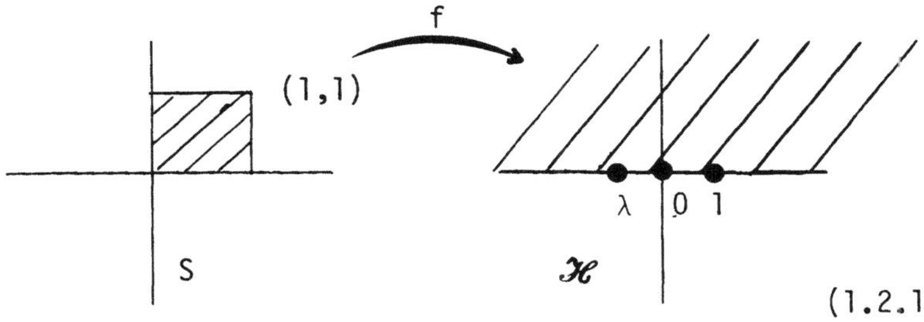

(1.2.1)

to points $0, 1, \lambda$ and the 4th corner to ∞. In particular f is meromorphic, as it should be, for f can be extended to a doubly periodic function on $\mathbb{C}$, by the Schwartz Reflection Principle. Actually, it is easier to construct a holomorphic map $g: \mathcal{H} \to S$. The Schwartz-Christoffel formula applies in this case to give the elliptic integral

$$g(P) = \int_{P_0}^{P} \frac{dz}{\sqrt{4z(1-z^2)}} \quad . \tag{1.2.2}$$

Such integrals had already been the subject of deep research by Fagano, Euler, Gauss, Abel, Jacobi and others, being first encountered in the computation of the arclength of an ellipse. In particular, Euler had shown that elliptic integrals, such as (1.2.2), satisfy addition formulae

$$g(P) + g(Q) = g(R) \tag{1.2.3}$$

where R is a rational function of P and Q, generalizing the familiar trigonometric addition formulae gotten from considering the lengths of arcs on a circle. Indeed, there are group theoretic ideas underlying (1.2.3) too.

That is, consider the meromorphic function f (cf. (1.2.1)) which inverts g. As we have noted f extends to a doubly periodic meromorphic function on $\mathbb{C}$ and hence to a meromorphic function on the torus, or more properly the elliptic curve,

$$\mathscr{E} = \mathbb{C}/\{n + im\}, \quad n, m \in \mathbb{Z},$$

gotten by identifying the (oriented) horizontal edges of the unit square and by identifying the (oriented) vertical edges. One therefore has a nontrivial meromorphic function,

$$f : \mathscr{E} \to \mathbb{C} \tag{1.2.4}$$

and a holomorphic 1-form on $\mathscr{E}$,

$$\frac{dz}{\sqrt{4z(1-z^2)}} \tag{1.2.5}$$

which turns out to be invariant under multiplication on the group $\mathscr{E}$. This can also be seen from the method of substitution applied to the integral (1.2.2). That is, substitute $y^2 = 4z(1-z^2)$ and consider integrating dz/y over paths defined on the algebraic curve, $y^2 = 4z(1-z^2)$. Homogenizing this curve we obtain

$$y^2x = 4z(x^2 - z^2), \tag{1.2.6}$$

and hence a cubic curve $X \subset \mathbf{P}^2$. One can see a beautiful geometric definition of the group law on X: choose 2 points P_1, P_2 on X and consider the line $\ell(\simeq\mathbf{P}^1)$ in $\mathbf{P}^2$ which they determine. By Bézout's Theorem, ℓ intersects X in a third point, $P_3 = (P_1 \cdot P_2)^{-1}$! Moreover, dz/y is an invariant holomorphic 1-form on X and from this one may obtain (1.2.3). However, more is true; X admits a non-constant meromorphic function derived from dz/y, viz. f. In fact, the field of meromorphic functions on X is easily seen to be $\mathbb{C}(y,z)$ where y and z are related as above. Again using the form dz/y and a formula relating the degree of X to the topology of X one may show that $X \simeq \mathcal{E}$ as complex manifolds!

Remark. As a sketch of the proof, one sees from the fact that X is a nonsingular cubic in $\mathbb{P}^2$ that X is not simply-connected and thus integrals $\int_\gamma dz/y$ where γ is a closed path on X are not necessarily 0, although the proper form of Cauchy's Theorem is still valid; viz., if $\gamma_1 \sim \gamma_2$ (are homologous) then the path integrals taken over γ_1, γ_2 are equal. And, although X is not simply-connected, one knows that there is a basis $\{\gamma_1, \gamma_2\}$ for the closed curves on X modulo homology. Thus there are two basic "periods" of dz/y,

$$\int_{\gamma_1} \frac{dz}{y} \quad \text{and} \quad \int_{\gamma_2} \frac{dz}{y} \quad . \tag{1.2.7}$$

Now, if $P_0 \in X$ is the identity (or any point) then one might consider the quantities

$$\int_{P_0}^{P} \frac{dz}{y} ,$$

for all $P \in X$. This quantity is not a well-defined complex number, as the integral depends on the choice of path. If γ and $\tilde{\gamma}$ are 2 paths joining P_0 and P, then

$$\int_\gamma \frac{dz}{y} - \int_{\tilde{\gamma}} \frac{dz}{y}$$

is an integral around a closed path, based at P_0, on X and is therefore (by Cauchy's Theorem) an integer combination of the periods (1.2.7). Thus, if Λ is the lattice in $\mathbb{C}$ generated by the periods (1.2.7) one has an isomorphism

$$X \to \mathbb{C}/\Lambda = \mathcal{E} , \qquad \text{defined via}$$

$$P \mapsto \left(\int_{P_0}^{P} \frac{dz}{y} \right) \quad . \tag{1.2.8}$$

To conclude the remark, if one instead considered integrals with a rational integrand (or, more generally, of the form $dz/\sqrt{z^2+az+b}$) the curve X in $\mathbb{P}^2$ turns out to be $\mathbb{P}^1$, as a conic in $\mathbb{P}^2$, which is simply connected, while the form dz/y is meromorphic and the usual residue calculus applies. This explains the ease with which rational integrals may be calculated as well as the relative difficulty involved in calculating elliptic integrals.

Summarizing, one has the interconnection between elliptic curves, complex tori, and certain fields of meromorphic functions. This is a special case of what has been properly referred [7] to as "the amazing synthesis." That is, one may consider three formally distinct classes of objects: non-singular projective curves (of any degree), complex compact manifolds of dimension 1, and fields of meromorphic functions. Then, the amazing synthesis is that any one of these objects determines the other two. Schematically,

$$\begin{array}{ccc} \left\{\begin{array}{l}\text{non-singular}\\ \text{algebraic curves}\end{array}\right\} & \longleftrightarrow & \left\{\begin{array}{l}\text{compact, complex}\\ \text{manifolds of dim 1, i.e.,}\\ \text{Riemann surfaces}\end{array}\right\} \\ \nwarrow\!\searrow & & \swarrow\!\nearrow \\ & \left\{\begin{array}{l}\text{fields of transcendence}\\ \text{degree 1 over } \mathbb{C}\end{array}\right\} . & \end{array} \qquad (1.2.9)$$

The deeper part of this correspondence is that from an abstract Riemann surface S one may recover the embedding of S into projective space and the equations defining this curve, or equivalently, that one may construct the field of meromorphic functions on S. Above, the meromorphic function f on the curve $\mathscr{E}$ was constructed via potential theory, i.e., in order to solve the Dirichlet problem. As Riemann demonstrated with liberal use of the "Dirichlet principle," such transcendental techniques can be used to construct non-trivial meromorphic functions on an arbitrary Riemann surface. Briefly, the intuition runs as follows.

First of all (and we will consider analytic equivalence in 1.3), a compact Riemann surface is topologically a sphere with g handles, where if the surface is given as a curve of degree d in $\mathbb{P}^2$ the genus g is given by $(d-1)(d-2)/2$. Thus, the elliptic curve $\mathscr{E}$ has genus $(3-1)(3-2)/2 = 1$ and is a sphere with 1 handle, i.e., a torus.

Next, a meromorphic function f on S has as many poles as zeroes. Where $f = u + iv$ is analytic, u and v satisfy Laplace's equation in light of the Cauchy-Riemann equations. Therefore, f gives rise to a time-invariant flow with inessential singularities on S where u = const. defines the equipotential curves and v = const. defines the lines of force. Conversely, Riemann's idea was to construct such an f by regarding, intuitively, S as a surface made of a conductive material and by placing the poles of a battery at each pole-zero pair of f. This can be made somewhat more precise by a much more careful description of S and an appeal to the Dirichlet principle. Indeed, the application of modern harmonic theory to the (Riemann-Roch) question of existence of meromorphic functions on S is one of the most beautiful sides of the "amazing synthesis." For a more detailed account of the intuitive discussion hinted at above, be sure to browse in F. Klein's book [ ].

In closing this section, we would like to make contact with what is perhaps a more familiar description of a Riemann surface, viz. as a branched cover of the extended complex plane $\mathbb{P}^1$. For example, at least the finite part of the Riemann surface of the relation $y^2 = 4z(1-z^2)$ can be obtained by forming the branch cuts between -1 and 0 and 1 and $+\infty$ and sewing two copies of the plan less these cuts together in the appropriate fashion. One can get at the whole Riemann surface more easily by considering the graph of the relation $y^2 = 4z(1-z^2)$. Explicitly, introduce homogeneous coordinators $([y,\tilde{y}],[z,\tilde{z}])$, and homogenize the relation, obtaining the curve $\mathscr{E}$

$$y^2\,\tilde{z}^3 - 4\tilde{y}^2 z(\tilde{z}^2 - z^2) \tag{1.2.10}$$

in $\mathbb{P}^1 \times \mathbb{P}^1$. However, we get more than just the curve $\mathscr{E} \subset \mathbb{P}^1 \times \mathbb{P}^1$, we also obtain 2 rational functions on $\mathscr{E}$

$$\mathrm{proj}_1\colon \mathscr{E} \to \mathbb{P}^1 , \qquad \mathrm{proj}_2\colon \mathscr{E} \to \mathbb{P}^1$$

which are, of course, the algebraic functions y and z, on the Riemann surface $\mathscr{E}$ of $y^2 = 4z(1-z^2)$. Notice that $y\colon \mathscr{E} \to \mathbb{P}^1$ exhibits $\mathscr{E}$ as a 2-fold cover of $\mathbb{P}^1$, branched at the 4 points $z = 0, \pm 1$, and ∞.

This is, at the very least, reminiscent of root loci. That is, for a scalar transfer function $T(s)$ one may regard, as in (1.1.7) etc., $T(s)$ as a branched cover of the Riemann spheres

$$T : \mathbb{P}^1 \to \mathbb{P}^1 ,$$

as depicted below

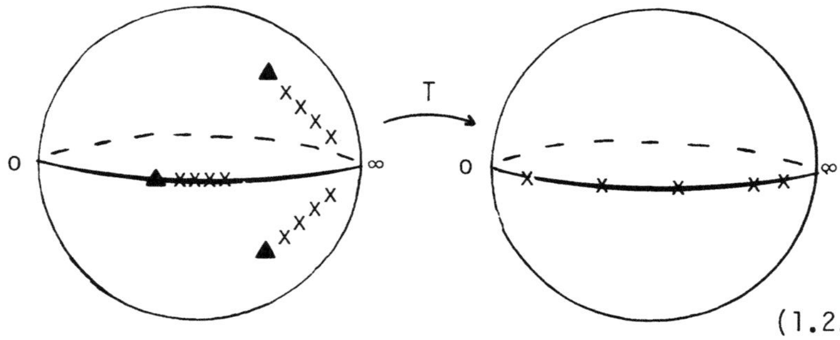

(1.2.11)

Here, the ▲'s denote $T^{-1}(\infty)$ and the x's denote the motion of the closed-loop root loci, as a function of $-1/k$ toward the open loop zeroes--one finite real zero and a branched zero at ∞. Indeed, for a scalar gain $K = kI$ and square multivariable tranfer function $T(s)$, an extension of these ideas has been given by A. MacFarlane and I. Postlethwaite.

Example [9]. Consider the transfer function

$$T(s) = \frac{1}{(1.25)(s+1)(s+2)} \begin{pmatrix} s-1 & s \\ -6 & s-2 \end{pmatrix}$$

and the scalar output gain

$$K = \begin{pmatrix} k & 0 \\ 0 & k \end{pmatrix}$$

In order to study the locus of roots of the closed-loop characteristic polynomial (see Professor Byrnes's lectures), it is enough to study the locus of roots of $\det(I + kT(s))$ or, setting $k = -1/g$, the Riemann surface X defined by

$$0 = \det(gI - G(s)) = g^2 - \mathrm{tr}G(s)g + \det G(s) .$$

Clearing denominators, one obtains

$$0 = f(s,g) = (1.25)(s+1)(s+2)g^2 - (2s-3)g + \frac{4}{5} = 0$$

leading to the algebraic functions

$$g\pm(s) = \frac{(2s-3) \pm \sqrt{1-24s}}{(2.5)(s+1)(s+2)} .$$

In this way, one has

$$\begin{array}{ccc} \mathbb{P}^1 & \times & \mathbb{P}^1 \\ & \cup & \\ & X & \\ {}^{s}\swarrow & & \searrow{}^{g} \\ \mathbb{P}^1 & & \mathbb{P}^1 \end{array}$$

where $s : X \to \mathbb{P}^1$ is a 2-sheeted cover of $\mathbb{P}^1$, branched at $x = 1/24$ and at $s = \infty$, as depicted below.

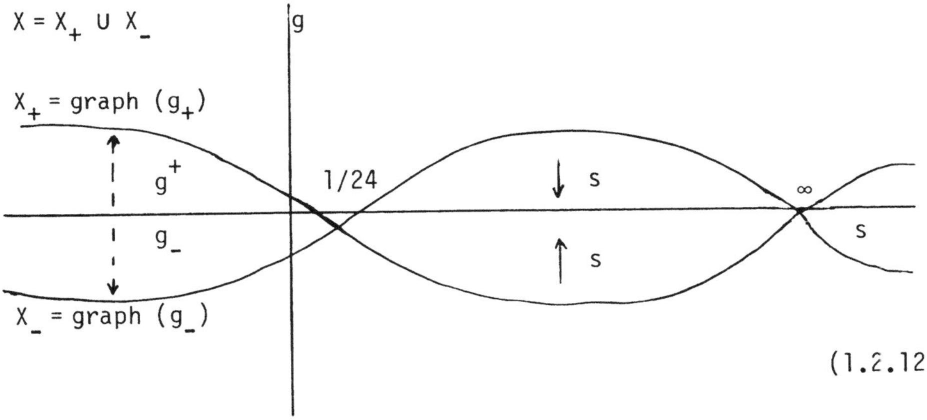

(1.2.12)

Now the study of root loci, is the study of the loci of s on X, for each fixed real positive gain k -- i.e., for each fixed real negative value of $g = -1/k$. Thus, the root locus is simply the arc on X given by g^{-1} (negative real axis) and to see this concretely it's perhaps easiest to study the pair of arcs γ_1, γ_2 given by $s(g^{-1}$ (negative real axis)). On the 2 copies X_+, X_- of the s-plane, branched at $s = 1/24$, $s = \infty$, one sees that these loci start at the open loop poles, $s = -1$, $s = -2$ and move to ∞, the only open loop zero, as follows (note $g_\pm$ is real iff s is real and $s < 1/24$.

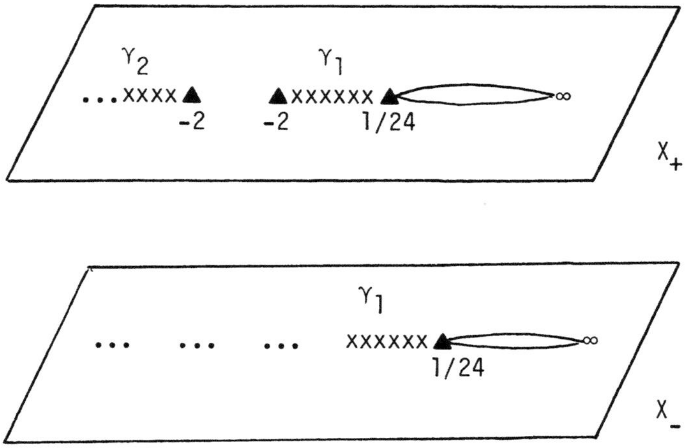

(1.2.13)′

Thus γ_2 moves, as $0 < g < \infty$, from the pole -2 on X_+ to the branch point ∞, while γ_1 moves from the pole -1 on X_+ to the branch point 1/24, where γ_1 changes sheets, moving to ∞ on X_-. We can describe $X = X_+ \cup X_-$ topologically as a sphere, for

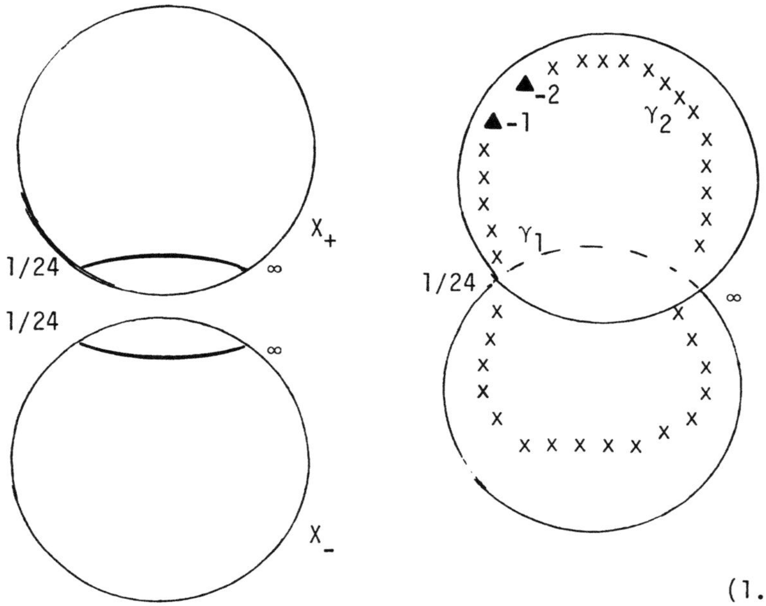

(1.2.14)

One now easily finds the regions of stability:

(a) for $0 \leq k < 1.25$, the closed loop system is asymptotically stable.

(b) for $1.25 \leq k \leq 2.5$, the system is unstable with one pole (on γ_2) in the left-belt plane.

(c) for $2.5 < k < \infty$, the system is stable.

Remark. Branched covers of $\mathbb{P}^n$ by complex manifolds of dimension n play a role in the study of root locus, when one allows arbitrary gains K; see Professor Byrnes's lectures.

Now, there is an alternate route to representing a plane curve as a branched cover of $\mathbb{P}^1$, recall that one may homogenize and projectivize, obtaining the algebraic curve X in $\mathbb{P}^2$ defined by $y^2x - 4z(x^2 - z^2) = 0$, as in (1.2.9). Then choosing any line $\mathbb{P}^1$ and a point P not on $\mathbb{P}^1$ or X, the branched cover of $\mathbb{P}^1$ is gotten by a "central projection" based at P. That is, by Bézout's Theorem any line ℓ through P intersects X in 3 points (counting multiplicity) and $\mathbb{P}^1$ in a single point and therefore defines a function,

$$f : X \to \mathbb{P}^1 \tag{1.2.15}$$

which sends these three points to the corresponding point on $\mathbb{P}^1$. One may calculate that there are 6 branch points on $\mathbb{P}^1$ for which multiplicities occur in $\ell \cap X$, where ℓ joins the branch point to P. [This is as it should be, for $\mathbb{P}^1 \simeq S^2$ is simply connected and therefore does not admit a non-trivial connected covering space.] Note that f has the form $f(x) = [q(x),p(x)]$ in homogeneous coordinates and thus corresponds to the coprime factorization of meromorphic function $f = q/p$ on X.

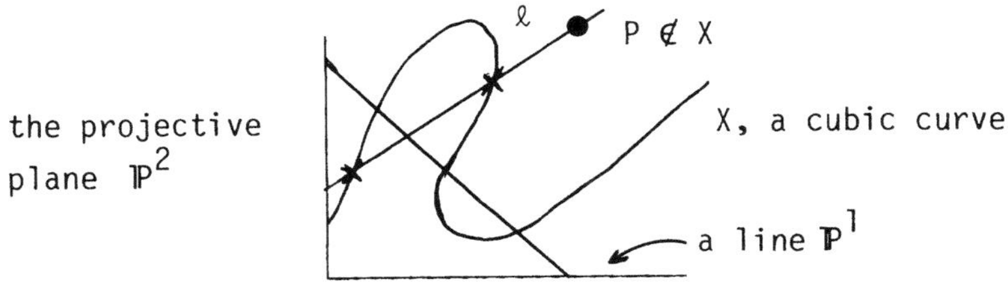

(1.2.16)

1.3 Invariants

In the final part of this section, we want to return to Cayley's Problem, especially the question of finite generation of the ring of invariants $R^{SL(n,\mathbb{C})}$. Set the notation: $V_{n,r}$ = {r-th degree forms in n-variables}, R is the ring of polynomials on $V_{n,r}$, and $SL(n,\mathbb{C})$ acts on $\mathbb{C}^n$ and therefore on $V_{n,r}$ by composition. $f \in R$ is said to be invariant under $SL(n,\mathbb{C})$ if, and only if, (1.1.2) holds and $S(n,r)$ denotes $R^{SL(n,\mathbb{C})}$. Now, for $n = 2$ the explicit structure of $S(2,r)$ is known for $r = 2,\ldots,8$, the case $r = 2$ being Gauss's Theorem, while the case $r = 8$ was only recently (1964) obtained by Shioda. Gordan, and later Clebsch and Gordan, was able to prove that the ring of $SL(2,\mathbb{C})$ invariants is finitely generated for all r.

Remark. Part of this problem is rather straightforward; i.e., if $R = \Sigma_{m\geq 0} R_m$ is grading of R into homogeneous polynomials of degree m, then since $SL(2,\mathbb{C})$ acts on $V_{2,r}$ by linear transformations $SL(2,\mathbb{C})$ acts on each R_m. In fact, this action is the symmetric tensor representation of $SL(2,\mathbb{C})$ on the space $\mathscr{S}^m(\mathscr{S}^2(\mathbb{C}^2))$ of symmetric tensors. The invariants in R_m correspond to the subspace of $\mathscr{S}^m(\mathscr{S}^2(\mathbb{C}^2))$ on which $SL(2,\mathbb{C})$ acts as the identity and this representation can be decomposed as in the Clebsch-Gordan formula. Moreover, the action of $SL(2,\mathbb{C})$ on $V_{2,r}$ is just the standard irreducible representation of dimension $r+1$. This explains, for example, the absence of any invariants of degree 1 in the ring $\mathbb{C}[\Delta]$. It is now, however, a proof that Δ generates $S(2,2)$. It should be remarked that for $n > 2$, the action of $SL(n,\mathbb{C})$ on R_m is the object of study in the "first main theorem of invariant theory" [2].

Now, in 1892 David Hilbert proved that $S(n,r)$ is finitely generated and, even better, gave a proof that revolutionized commutative algebra. Before sketching a proof, we would like to point out the connection with the construction of moduli (or parameter) spaces--in this case, the moduli space of homogeneous forms. That is, one is interested (as in the case of constructing the space of systems) in regarding $V_{n,r}/SL(n,\mathbb{C})$, the set of equivalence classes of forms modulo a special linear change of coordinates, as a variety or as a manifold in a natural way, viz., so that the map

$$\pi: V_{n,r} \to V_{n,r}/SL(n,\mathbb{C}) \qquad (1.3.1)$$

is algebraic or smooth. First of all, the orbits must be closed in $V_{n,r}$ as they are the inverse images of the closed points of $V_{n,r}/SL(n,\mathbb{C})$. Second, if $V_{n,r}/SL(n,\mathbb{C})$ is an affine variety, there must be enough functions

$$f:V_{n,r}/SL(n,\mathbb{C}) \rightarrow \mathbb{C} \tag{1.3.2}$$

to separate points and, moreover, this algebra of such f's must be finitely-generated. Such f's are, however, invariant polynomials on $V_{n,r}$ since π in (1.3.1) is assumed to be algebraic. Thus, two necessary conditions for an affine quotient to exist are:

(a) all orbits are closed

(b) $R^{SL(n,\mathbb{C})} = S(n,r)$ is finitely-generated.

Notice that if one had, instead, a compact group G acting on a vector space V, then (a) would be trivial, whereas by "averaging over G" one can always construct enough G-invariant functions to separate orbits. In fact, the existence of a process for averaging over $SL(n,\mathbb{C})$ underlies the validity of Hilbert's Theorem. This fact was brought out quite clearly by Nagata, who gave satisfactory answers to Hilbert's 14th Problem, which is a natural generalization of Cayley's Problem.

1.3.3. Theorem (Hilbert). *S(n,r) is finitely-generated for all n and all r.*

Sketch of Proof (from [8]). One first of all has the Hilbert basis theorem: each ideal of R, the ring of polynomials on $V_{n,r}$, is finitely generated--for a proof of this fact, one may refer to Chapter 2, Theorem 2.9. Next, one introduces the Reynolds operators (i.e., averaging over $SL(n,\mathbb{C})$): if V is an $SL(n,\mathbb{C})$-module, then the submodule $V^{SL(n,\mathbb{C})}$ of invariants has a unique $SL(n,\mathbb{C})$-invariant complement $V_{SL(n,\mathbb{C})}$. Alternatively, one has a projection

$$R:V \rightarrow V^{SL(n,\mathbb{C})} \tag{1.3.4}$$

commuting with the action of $SL(n,\mathbb{C})$. R is called the Reynolds operator, and could be represented symbolically in a seductive (but formal) way,

$$RV = \int_{SL(n,\mathbb{C})} gv \, dg.$$

By uniqueness, one may deduce that, for I an ideal of $R^{SL(n,\mathbb{C})} = S(n,r)$,

$$(R/IR)^{SL(n,\mathbb{C})} \simeq R^{SL(n,\mathbb{C})}/I \quad . \qquad (1.3.5)$$

That is, $I \to IR$ is an injection of the lattice of ideals of $R^{SL(n,\mathbb{C})}$ into the lattice of ideals of R. Hence, $R^{SL(n,\mathbb{C})}$ is Noetherian, by the Hilbert Basis Theorem. In particular, the ideal $\Sigma_{m>0} R_m^{SL(n,\mathbb{C})}$ of $R^{SL(n,\mathbb{C})}$ is finitely generated, say by $x_1,\ldots,x_\ell$. One next proves by induction that monomials in the x_i's generate each homogeneous piece $R_m^{SL(n,\mathbb{C})}$ and therefore $R^{SL(n,\mathbb{C})}$ is finitely generated over $\mathbb{C}$.

It should be emphasized that Hilbert's proof preceded and to a large extent motivated the introduction of chain conditions into ring theory and it should be remarked that the detailed structure of $S(2,6)$ was the subject of E. Noether's thesis.

Finally, one rather interesting and tractable case is $n = 2$, $r = 4$. Here, it is known [8] that $S(2,4) = \mathbb{C}[P,Q]$, where $\deg P = 2$, $\deg Q = 3$. In fact, if $f(x,y) = a_0x^4 + a_1x^3y + a_2x^2y^2 + a_3xy^3 + a_4y^4 \in V_{2,4}$, then Q is defined via

$$Q(f) = \det \begin{pmatrix} a_0 & a_{1/4} & a_{2/6} \\ a_{1/4} & a_{2/6} & a_{3/4} \\ a_{2/6} & a_{3/4} & a_4 \end{pmatrix} \quad . \qquad (1.3.6)$$

viz. as the determinant of a Hankel matrix! Moreover, the $SL(2,\mathbb{C})$ action on the space of 3×3 non-singular Hankels can be obtained in terms of control-theoretic scaling actions on the space of Hankel, as in Professor Brockett's lecture. Now, the structure of $S(2,4)$ (indeed of $S(2,2g+2)$) is also of interest in Riemann surface theory. That is, any elliptic curve $\mathscr{E}$ is a 2-sheeted branched cover, with 4 branch points of $\mathbb{P}^1$. In this way, the moduli space of elliptic curves can be represented as the moduli space of 4 unordered points on $\mathbb{P}^1$, up to equivalence under projective automorphisms, i.e., the group $GL(2,C)/\{\alpha I\}$ acting on lines in $\mathbb{C}^2$. Notice, however, that 4 lines in $\mathbb{C}^2$ determine, up to a multiplicative constant, a homogeneous quartic polynomial $f(x,y)$, i.e., a point $f \in V_{2,4}$, while projective equivalence corresponds to equivalence modulo $GL(2,\mathbb{C}) \supset SL(2,\mathbb{C})$.

In this way, an analysis of the $\mathbb{C}^* = \{\lambda \mathrm{Id}\}$ action on $V_{2,4}/SL(2,\mathbb{C})$ leads to the construction of the (moduli) space of all elliptic curves. By counting dimensions one sees that such a space must have dimension 1, for $\dim \mathbb{C}^* = 1$ and $\text{tr.deg. } \mathbb{C}[P,Q] = 1$. This existence of this one parameter family of elliptic curves (these turn out to be points in $\mathbb{C}$) illustrates the fact that there are too many conformally distinct yet topologically equivalent Riemann surfaces. In fact, Riemann asserted that there are 3g-3 parameters which describe all Riemann surfaces of genus $g > 1$. Another nice extension by Mumford of the work of Hilbert and Nagata enables one, for example, to construct such moduli spaces and therefore to speak about their dimension.

We remark that such problems arise frequently in control theory; for example, in the construction of moduli and canonical forms for linear dynamical systems. Here, one might ask, for fixed n, m, and p and for an arbitrary minimal triple (F,G,H) of these dimensions: do there exist canonical forms (F_C,G_C,H_C) for the action of $G\ell(n)$, via change of basis in the state space, such that the entries of (F_C,G_C,H_C) are algebraic in (F,G,H)? Since the entries of (F_C,G_C,H_C), as it were, are invariant functions (for this $G\ell(n)$ action) one might ask in particular for an explicit description of the ring of invariants. A description of the functions $f(F,G,H)$, invariant under the $G\ell(n)$ action on mixed tensors (F), vectors (G), and co-vectors (H), is well-known classically [5]; viz. the ring of such f's is generated by the entries of the matrices, HF^iG. However, it turns out that, because of the geometry of the moduli space $\{(F,G,H)\}/G\ell(n)$--or, equivalently the geometry of the corresponding space of Hankel matrices--neither algebraic, nor even continuous canonical forms exist (see Professor Hazewinkel's lectures).

REFERENCES

[1] Birkoff, G.: *A source Book in Classical Analysis*, Harvard University Press.

[2] Dieudonné, J. A.: *La theorie des invariants au XIXe siecle*, Sem. Bourbaki, no. 395.

[3] Dieudonné, J. A.: 1974, *The historical development of algebraic geometry*, Amer. Math. Monthly, October, 827-866.

[4] Gauss, K.: 1801, *Disquisitions Arithmeticae*.

[5] Gurevich, G. B.: 1964, *Foundations of the Theory of Algebraic Invariants*, P. Noordhoff, Groningen, The Netherlands.

[6] Klein, F.: *On Riemann's Theory of Algebraic Functions and Their Integrals*, Dover.

[7] Mumford, D.: 1975, *Curves and Their Jacobians*, Univ. of Michigan Press.

[8] Mumford, D., and Suominen, K.: 1970, *Introduction to the Theory of Moduli, Algebraic Geometry*, Edited by F. Oort.

[9] Postlethwaite, J., and MacFarlane, A. G.: 1979, *A complex variable approach to the analysis of linear multivariable feedback systems*, Lecture Notes in Control and Information Sciences, Vol. 12. New York: Springer-Verlag.

[10] Walker, R.: 1978, *Algebraic Curves*, Graduate Texts in Mathematics, Springer-Verlag.

2. MODULES OVER NOETHERIAN RINGS AND PRINCIPAL IDEAL DOMAINS

One of the fundamental steps in the study of automata and linear system theory is the introduction of reduced, or minimal, realizations. These are doubly interesting, since they are unique up to isomorphism and lead to an implementation of the system using a minimum number of certain components.

We all know that in order to carry out such a reduction one must take the subset of the state space consisting of the reachable states. As we shall see when we study the realization theory of linear systems, the size of the realization is directly related to the number of generators of the state module. If we, therefore, believe in the interest and applicability of linear models with coefficients belonging to a ring--and there is good reason to do so--it is vital to know over which rings this reduction process will lead to a physically realizable system (i.e., one with a finitely generated state module) or, even better, to a smaller system (i.e., one with a state module having fewer generators).

2.1 Noetherian Rings and Modules: Fundamental Results

Let R be a commutative ring.

2.1.1 Definition. *A module* M *is Noetherian if every submodule of* M *is finitely generated.*

It follows, of course, that M itself is finitely generated. Since a ring can be viewed as module over itself, its submodules being the ideals, (2.1) subsumes the following

2.2.2 Definition. *A ring* R *is Noetherian if every ideal is finitely generated.*

We shall first prove some elementary properties of Noetherian modules, then show how they relate to Noetherian rings; we shall afterwards prove that a lot of the state modules we shall find in system theory fall into this category.

First of all, there is a characterization available for Noetherian modules.

2.1.3 Theorem. *A module* M *is Noetherian if, and only if, every strictly increasing sequence of submodules*

$$N_1 \subset N_2 \subset \dots \subset N_i \subset M,$$

is finite.

Proof. Assume first of all that M is Noetherian; then the submodule $N = \bigcup_i N_i$ of M is finitely generated, and these generators lie in one of the N_i's, which is therefore equal to all of N.

Conversely, let N be a submodule of M, S the set of its finitely generated submodules; S is not empty, since it contains {0}. Let us show that it has a maximal element: indeed, since S is non-empty, we can choose a submodule N_0 in S; if it is not maximal, it is contained in a strictly larger submodule N_1, which is itself either maximal or contained in a strictly larger submodule N_2, etc., ...; the chain thus constructed being finite by assumption, S contains a maximal element. But this maximal element must be N itself, for otherwise we would add another generator of N to it, thereby constructing a larger finitely generated submodule of N. N is, therefore, in S, hence finitely generated.

This property, very useful in practice, is called the *Ascending Chain Condition* (or *A.C.C.*).

2.1.4 Lemma. *The submodules and quotient modules of a Noetherian module are themselves Noetherian.*

Proof. The relation $N \subset M$ for submodules being transitive, the case for submodules follows directly from the definition of Noetherian modules (2.1.1).

Let $L = M/N$, and $L_0 \subset L_1 \subset \dots$ be a strictly increasing sequence of submodules of L; let $M_0 \subset M_1 \dots$ be a sequence of representative elements of the equivalence classes in M (i.e., $L_i = M_i/N$); it is strictly increasing. M being Noetherian by assumption, (2.1.3) implies that the sequence is finite. So the original sequence $\{L_i\}$ is finite too, and L is Noetherian.

This lemma has a converse:

2.1.5 Lemma. *Suppose we have three modules and module homomorphisms* $L \xrightarrow{g} M \xrightarrow{h} N$ *such that* $\operatorname{im} g = \ker h$ *(we thus have what is called an exact sequence). Then if both* L *and* N *are Noetherian, so is* M.

Proof. Let $L' = \operatorname{im} g$, $N' = \operatorname{im} h$. (1.4) implies that L', isomorphic to a quotient module of L, and N', being a submodule of N, are both Noetherian. We can write an exact sequence

$$0 \to L' \to M \to N' \to 0 \quad .$$

Let $M' \subset M$ be a submodule of M; we must show that M' is finitely generated. We have an exact sequence

$$0 \to L' \cap M' \to M' \to M'/M' \cap L' \to 0$$

and $L' \cap M'$ and $M'/M' \cap L'$, submodules respectively of L' and N', are Noetherian; they are, therefore, finitely generated, by say $\{\ell_i\}$ and $\{n_j\}$ respectively.

Let x be an element of M'; its image in $M'/M' \cap L'$ is $\sum_j x_j n_j$, $x_j \in R$. If $\{\bar{n}_j\}$ designates a set of pre-images of $\{n_j\}$ in M', then the element $x - \Sigma x_j \bar{n}_j$ of M' is in the kernel of the projection $M' \to M'/M' \cap L'$.

Since the sequence is exact, it is in the image of the injection $L' \cap M' \to M'$, so $x - \Sigma x_j \bar{n}_j = \Sigma y_i \ell_i$, and

$$x = \Sigma y_i \ell_i + \Sigma x_j \bar{n}_j \quad .$$

M' is, therefore, generated by $\{\ell_i, \bar{n}_j\}$, which is a finite set.

2.1.7 Corollary. *A finite direct sum of Noetherian modules is Noetherian.*

Proof. It follows directly from (2.1.5) by induction on the number of direct summands.

We are now in a position to prove the important

2.1.8 Theorem. *Let* R *be a Noetherian ring. Then a module* M *is Noetherian if, and only if, it is finitely generated.*

Proof. It follows directly from definition 2.1.1 that a Noetherian module is finitely generated. Conversely, let M be a finitely generated module; we have an exact sequence

$$0 \to N \to F \to M \to 0$$

where F is the free R-module built on a set of generators of M; F is therefore a finitely generated free module over a Noetherian ring, hence, by (2.1.7), a Noetherian module. M, being a quotient of a Noetherian module, is itself Noetherian by (2.1.4) (note that this also implies that N is Noetherian, hence finitely generated).

2.2 Examples of Noetherian Rings

In particular, the state module of a finite dimensional linear system, being finitely generated, will be Noetherian whenever the ring is Noetherian. It becomes now urgent to exhibit some Noetherian rings, and to show that a large number of the rings we encounter in system-theoretic applications fall indeed in that category.

2.2.1 Definition. *A Principal Ideal Domain (P.I.D.) is an integral domain in which each ideal is principal (i.e. is generated by a single element).*

Since each ideal in a P.I.D. has a single generator, a P.I.D. is an example of a Noetherian ring; so $\mathbb{Z}$, for example, is Noetherian; and so is a field, of course. We can greatly enlarge the class by using the following:

2.2.2 Hilbert Basis Theorem. *A polynomial ring in finitely many unknowns over a Noetherian ring is also Noetherian.*

Proof. Since $R[x_1,\ldots,x_n] = R[x_1,\ldots,x_{n-1}]\,[x_n]$, it is clear by induction that we need only consider the case of polynomial ring in a single indeterminate $R[x]$.

Let I be an ideal in $R[x]$, and A_i the set of leading coefficients of polynomials of degree i in I. Since I is an ideal, A_i is an ideal too; furthermore, $f(x) \in I \Rightarrow xf(x) \in I$, so

$$A_0 \subset A_1 \subset \ldots$$

R being Noetherian, this sequence of ideals becomes eventually constant; let A_n be the maximal element for this chain. By a second application of the Noetherian assumption, we get that each A_i is generated by a finite set of generators $\{a_{ij}\}$, leading coefficients of a set of polynomials $\{f_{ij}\}$ of degree i in I.

Let us show that these polynomials $\{f_{ij}\}$ generate I, and consider an arbitrary polynomial $g(x)$ in I:

$$g(x) = g_m x^m + \ldots + g_1 x + g_0$$

We shall see that there exists a linear combination $h(x)$ of the $f_{ij}(x)$ such that $g(x) - h(x)$ be of strictly lower degree than $g(x)$, thereby establishing the desired result by induction: Since A_n is the maximal ideal in the chain, $g_m \in A_m \subset A_n$, so

$$g_m = \sum_j r_j d_{nj} \quad , \qquad r_j \in R$$

$$h(x) = \begin{cases} x^{m-n}(\sum r_j f_{nj}(x)) & \text{if } m > n \\ \sum f_j f_{mj}(x) & \text{otherwise} \end{cases}$$

is of degree m and has leading coefficient g_m. Thus $h(x)$ is a linear combination over $R[x]$ of the $f_{ij}(x)$'s, and $g(x) - h(x)$ has strictly lower degree.

Polynomial rings over fields or over the integers are therefore Noetherian.

2.3 On Duality and the Structure of Modules over Noetherian Rings

As an exercise in making use of the Noetherian assumption, let us establish two interesting results about Noetherian modules. The first one will be a structure theorem analogous in spirit to the Jordan-Hölder theorem for groups (after all, groups are but modules over the Noetherian ring $\mathbb{Z}$!) The second will give us an introduction to duality theory useful for future lectures in system theory.

2.3.1 Theorem. *Let* R *be a Noetherian ring and* M *a finitely generated* R*-module. Then there is a sequence of submodules*

$$0 = M_0 \subset M_1 \subset \ldots \subset M$$

such that for each i *the module* M_i/M_{i-1} *is isomorphic to* R/p_i, *where* p_i *is a prime ideal of* R.

Proof. Let S be the set of submodules of M for which the theorem holds. We can select out of S a sequence of strictly increasing elements

$$0 = M_0 \subset M_1 \subset \ldots \subset M_r \subset \ldots$$

By the Noetherian assumption, such a sequence is necessarily finite and ends with a maximal element, say M_r. If $M_r = M$, then we are done. Otherwise, let us show that $N = M/M_r$ contains a submodule isomorphic to R/p for some prime ideal p.

This will be achieved by a second use of the Noetherian

assumption. But first note that for a module to contain a submodule isomorphic to R/p is equivalent to saying that p is the annihilator of some element x of the module (i.e., $p = \{r \in R | rx = 0\}$): the map

$$r \mapsto rx$$

which sends R onto the cyclic submodule generated by x has p for kernel, hence that cyclic submodule is isomorphic to R/p; conversely, if the module contains a copy of R/p, then p is the annihilator of its generator.

Let, therefore, F be the family of ideals other than R which annihilate elements of N, and, once again, let I be a maximal element of F. Let us prove that I is prime. Say x is the element in N annihilated by I; then, if $ab \in I$ but $b \notin I$, $bx \neq 0$; any element in I annihilates x, hence bx too, so I contains the annihilator of bx and is equal to it, being a maximal element in F. But $ab \in I \Rightarrow abx = 0$: a annihilates bx, hence is in I. I is therefore prime.

We can now return to the main line of the argument: $N = M/M_r$ contains a submodule isomorphic to R/p, which corresponds to a submodule N' of M containing M_r and such that N'/M_r be isomorphic to R/p; the sequence

$$0 = M_0 \subset M_1 \subset \dots \subset M_r \subset N'$$

is therefore a strictly increasing sequence in S, contradicting the maximality of M_r.

2.3.2 <u>Definition</u>. *The dual* M *of a module* M *over a ring is the set of module-homomorphisms from* M *into* R.

As long as we limit ourselves to free modules over an integral domain, the theory remains the same as that of vector-space duality: the dual of a finitely generated free module is a finitely generated free module of same rank, and the proofs are the same. When the module is not free anymore, the issue, of course, becomes different; we however still have:

2.3.3 <u>Theorem</u>. *Let* R *be a Noetherian integral domain,* M *a finitely generated* R*-module. Then* M^* *is finitely generated as an* R*-module too.*

<u>Proof</u>. M, being finitely generated, is a quotient of a finitely generated free module L. But Hom $(\cdot,R)$ is a contravariant left exact functor. Hence

$$L \to M \to 0 \Rightarrow 0 \to M^* \to L^*$$

and M^* is a sub-module of a finitely generated free module. It follows from the Noetherian assumption and (2.1.8) that M^* is a Noetherian module, hence finitely generated.

2.3.4 Definition. *The Krull dimension of a ring* R *is the length* n *of the longest chain*

$$p_0 \subset p_1 \subset \ldots \subset p_n \neq R$$

of prime ideals in R *(infinite if there is no maximal chain).*

2.3.5 Definition. *Let* R *be a finitely generated algebra over a field* k *(i.e., a quotient of a polynomial ring over* k*), which is itself an integral domain. Let* K *be its quotient field. Then the transcendence degree of* R *is the dimension of* K *as a vector-space over* k.

2.3.6 Definition. *Let* M *be an* R-*module. We shall say that* $hd(M) \leq n$ *if there exists a projective resolution of* M

$$0 \to P_n \to \ldots \to P_1 \to P_0 \to M \to 0$$

Then the global dimension of R *is* $\sup\{hd(M) | M : \text{module over } R\}$.

By a projective module is meant a module P which is complemented in a free module, i.e., one for which there exists a splitting

$$M \simeq P \oplus Q$$

where M is free. Note that a module is projective if, and only if, it may be realized as the image of a projection operator

$$M \to P$$

defined on a free module. One can check that if P is itself finitely generated, then M may be taken to be finitely generated and free.

2.3.7 Theorem. *In the case of polynomial rings over a field, all three notions of dimension are equivalent.*

Remark. In particular, a polynomial ring has finite Krull dimension. However, Negata has given an example of a Noetherian ring with infinite Krull dimension.

2.4 Modules Over a Principal Ideal Domain

If the ring not only is Noetherian but is a P.I.D., then one can say even more about the generators of a submodule:

2.4.1 Theorem. *Every submodule of a finitely generated free module over a P.I.D. R is a free R-module. (The finitely generated assumption is not necessary but makes the proof shorter).*

Proof. Let L be a free module, $\{e_i\}$ a basis for L, $\{p_i\}$ the corresponding coordinate functions. Let M be a submodule of L. The image of M in R by projection p_i is an ideal of R, which is principal by assumption, say Ra_i. Let m_i be an element of M such that $p_i(m_i) = a_i$ (if $a_i = 0$ take $m_i = 0$).

Let us show that $\{m_i\}$ generates M: if $m \in M$ and $p_i(m) = r_i a_i$ let $m' = \Sigma r_i m_i$; then $m - m'$ projects to 0 on every coordinate, hence is 0.

Furthermore, the $\{m_i\}$'s are free:

$$\sum r_i m_i = 0 \Rightarrow p_i(\sum r_i m_i) = 0 \Rightarrow r_i a_i = 0, \forall i$$

Since a_i is different from 0 for $m_i \neq 0$, it follows that $r_i = 0_i$.

2.4.2 Definition. *An element* $m \neq 0$ *in* M *is said to have torsion if there exists* $r \neq 0$ *in* R *such that* $rm = 0$. *If no element in* M *has torsion, then* M *is called torsion-free.*

2.4.3 Lemma. *A finitely generated torsion-free module* M *over an integral domain* R *can be embedded in a finitely generated free module.*

Proof. Let K be the quotient field of R, and let $\{a_1,\ldots,a_n\}$ be generators of M. Let $\{b_1,\ldots,b_\ell\}$ be a basis for the vector space $M \otimes_R K$ over K. Then

$$a_i = \sum (r_{ij}/s_{ij}) b_j \quad , \qquad r_{ij}, s_{ij} \in R$$

Let s be a common multiple of the s_{ij}'s. Then

$$\{b_1/s, \ldots, b_\ell/s\}$$

being linearly independent over K generates a free R-module which contains M.

2.4.4 Corollary. *Every finitely generated torsion-free module over a P.I.D. is free.*

Proof. This is a direct consequence of (2.4.3) and (2.4.1).

Remark. The finitely generated assumption is crucial. Indeed, the quotient field K of R is a torsion free R-module, but is not free.

The structure theorem we established for Noetherian rings also takes a more powerful form:

2.4.6 Theorem. *Let* R *be a P.I.D. Then any finitely generated module* M *over* R *is isomorphic to a direct sum of sub-modules* R/p^n, p *prime.*

The structure theorem for finitely-generated modules over a P.I.D. is very powerful in studying the algebra of linear maps and, of course, linear systems defined on such modules. Suppose, for example, that we wish to study the linear system

$$x(t+1) = Ax(t) + Bu(t)$$

$$y(t) = Cx(t) \tag{2.4.7}$$

where $u \in \mathbb{Z}^{(m)}$, $y \in \mathbb{Z}^{(p)}$, $x \in M$ a finitely-generated module over $\mathbb{Z}$. Of course, (A,B,C) we assumed to be $\mathbb{Z}$-linear maps. If (2.4.7) is observable, then M is necessarily free. For, observability implies that M may be imbedded, by successive observations, in the direct sum

$$\bigoplus_{i=1}^{\infty} \mathbb{Z}^{(p)}$$

and, therefore, has no non-zero torsion elements.

As a second illustration, consider an R-linear map

$$P : R^{(n)} \to R^{(n)}$$

which is a projection, i.e. $P^2 = P$. If R is a field, then it is a significant fact in linear algebra that one may choose a basis of $R^{(n)}$ so that the matrix of P, with respect to this basis, has only 1's and 0's on the diagonal and 0's elsewhere. This is not true for all R, but we can give a proof if R is a P.I.D. For, consider $M = P(R^{(n)}) \subset R^{(n)}$ it is finitely generated, as the image of $R^{(n)}$, and has no torsion, so M must be free and one can choose a basis:

$$\text{span } \{x_1,\ldots,x_r\} = M, \qquad \text{over } \mathbb{R}$$

Fortunately, one can actually extend this basis, since the same statements are valid for $N = \text{image } (I-P)$ and $N \cap M = (0)$, $N + M = R^{(n)}$.

This basic result is also true for polynomial rings over a field but is a much deeper result than one might first suspect--it used to be known as the Serre conjecture, and has been proven by Quillen and Suslin.

As a final observation, we suppose given a R-linear map

$$T : R^{(n)} \to R^{(\ell)}$$

and ask whether T is injective. Passing to the fraction field K of R, one has an extended K-linear map

$$T_K : R^{(n)} \otimes_R K \to R^{(\ell)} \otimes_R K \quad , \quad \text{or simply}$$

$$T_K : K^{(n)} \to K^{(\ell)}$$

a K-linear map of K-vector spaces, where the question is answered quite easily. Since a non-zero element of M is zero in $M \otimes_R K$ only if it is a torsion element, for a map of free modules over a P.I.D., T is injective if, and only if, T_K is injective.

2.4.8 Example. Consider $T : \mathbb{Z} \to \mathbb{Z}$, $T(z) = 2z$. Then $T : \mathbb{Q} \to \mathbb{Q}$ is both injective and surjective, while T itself is only injective.

3. DIFFERENTIABLE MANIFOLDS, VECTOR BUNDLES, AND GRASSMANNIANS

This section discusses first some of the elements of the theory of differentiable manifolds, then discusses that powerful tool "partitions of unity" and then proceeds to say a few things about vector bundles. One particular family of manifolds, the Grassmann manifolds, have proved to be very important in linear system theory and one particular vector bundle over the manifolds enjoys a similar status. The last two almost telegraphic subsections are intended to indicate that this phenomenon is not peculiar to system theory: these manifolds and bundles play an equally distinguished role in the general theory of vector bundles itself, a feat which may help to understand the role they play in system theory.

There are many books and and lecture notes in which the theory of manifolds and vector bundles is clearly explained. Some of the present writer's favorites are:

M. F. Atiyah, *K-Theory*, Harvard Lecture Notes, Fall 1964. (Published by Benjamin)

S. Helgason, *Differential geometry, Lie groups and symmetric spaces*, Acad. Pr., 1978. (on press)

F. Hirzebruch, *Introduction to the theory of vector bundles and K-theory*, Lectures at the University of Amsterdam and Bonn, University of Amsterdam, 1965.

D. Husemoller, *Fibre bundles*, McGraw-Hill, 1966.

J. W. Milnor, J. D. Stasheff, *Characteristic classes*, Princeton University Press, 1974.

The last one named is especially recommended. Finally at a more introductory level recommended

L. Auslander, R. E. MacKenzie, *Introduction to differentiable manifolds*, Dover (reprint) 1977.

3.1 Differentiable Manifolds

3.1.1 <u>Definition</u> (Differentiable maps). Let $U \subset \mathbb{R}^m$ and $V \subset \mathbb{R}^n$ be open subsets. A mapping $\phi : U \to V$ is differentiable if the coordinates $y_i(\phi(x))$ of $\phi(x)$ are differentiable functions of $x = (x_1,\dots,x_m) \in U \subset \mathbb{R}^m$, $i = 1,\dots,n$. Here a function is said to be differentiable if all partial derivatives of all orders exist and are continuous. The differentiable mapping ϕ if it is 1-1, onto and if ϕ^{-1} is also differentiable.

3.1.2 Definition (Charts). Let M be a Hausdorff topological space. An *open chart* on M is a pair (U,ϕ) consisting of an open subset U of M and a homeomorphism ϕ of U onto some open subset of an $\mathbb{R}^m$; the number m is called the *dimension* of the chart.

3.1.3 Definition (Differentiable manifolds). Let M be a Hausdorff space. A *differentiable structure* on M consists of a collection of open charts (U_i,ϕ_i), $i \in I$ such that the following conditions are satisfied

$$\bigcup_{i \in I} U_i = M \tag{3.1.4}$$

for all $i,j \in I$ the mapping $\phi_j \cdot \phi_i^{-1}:\phi_i(U_i \cap U_j) \to \phi_j(U_i \cap U_j)$ is a diffeomorphism. (3.1.5)

The collection (U_i,ϕ_i), $i \in I$ is maximal with respect to properties (3.1.4) and (3.1.5) (3.1.6)

A *differentiable manifold* is a Hausdorff topological space together with a differentiable structure.

Locally it is just like $\mathbb{R}^n$, but globally not. The charts permit us to do (locally) calculus and analysis as usual. It is possible that one and the same topological space admits several inequivalent differentiable structures where inequivalent means "non diffeomorphic"--a notion which is defined below in 3.2.

If M is connected as a topological space, all the charts (U_i,ϕ_i) necessarily have the same dimension which is then also (by definition) the dimension of the differentiable manifold M.

Often a differentiable structure is defined by giving a collection of charts (U_i,ϕ_i) such that only (3.1.4) and (3.1.5) are satisfied. Then there is a unique larger collection of charts such that also (3.1.6) holds. (Easy exercise.)

3.1.7 Example: The circle. Consider the subset of $\mathbb{R}^2$ defined by

$$S^1 = \{(x_1,x_2)\ x_3^2 + x_2^2 = 1\} \tag{3.1.8}$$

Let $U_1 = \{x \in S^1 | x \neq (0,-1)\}$, $U_2 = \{x \in S^1 | x \neq (0,1)\}$.

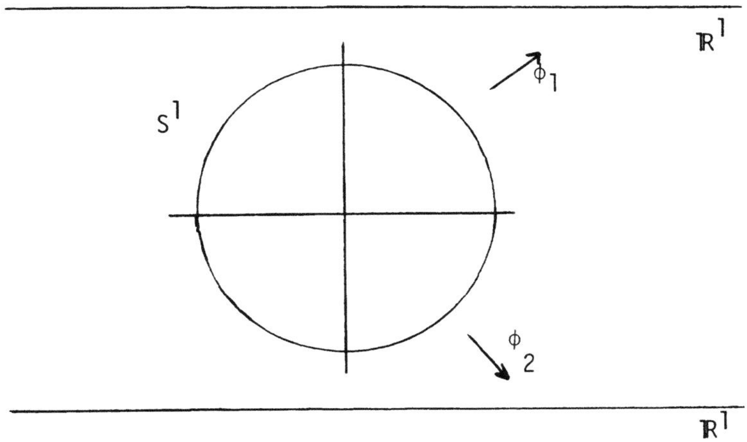

Now define $\phi_1 : U_1 \to \mathbb{R}^1$ by $\phi_1(x_1,x_2) = \dfrac{x_1}{1+x_2}$ and $\phi_2 : U_2 \to \mathbb{R}^1$ by $\phi_2(x_1,x_2) = \dfrac{x_1}{1-x_2}$. These are both homeomorphisms. The inverse of ϕ_1 is given by the formula $x \mapsto (x_1,x_2)$ with

$x_1 = \dfrac{2x}{x^2+1}$, $x_2 = \dfrac{1-x^2}{1+x^2}$ and the inverse of ϕ_2 by the very similar formula: $x \mapsto (x_1,x_2)$, $x_1 = \dfrac{2x}{1+x^2}$, $x_2 = \dfrac{x^2-1}{1+x^2}$

The map $\phi_2 \cdot \phi_1^{-1} : \phi_2(U_1 \cap U_2) = \mathbb{R}^1 \backslash \{0\} \to \phi_2(U_1 \cap U_2) = \mathbb{R}^1 \backslash \{0\}$ is given by $x \to x^{-1}$ and hence is a diffeomorphism, so that (U_1,ϕ_1) and (U_2,ϕ_2) do indeed define a differentiable structure on S^1.

3.1.9 <u>Trivial Example: $\mathbb{R}^n$ Itself</u>. Let $M = \mathbb{R}^n$ and define a chart (U,ϕ) by $U = M = \mathbb{R}^n$, ϕ = identity map : $U \to \mathbb{R}^n$. This one element collection of charts satisfies, of course, (3.1.4) and (3.1.5), and hence defines a differentiable structure on $\mathbb{R}^n$.

3.1.10 Constructing differentiable manifolds 1. Embedded manifolds. The example above illustrates one way in which differentiable manifolds often arise. Namely the topological space M is given as a "smooth" subset of some $\mathbb{R}^n$ and the differentiable structure is induced from the natural differentiable structure of $\mathbb{R}^n$. Indeed, apart from a factor 2 the maps ϕ_1 and ϕ_2 of example 3.1.7 above arise by projecting the circle from $(0,-1)$ onto the line $x_2 = 1$ in $\mathbb{R}^2$ and by projecting the circle from $(0,1)$ onto the line $x_2 = 1$ in $\mathbb{R}^2$.

Abstractly a smoothly embedded differentiable manifold of dimension m is a subset $M \subset \mathbb{R}^n$ (for some n) such that for each $x \in M$ there is a differentiable map $\psi: V \to \mathbb{R}^n$ defined on some open subset $V \subset \mathbb{R}^m$ such that

ψ maps V homeomorphically onto some open neighborhood U of x in M (3.1.11)

for each $y \in V$ the matrix $\left(\frac{\partial \psi_i}{\partial y_j}(y)\right)$, $i = 1,\dots,n$, $j = 1,\dots,m$ has rank m. (3.1.12)

It is not difficult to prove (using the implicit function theorem) that the pairs (U,ψ^{-1}) for varying x now define a differentiable structure on M; i.e., that these pairs constitute a collection of charts which satisfy (3.1.4) (3.1.5). Inversely it is a theorem (Whitney) that every differentiable manifold with a countable basis arises in this way (up to diffeomorphism).

3.1.13 Constructing differentiable manifolds 2. Gluing. A second very frequently used method of obtaining a differentiable manifold is by a gluing procedure. Suppose that we have for each i in some index set I (often a finite set) some open set $U_i \subset \mathbb{R}^m$. Suppose moreover that for each $i,j \in I$, $i \neq j$, there are defined open subsets $U_{ij} \subset U_i$ and $U_{ji} \subset U_j$ and a diffeomorphism $\phi_{ij} : U_{ij} \to U_{ji}$. Suppose more over that the following compatibility conditions hold

$$U_{ii} = U_i, \quad \phi_{ii} = \text{identity for all } i \in I$$

and for all $i,j,k \in I$

$$U_{ij} \cap \phi_{ij}^{-1}(U_{jk}) \subset U_{ik} \quad \text{and} \quad \phi_{jk} \cdot \phi_{ij} = \phi_{ik} \quad \text{on} \quad U_{ij} \cap \phi_{ij}^{-1}(U_{jk}). \tag{3.1.14}$$

(Note that this implies that $\phi_{ij} = \phi_{ji}^{-1}$). Then we define a topological space M by taking the disjoint union $\amalg U_i$ and then identifying $x \in U_i$ with $y \in U_j$ iff $y = \phi_{ij}(x)$, $x \in U_{ij}$, $y \in U_{ji}$. This is an equivalence relation because of (3.1.14). Let M be the topological space $\amalg U_i/\sim$ with the quotient topology, where $\sim$ denotes the equivalence relation just defined.

Let $\phi_i : U_i \to \amalg U_i \to \amalg U_i/\sim$ be the obvious map. Suppose that M is Hausdorff (this is not automatically the case), then the (U_i, ϕ_i) are a collection of charts satisfying (3.1.4) and (3.1.5) so that they define a differentiable structure on M.

3.1.15 <u>Example: real n-dimensional projective space</u>. Let $I = \{0,1,\ldots,n\}$ and for each $i \in I$ let $U_i = \mathbb{R}^n$, and for each $i \in I$ let $\alpha_i : U_i \to \mathbb{R}^{n+1}$ be the embedding $\alpha_i(x_1,\ldots,x_n) = (x_1,\ldots,x_i,1,x_{i+1},\ldots,x_n)$. Label the coordinates of $\mathbb{R}^{n+1}$ by $0,1,\ldots,n$. Thus $\alpha_i(U_i) = \{y = (y_0,\ldots,y_n) \in \mathbb{R}^{n+1} \mid y_i = 1\}$. Let $i,j \in I$, $i \neq j$ and define U_{ij} as $\alpha_i^{-1}V_{ij}$ where $V_{ij} = \{y \in \mathbb{R}^{n+1} \mid y_i = 1,\ y_j \neq 0\}$, and define $\phi_{ij} : U_{ij} \to U_{ji}$ as the composite $\alpha_j^{-1} \cdot \psi_{ij} \cdot \phi_i$, where $\psi_{ij} : V_{ij} \to V_{ji}$ is defined by $\psi_{ij}(y_0,\ldots,y_n) = (y_j^{-1}y_0,\ldots,y_j^{-1}y_n)$. (Note that indeed $\psi_{ij}(V_{ij}) = V_{ji}$, so that $\phi_{ij}(U_{ij}) = U_{ji}$.

The compatibility conditions (3.1.14) hold and the topological space M is Hausdorff. Thus then gluing data define a differentiable manifold which is denoted $\mathbb{P}^n(\mathbb{R})$ and called real n-dimensional projective space.

Consider the differentiable manifold $X = \mathbb{R}^{n+1}\setminus\{0\}$. For each $y \in X$, $y = (y_0,\ldots,y_n)$ choose an i such that $y_i \neq 0$. Now define $\pi : X \to \mathbb{P}^n(\mathbb{R})$ by assigning to y the equivalence class of $\alpha_i^{-1}(y_0y_i^{-1},\ldots,y_{i-1}y_i^{-1},1,y_{i+1}y_i^{-1},\ldots,y_ny_i^{-1})$. Note that

$\pi(y)$ does not depend on the choice of i. It is now an easy exercise to check that $\pi(y) = \pi(y')$ if and only if there is an $\lambda \neq 0$ such that $y'_i = \lambda y_i$, $i = 0,\ldots,n$. Thus, the construction above defines as a differentiable manifold structure on the set of all lines through the origin of $\mathbb{R}^{n+1}$.

3.1.16 Grassmann manifolds. Let $1 \leq k < n$, $k,n \in \mathbb{N}$. Then $\mathcal{G}_{k,n}$ is by definition the set of all k-dimensional subspaces of $\mathbb{R}^n$. This set can be given a differentiable manifold structure in a manner rather similar to the one used above in (3.1.15). For explicit details see section 4 of this chapter.

3.1.17 Morphisms of manifolds: differentiable mappings. Let M and N be two differentiable manifolds. Let (U_i,ϕ_i) and (V_j,ψ_j) be collections of charts for M and N respectively such that (3.1.4) and (3.1.5) hold. A map $\phi : M \to N$ is a *morphism of differentiable manifolds* or a *differentiable map(ping)* if for all $i \in I$, $j \in J$ the map

$$\phi_j \circ \phi \circ \phi_i^{-1} : \phi_i(U_i \cap \phi^{-1}(V_j)) \to \phi_j(V_j)$$

is a differentiable map in the sense of 3.1.1 above. A differentiable mapping ϕ which is 1-1 and onto and such that ϕ^{-1} is also a differentiable mapping is called a *diffeomorphism*.

3.1.18 Example. Give $X = \mathbb{R}^{n+1}\backslash\{0\}$ the differentiable structure defined by the one element collection of charts $U = X$, ϕ = identity. Then $\pi : X \to \mathbb{P}^n(\mathbb{R})$ as defined in example 3.1.15 above is a differentiable mapping.

3.1.19. Differentiable map and gluing data. Suppose the two differentiable manifolds M and N have been obtained by means of the procedure discussed above in 3.1.13 from the local pieces U_i and patching data ϕ_{ij} (resp. local pieces V_k and patching data $\psi_{k\ell}$). Then a frequently used method of specifying a differentiable map $\alpha: M \to N$ is as follows. For each i and k let there be given an open subset $U_{ik} \subset U_i$ and a differentiable map (in the sense of 3.1.1)

$$\alpha_{ik} : U_{ik} \to V_k, \qquad \bigcup_k U_{ik} = U_i$$

Suppose that the following compatibility condition holds where appropriate

$$\psi_{k\ell} \circ \alpha_{ik} = \alpha_{j\ell} \circ \phi_{ij} \tag{3.1.20}$$

i.e. if $n \in U_{ik}$ and $y \in U_{j\ell}$ and $\phi_{ij}(n) = y$, then $\alpha_{ik}(n) \in V_k$, $\alpha_{j\ell}(y) \in V_\ell$ and $\psi_{k\ell}(\alpha_{ik}(n)) = \alpha_{j\ell}(k)$. Then the α_{ik} combine to define a differentiable map $\alpha : M \to N$ as is easily checked.

3.1.21 Example. Consider $\mathbb{P}^1(\mathbb{R})$ as defined in 3.1.15 above. Now define $\alpha : \mathbb{P}^1(\mathbb{R}) \to \mathbb{R}^2$ as follows

$$\alpha_1 : U_1 = \mathbb{R} \to \mathbb{R}^2, \quad x_0 \to \left(\frac{2x_0}{x_0^2+1}, \frac{1-x_0^2}{x_0^2+1}\right)$$

$$\alpha_0 : U_0 = \mathbb{R} \to \mathbb{R}^2, \quad x_1 \to \left(\frac{2x_2}{x_2^2+1}, \frac{x_1^2-1}{x_1^2+1}\right)$$

Recall that $U_{10} = \{x_0 \in \mathbb{R} | x_0 \neq 0\}$, $U_{01} = \{x_1 \in \mathbb{R} | x_1 \neq 0\}$ and that the gluing map ϕ_{10} is given by $\phi_{10}(x_0) = x_0^{-1}$. And we check that on U_{10}

$$\alpha_0\phi_{10}(x_0) = \alpha_0(x_0^{-1}) = \left(\frac{2x_0^{-1}}{x_0^{-2}+1}, \frac{x_0^{-2}-1}{x_0^{-2}+1}\right)$$

$$= \left(\frac{2x_0}{1+x_0^2}, \frac{1-x_0^2}{1+x_0^2}\right) = \alpha_1(x_0)$$

so that the compatibility condition (3.1.20) is fulfilled, and the α_0, α_1 do indeed combine to define a differentiable map $\alpha : \mathbb{P}^1(\mathbb{R}) \to \mathbb{R}^2$. Note that $\alpha(\mathbb{P}^1(\mathbb{R})) = S^1 = \{(x_1, x_2) \in \mathbb{R}^2 | x_1^2 + x_2^2 = 1\}$. The map α is also 1-1 and surjective onto S^1 and the inverse map $\alpha^{-1} : S^1 \to \mathbb{P}^1(\mathbb{R})$ is also differentiable. Thus α induces a diffeomorphism of $\mathbb{P}^1(\mathbb{R})$ with the circle S^1.

3.1.22. Products. Let M and N be differentiable manifolds of dimension m and n respectively. Then the cartesian product $M \times N$ has a natural differentiable structure defined as follows. Let (U_i, ϕ_i), $i \in I$ be a collection of open charts for M such that (3.1.4) and (3.1.5) hold; and let (V_j, ψ_j), $j \in J$ be a similar collection for N. Then the open sets $U_i \times V_j$, $i \in I$, $j \in J$ cover the topological space $M \times N$ and the maps ϕ_i and

ψ_j combine to define a homeomorphism $\phi_i \times \psi_j : U_i \times V_j \to \phi_i(U_i) \times \psi_j(V_j) \subset \mathbb{R}^m \times \mathbb{R}^n$. This defines a collection of charts $(U_i \times V_j, \phi_i \times \psi_j)$, $i \in I$, $j \in J$ which satisfies (3.1.4) and (3.1.5) and hence defines a differentiable structure on $M \times N$.

If both M and N are embedded manifolds, cf. 3.1.10 above, say, $M \subset \mathbb{R}^r$, $N \subset \mathbb{R}^s$, then $M \times N \subset \mathbb{R}^r \times \mathbb{R}^s = \mathbb{R}^{r+s}$ is naturally again an embedded manifold.

If both M and N are obtained by a local pieces and gluing data construction $M \times N$ can be described in a similar way. Indeed if (U_i, U_{ij}, ϕ_{ij}) describe M and $(V_k, V_{k\ell}, \psi_{k\ell})$ the manifold N then M N is described by the local pieces and gluing data $(U_i \times V_k,\ U_{ij} \times V_{k\ell},\ \phi_{ij} \times \psi_{k\ell})$.

3.1.23. <u>Example</u>. $\mathbb{P}^1(\mathbb{R}) \times \mathbb{P}^1(\mathbb{R})$. According to the recipe above $\mathbb{P}^1(\mathbb{R}) \times \mathbb{P}^1(\mathbb{R})$ is obtained by gluing together four local pieces

$$U_1 \times V_1 = \mathbb{R} \times \mathbb{R},\quad U_1 \times V_0 = \mathbb{R} \times \mathbb{R},\quad U_0 \times V_1 = \mathbb{R} \times \mathbb{R},$$
$$U_0 \times V_0 = \mathbb{R} \times \mathbb{R}$$

by means of the following six diffeomorphsims (and their inverses)

$$\mathrm{id} \times \psi_{10} : U_1 \times V_{10} \to U_1 \times V_{01},\quad (x_0, y_0) \to (x_0, y_0^{-1})$$

$$\phi_{10} \times \mathrm{id} : U_{10} \times V_1 \to U_{01} \times V_1,\quad (x_0, y_0) \to (x_0^{-1}, y_0)$$

$$\phi_{10} \times \psi_{10} : U_{10} \times V_{10} \to U_{01} \times V_{01},\quad (x_0, y_0) \to (x_0^{-1}, y_0^{-1})$$

$$\phi_{10} \times \psi_{01} : U_{10} \times V_{01} \to U_{01} \times V_{10},\quad (x_0, y_1) \to (x_0^{-1}, y_1^{-1})$$

$$\phi_{10} \times \mathrm{id} : U_{10} \times V_0 \to U_{01} \times V_0,\quad (x_0, y_1) \to (x_0^{-1}, y_1)$$

$$\mathrm{id} \times \psi_{10} : U_0 \times V_{10} \to U_0 \times V_{01},\quad (x_1, y_0) \to (x_1, y_0^{-1})$$

Let us use this description to define a morphism $\alpha : \mathbb{P}^1(\mathbb{R}) \times \mathbb{P}^1(\mathbb{R}) \to \mathbb{P}^3(\mathbb{R})$ as follows. Recall that $\mathbb{P}^3(\mathbb{R})$ is built out of four pieces $W_i = \mathbb{R}^3$, $i = 0,1,2,3$; cf. 3.1.15. We define α by means of the maps

$$\alpha_1 : U_1 \times V_1 \to W_3,\ (x_o,y_o) \to (x_oy_o,x_o,y_o)$$

$$\alpha_2 : U_1 \times V_o \to W_2,\ (x_o,y_1) \to (x_o,x_oy_1,y_1)$$

$$\alpha_3 : U_o \times V_1 \to W_1,\ (x_1,y_o) \to (y_o,x_1y_o,x_1)$$

$$\alpha_4 : U_o \times V_o \to W_o,\ (x_1,y_1) \to (y_1,x_1,x_1y_1)$$

It is now easy to check that the compatibility conditions 3.1.20 are satisfied. For example that $\alpha_2 \cdot (\mathrm{id} \times \psi_{10}) = \chi_{32} \cdot \alpha_1$ is illustrated by the diagram below (there χ_{32} is the gluing diffeomorphism $W_{32} \to W_{23}$ of 3.1.15 above and we use (for convenience) the embedding $W_i \to \mathbb{R}^4$ which we also used in 3.1.15).

$$\begin{array}{ccccc} (x_o,y_o) & \overset{\alpha_1}{\mapsto} & (x_oy_o,x_o,y_o) & \leftrightarrow & (x_oy_o,x_o,y_o,1) \\ \Big\downarrow{\scriptstyle \mathrm{id}\times\psi_{10}} & & \Big\downarrow{\scriptstyle \chi_{32}} & & \Big\downarrow \\ (x_o,y_o^{-1}) & \overset{\alpha_2}{\mapsto} & (x_o,x_oy_o^{-1},y_o^{-1}) & \leftrightarrow & (x_o,x_oy_o^{-1},1,y_o^{-1}) \end{array}$$

The morphism constructed above in such painful detail is a very well known one. If we view $\mathbb{P}^n(\mathbb{R})$ as the set of all lines through the origin in $\mathbb{R}^{n+1}$, i.e. as equivalence classes of points in $\mathbb{R}^{n+1}$ under the equivalence relation $(x_o,\ldots,x_n) \sim (x_o^1,\ldots,x_n^1)$ iff $\exists\lambda \neq 0$ such that $x_i^1 = \lambda x_i$, $i = 0,\ldots,n$, then $\alpha : \mathbb{P}^1(\mathbb{R}) \times \mathbb{R}^1(\mathbb{R}) \to \mathbb{P}^3(\mathbb{R})$ is induced by $((x_o,x_1),(y_o,y_1)) \to (x_oy_o,x_oy_1,x_1y_o,x_1y_1)$ and from this the explicit local pieces description above is easily deduced.

3.1.24. <u>Submanifolds.</u> Let M be a differentiable manifold of dimension n. A subset $N \subset M$ is a submanifold of dimension $p \le n$ if there exists for every $x \in N$ an open chart $\phi : U \to \mathbb{R}^n$ such that $\phi(x) = 0$ and the $V = \{x \in U \mid \phi_{p+1}(x) = \ldots = \phi_n(x) = 0\}$ together with the restriction of ϕ to V (as a map to $\mathbb{R}^p$) form a system of open charts for N. The differentiable manifold N is said to be a *regular submanifold* of M if for every $x \in M$ there is a U as above such that moreover $V = N \cap U$. (V as above).

An example is $S^1 = \{(x_1 x_2) | x_1^2 + x_2^2 = 1\} \subset \mathbb{R}^2$. This is a regular submanifold (Exercise: prove this.) This winding line (with irrational winding angle) on a torus is an example of a nonregular submanifold. (The torus T^2 is the differentiable manifold $S^1 \times S^1$ and $\mathbb{R}$ can be seen as a subset of $S^1 \times S^1$ by mapping t to $(e^{2\pi i t}, e^{2\pi i \alpha t})$, α irrational; note that the induced topology on $\mathbb{R}$ from this injection into T^2 is not the original topology of $\mathbb{R}$. $N \subset M$ is a regular submanifold the induced topology on N is indeed original topology (belonging to the differentiable structure) of N.

3.1.25. Analytic manifolds. Similarly to differentiable manifolds one can define *analytic manifolds* by replacing everywhere differentiable map by analytic map. Thus an analytic manifold is locally like $\mathbb{R}^m$ and the local coordinate transition mapping $\psi_j \cdot \phi_i^{-1}$, cf. 3.1.3, are analytic, i.e. they admit (locally) convergent power series expansions.

To define *complex manifolds* replace $\mathbb{R}$ by $\mathbb{C}$ everywhere and require that the coordinate transition mappings $\phi_j \cdot \phi_i^{-1}$ are holomorphic.

3.2. Partitions of unity

A powerful and often used tool in differential topology are partitions of unity.

3.2.1 Some definitions and facts from general topology. Recall that a covering $\{U_i, i \in I\}$ of a topological space X is said to be locally finite if for every $x \in X$ there is an open neighbourhood V containing x such that $U_i \cap V \neq \phi$ for only finitely many i. Recall also that a topological space is *paracompact* if every covering admits a locally finite refinement. A space is *normal* if for all closed $A, B \subset X$ there are open $U, V \subset X$ such that $A \subset U$, $B \subset V$ and $U \cap V = \phi$. A locally compact Hausdorff space with countable base is paracompact and every paracompact space is normal.

3.2.2. Convention. We shall assume from now on that every differentiable manifold is paracompact. This is not automatically the case, though it is not easy to construct counterexamples. If M is built up out of countably many $U_i \subset \mathbb{R}^m$ by a local pieces and gluing data procedure as in 3.1.13 above it is automatically paracompact (by the remarks made above). Thus manifolds like spheres, projective space, Grassmannians are all paracompact.

3.2.3 Theorem. *Let* M *be a paracompact differentiable manifold and let* $\{U_i \mid i \in I\}$ *be a locally finite open covering of* M. *Assume that all* $\bar{U}_i$ *are compact. Then there exists a collection* $\{\phi_i \mid i \in I\}$ *of differentiable functions on* M *such that*

$$\operatorname{Supp}(\phi_i) \subset U_i \qquad (3.2.4)$$

$$\phi_i(x) \geq 0 \quad \text{for all } x \in M \qquad (3.2.5)$$

$$\sum_i \phi_i(x) = 1 \quad \text{for all } x \in M \qquad (3.2.6)$$

Here $\operatorname{Supp}(\phi)$ is the closure of the set of all $x \in M$ such that $\phi(x) \neq 0$. Note that in the sum (3.2.6) for all x there are only finitely many i such that $\phi_i(x) \neq 0$ (because the covering is locally finite and because of (3.2.6)) so that this sum makes sense.

3.3 Vectorbundles

3.3.1 Definition (real vector bundles). An n-dimensional real vector bundle over a topological space X is a topological space E together with a continuous map $\pi : E \to X$ (called the projection on X) such that

For each $x \in X$, $\pi^{-1}(x)$ is (equipped with a structure of) a real n-dimensional vector space (3.3.2)

For every $x \in X$ there is an open neighborhood U of x such that $\pi^{-1}(U)$ is isomorphic to $U \times \mathbb{R}^n$, (3.3.3)

where with this last phrase we mean that there is a homeomorphism $\phi : \pi^{-1}(U) \to U \times \mathbb{R}^n$ such that the following diagram commutes

$$\begin{array}{ccc} \pi^{-1}(U) & \xrightarrow{\phi} & U \times \mathbb{R}^n \\ \downarrow \pi & & \downarrow p_1 \\ U & \xrightarrow{\text{id}} & U \end{array}$$

where p_1 is the projection onto the first factor, and such that moreover $\phi : \pi^{-1}(x) \to \{x\} \times \mathbb{R}^n$ is an isomorphism of vector $x \in U$, (where, of course, $\{x\} \times \mathbb{R}^n$ is given the vectorspace structure arising from identifying $\{x\} \times \mathbb{R}^n$ with $\mathbb{R}^n$ in the obvious way).

The vectorspace $\pi^{-1}(x) \subset E$ is called the *fibre* of the vector bundle over x and is often denoted E_x.

3.3.4 Example (trivial bundle). $E = X \times \mathbb{R}^n \overset{\pi}{\to} X$, where π is projection on the first vector.

3.3.5 Example (Tangent bundle of S^2). Consider $S^2 = \{(x_1,x_2,x_3) | x_1^2+x_2^2+x_3^2 = 1\}$ and consider in $S^2 \times \mathbb{R}^3$ the subspace E defined by

$$E = \{(x,v) \in S^2 \times \mathbb{R}^3 | x_1v_1 + x_2v_2 + x_3v_3 = 0\} \qquad (3.3.6)$$

and define $\pi : E \to S^2$ by $(x,v) \mapsto x$. For each fixed $x \in E$ the set $\pi^{-1}(x) = E_x$ consists of all v satisfying the equation $x_1v_1 + x_2v_2 + x_3v_3 = 0$. Now give E_x the vectorspace structure of this subspace of $\mathbb{R}^3$. We check that property (3.3.3) holds. Let $x \in S^2$, then at least one of the x_i is $\neq 0$, say, x_1. Let $U = \{x \in S^2 | x_1 \neq 0\}$. Now define $\phi : U \times \mathbb{R}^2 \to \pi^{-1}(X)$ by $(x,(w_1,w_2)) \mapsto (x,(-x_1(x_2,w_1 + x_3w_2),w_1w_2))$. This ϕ is an isomorphism as required in (3.3.3).

3.3.7 Homomorphisms of vector bundles. Let $\pi : E \to X$, $\pi' : E' \to X$ be two vector bundles over X. A *homomorphism of vector bundles* is a continuous map $\phi : E \to E'$ such that the following diagram commutes

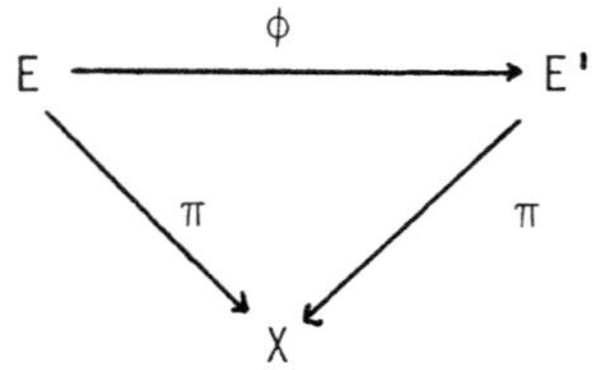

and such that the induced map $\phi_x : E_x \to E'_x$ are homomorphisms of vector spaces. The homomorphism ϕ is called an *isomorphism* if the maps $E_x \to E'_x$ are all isomorphisms.

Thus, for example, the map ϕ in (3.3.3) above is an isomorphism of the vector bundle $\pi : \pi^{-1}(U) \to U$ with the bundle $p_1 : U \times \mathbb{R}^n \to U$. A vector bundle which is isomorphic to one as in example 3.3.4 is called *trivial*.

3.3.8 Constructing vector bundles 1: local pieces and gluing data. Property (3.3.3) shows that every vector bundle can be obtained (up to isomorphism) by gluing trivial bundles together. In detail this goes as follows. Let X be a topological space and $\{U_i\}$, $i \in I$ an open covering of X. Suppose we have for each $i,j \in I$ a continuous map

$$\phi_{ij} : U_i \cap U_j \to G\ell_n(\mathbb{R}) \tag{3.3.9}$$

where $G\ell_n(\mathbb{R})$ is the (Lie) group of all invertible real $n \times n$ matrices.

We now require the ϕ_{ij} to be compatible in the following sense

$$\phi_{ii}(x) = I_n, \text{ the } n \times n \text{ unit matrix for all } n \in U_i$$
$$\phi_{jk}(x)\phi_{ij}(x) = \phi_{ik}(x) \quad \text{for all } x \in U_i \cap U_j \cap U_k \tag{3.3.10}$$

From these data we can construct a vector bundle E over X as follows. Take the disjoint union $\cup ! U_i \times \mathbb{R}^n$. Now define an equivalence relation $\sim$ as follows. The element $(x,v) \in U_i \times \mathbb{R}^n$ is equivalent to $(y,w) \in U_j \times \mathbb{R}^n$ if $x = y$ in X and $\phi_{ij}(x)v = w$. Let $E = \cup ! U_i \times \mathbb{R}^n/\sim$ and let π be induced by $(x,v) \to x$. The local trivialization maps required in (3.3.3) are given by $U_i \times \mathbb{R}^n \subset \cup ! U_i \times \mathbb{R}^n \to E$, and these also define the vectorspace structures on the fibres.

3.3.12 Example. Consider $\mathbb{P}^1(\mathbb{R})$ as the set of all lines through zero in $\mathbb{R}^2$, i.e. as the set of all ratios $(x_0:x_1)$, $x_0,x_1 \in \mathbb{R}$ $(x_0,x_1) \neq (0,0)$. Let $U_0 = \{x \in \mathbb{P}^1(\mathbb{R}) | x_0 \neq 0\}$, $U_1 = \{x \in \mathbb{P}^1(\mathbb{R}) | x_1 \neq 0\}$. Define $\phi_{01} : U_0 \cap U_1 \to G\ell_1(\mathbb{R})$ by $\phi(x_0:x_1) = x_1^{-1}x_0^1$. Set $\phi_{10} = \phi_{01}^{-1}$ and the compatibility conditions (3.3.11) hold. Let E be the resulting vector bundle. We claim that E is nontrivial. Indeed suppose E were trivial, then there would be an isomorphism $\phi: E \to \mathbb{P}^1(\mathbb{R}) \times \mathbb{R}$ compatible with the projections and hence there would be a map $s : \mathbb{P}^1(\mathbb{R}) \to E$ defined by $s(x) = \phi^{-1}(x,1)$ which satisfies

$$\pi \cdot s = id \tag{3.3.13}$$

and which is moreover such that $s(x) \neq 0 \in E_x$ for all x. Now $U_0 = \{(1:x_1) | x_1 \in \mathbb{R}\}$, $U_1 = \{(x_0:1) | x_0 \in \mathbb{R}\}$. From the construction of E we know that a map s satisfying (3.3.13) is given by two functions $f_1 : x_1 \to f_2(x_1)$, $f_0 : x_0 \to f_0(x_0)$ such that moreover $x_1^{-1} f_1(x_1) = f_0(x_1^{-1})$ for $x_1 \neq 0$. The requirement $s(x) \neq 0$ $\forall x$ means that $f_i(x_i) \neq 0$ $\forall x_i$. Hence by continuity $f_1(x_1)$ has the same sign for all x_1 and $f_0(x_0)$ has the same sign for all x_0. This, however, is incompatible with $x_1^{-1} f_1(x_1) = f_0(x_1^{-1})$.

A picture of this bundle is the so-called Möbius band

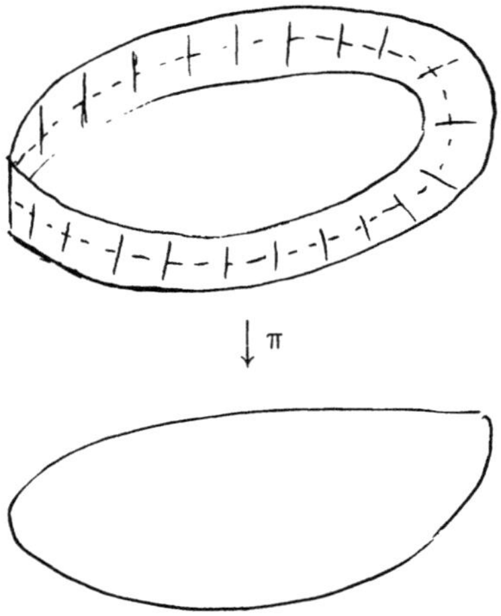

where π is the projection on the central circle. (The Möbius band is obtained by taking a rectangular strip of paper twisting it around once and gluing the ends together (as indicated below).

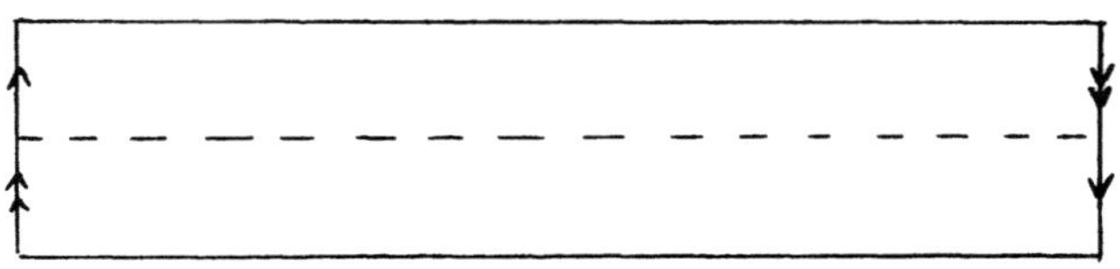

3.3.14 Linear constructions. Linear algebra or more precisely the category of finite dimensional vector spaces has many constructions which assign a new vector space to a set of one or more old vector spaces. Such a functor T is called continuous if the associated map $T : \mathrm{Hom}(V,W) \to \mathrm{Hom}(T(V),T(W))$ is continuous, where for simplicity we have taken a covariant functor in one variable. These constructions extend to constructions for vector bundles by simply performing the construction pointwise for every fibre. Thus given two vector bundles E,F over X one has e.g. the new vector bundles

$E \oplus F$, the direct sum of E and F

$E \otimes F$, the tensor product of E and F

$\mathrm{Hom}(E,F)$, the bundle over X where fibre over x is $\mathrm{Hom}(E_x,F_x)$

E^*, the dual bundle over X whose fibre over x is $\mathrm{Hom}(E_x,\mathbb{R})$

$\lambda^i(E)$, the i-th exterior power of X

A similar remark holds with respect to the natural isomorphisms of linear algebra. So one has e.g. $\mathrm{Hom}(E,F) \simeq E^* \otimes F$.

3.3.15 Sections. Let E be vector bundle over X. A continuous *section* of E is a continuous map $s : X \to E$ such that $\pi \cdot s = id_X$. The set of sections forms a vector space (pointwise addition and scalar multiplication) which is denoted ΓE or $\Gamma(E;X)$.

In example 3.3.12 we showed that for every section s of the Möbius band bundle there is an $x \in \mathbb{P}^1(\mathbb{R})$ such that $s(x) = 0$ thus proving that this bundle is nontrivial. (A trivial bundle clearly has sections which are everywhere nonzero. Exercise: Let E be an n-dimensional vector bundle over X. Suppose that there are n continuous sections $s_1,\ldots,s_n$ such that $s_1(x),\ldots,s_n(x)$ are linearly independent vectors in E_x for all $x \in X$. Prove that E is trivial.)

It is worth noting that $\Gamma\mathrm{Hom}(E,F)$ is the vector space of vector bundle homomorphisms $E \to F$ (cf. 3.3.14 and 3.3.7; Exercise: Prove this.)

3.3.16 Example. Tangent bundle of a manifold. Let M be an m-dimensional differentiable manifold. Let (U_i,ϕ_i), $i \in I$ be a collection of charts such that (3.1.4),(3.1.5) hold. We now construct a bundle over M by the local pieces and patching data descriptions of 3.3 above. To this end define

$$\phi_{ij} : U_i \cap U_j \to G\ell_n(\mathbb{R})$$

by the formula

$$\phi_{ij}(x) = \mathcal{J}(\phi_j \cdot \phi_i^{-1})(\phi_i(x))$$

where the symbol on the right is the Jacobian matrix of the diffeomorphism $\phi_j \cdot \phi_i^{-1}$ evaluated at $\phi_i(x)$. Note that the compatibility condition (3.3.11) follows from the chain rule.

The fibre of this bundle over $x \in M$ is called the tangent space of M and x and is denoted T_xM.

The bundle itself is denoted $TM \to M$, or simply TM. We can view the whole bundle $TM \to M$ as obtained by a local pieces and gluing data procedure as follows.

Consider the open pieces $\phi(U_i)$, $i \in I$ (where the U_i are as above). Now consider the pieces

$$\phi_i(U_i) \times \mathbb{R}^n \ , \quad i \in I$$

and we write an element of this set as a 2n-tuple

$$(x_1,\ldots,x_n,a_1,\ldots,a_n)^T$$

The total space TM of the tangent bundle of M is now obtained by gluing together the $\phi(U_i) \times \mathbb{R}^n$ by means of the isomorphisms

$$\phi_{ij} : \phi_i(U_i \cap U_j) \times \mathbb{R}^n \to \phi_j(U_i \cap U_j) \times \mathbb{R}^n$$

$$(x,a) \to ((\phi_i \cdot \phi_j^{-1})(x), \quad \mathcal{J}(\phi_j \cdot \phi_i^{-1})(x)(a))$$

These identifications are compatible with the projections

$$\phi_i(U_i) \times \mathbb{R}^n \to \phi_i(U_i) \ , \quad (x,a) \to x \ .$$

and thus the whole bundle $TM \to M$ is described.

Note that these considerations make it clear that TM is itself a differentiable manifold and that $\pi : TM \to M$ is a differentiable map. We can thus speak of differentiable sections.

3.3.17 Vector fields. Let $TM \to M$ be the tangent bundle of a differentiable manifold M. A differentiable section (cf. 3.3.15 above) of this bundle is called a vector field. In terms of local pieces and gluing data such a section thus is given by differentiable functions

$$a(i) : \phi_i(U_i) \to \mathbb{R}^n$$

(the local pieces of the section are then given by $\phi_i(U_i) \to \phi_i(U_i) \times \mathbb{R}^n$, $x \to (x, a(i)(x))$. Then functions must then satisfy the compatibility condition

$$\mathcal{J}(\phi_j \cdot \phi_i^{-1})(x)(a(i)(x)) = a(j)(\phi_j \cdot \phi_i^{-1}(x)) \; . \qquad (3.3.18)$$

3.3.20 Derivations. Let A be an algebra over a field K. (Take $K = \mathbb{R}$ or $\mathbb{C}$ if desired.) A *derivation* of A is a K-linear map $D : A \to A$, such that $D(fg) = f(Dg) + (Df)g$.

3.3.21 Vector fields as derivations. Let M be a differentiable manifold and let S(M) be the ring of differentiable functions on M. Let s be a differentiable section of the tangent bundle $TM \to M$. We claim that s defines a derivation of S(M). Indeed let s be given by the function $s(i) : \phi_i(U_i) \to \mathbb{R}^n$. A differentiable function on M can be viewed as a collection of functions $f(i) : \phi(U_i) \to \mathbb{R}$, $f(i) = f \cdot \phi_i^{-1}$, satisfying the compatibility condition

$$f(j)(\phi_j \cdot \phi_i^{-1}(x)) = f_i(x), \quad x \in \phi_i(U_i \cap U_j) \qquad (3.3.22)$$

Now define the collection of functions $g(i) : \phi(U_i) \to \mathbb{R}$ by the formula

$$g(i)(x) = \sum_{k=1}^{n} s(i)(x)_k \frac{\partial f(i)}{\partial x_k}(x) = \frac{\partial f(i)}{\partial x} s(i)(x)$$

where $s(i)(x)_k$ is the k-th component of column vector $s(i)(x)$, and $\frac{\partial f(i)}{\partial x}$ is the row vector $\left(\frac{\partial f(i)}{\partial x_i}(x), \ldots, \frac{\partial f(i)}{\partial x_n}(x)\right)$. We now claim that the $g(i)$ satisfy the compatibility condition (3.3.22). Indeed from (3.3.22) we find by the chain rule that (writing y for $(\phi_j \cdot \phi_i^{-1})(x))$

$$\frac{\partial f(j)}{\partial y}(y) = \frac{\partial f(i)}{\partial x}(x) \; \mathcal{J}(\phi_i \cdot \phi_j^{-1})(y)$$

Therefore

$$g(j)(y) = \frac{\partial f(j)}{\partial y} s(j)(y)$$

$$= \frac{\partial f(i)}{\partial x}(x)\ \mathcal{J}(\phi_i \cdot \phi_j^{-1})(y)\ \mathcal{J}(\phi_j \cdot \phi_i^{-1})(x)\ s(i)(x)$$

$$= g(i)(x)$$

because $\mathcal{J}(\phi_i \cdot \phi_j^{-1})(y)\ \mathcal{J}(\phi_j \cdot \phi_i^{-1})(x) = I_n$ and the compatibility relation (3.3.18). Inversely every derivation defines a vector field.

3.3.24 <u>The Lie bracket.</u> Let D_1, D_2 be two derivations of an algebra over $\mathbb{R}$ (or any other field). Then (as easily checked)

$$[D_1, D_2] = D_1 D_2 - D_2 D_1$$

is again a derivation. Now let s_1, s_2 be vector fields on a differentiable manifold M, with corresponding derivations D_1, D_2. Then the vector field corresponding to the derivation $[D_1, D_2]$ is denoted by $[s_1, s_2]$ is called the *Lie bracket* of the vector fields s_1 and s_2. The vector field $[s_1, s_2]$ can be calculated in terms of local pieces as follows. Let s_1 and s_2 be given locally by the functions $s_2(i), s_1(i) : \phi_i(U_i) \to \mathbb{R}^n$. Then $[s_1, s_2]$ is given by the functions

$$a(i) : {}_i(U_i) \quad \mathbb{R}^n,$$

$$a(i)(x) = (\mathcal{J}s_1(x)(x)s_2(i)(x)) - (\mathcal{J}s_2(i)(x)s_1(i)(x))$$

which in slightly less precise notation can be written

$$\frac{\partial s_1(i)}{\partial x} s_2(i) - \frac{\partial s_2(i)}{\partial x} s_1(i)$$

3.3.26 <u>Exercise.</u> Check that the $a(i)$ of (3.3.25) satisfy the compatibility relation (3.3.18) and that the derivation operator defined by these $a(i)$ according to (3.3.23) is indeed the derivation $D_1 D_2 - D_2 D_1$.

3.3.27 <u>Constructing homomorphisms by local pieces and patching data.</u> Let E and F be two vector bundles over a topological space X, both given in terms of local pieces and gluing data. Then often, a homomorphism $E \to F$ is easiest described in terms of local pieces too. Suppose for simplicity that the local pieces describing E and F are with respect to the same covering U_i.

(This can always be assured by taking a common refinement of the coverings defining E and F.) Let $\mathbb{R}^{m\times n}$ be the space of $m \times n$ matrices, let E and F be described by

$$\phi_{ij} : U_i \cap U_j \to G\ell_n(\mathbb{R}),\ \psi_{ij} : U_i \cap U_j \to G\ell_m(\mathbb{R})$$

then a homomorphism $\alpha : E \to F$ is unique described by a family of maps

$$\alpha_i : U_i \to \mathbb{R}^{m\times n}$$

such that for all $i,j \in I$ and $x \in U_i \cap U_j$

$$\psi_{ij}(x)\alpha_i(x) = \alpha_j(x)\phi_{ij}(x) \tag{3.3.28}$$

3.3.29 Metrics. If V is a vector space, let $Q(v)$ be the vector space of all quadratic forms on V. This is an example of a continuous functor in the sense of 3.3.14 above. Thus given a vector bundle E over X there is an associated vector bundle $Q(E)$ whose fibre over x is the space $Q(E_x)$ of all quadratic forms on E_x. A *metric* on E is now a section s of $Q(E)$ such that $s(x)$ is positive definite for all $x \in X$.

In more down to earth terms this means the following. Let E be built out of trivial pieces with respect to the covering U_i. Let $\phi_{ij} : U_i \cap U_j \to G\ell_n(\mathbb{R})$ be the gluing maps. Then a metric on E consists of continuous maps

$$s_i : U_i \to P_n$$

where P_n is the space of all positive definite quadratic forms on $\mathbb{R}^n$ such that

$$\phi_{ij}^T(x)s_j(x)\phi_{ij}(x) = s_i(x) \tag{3.3.30}$$

for all $x \in U_i \cap U_j$, where the upper T denotes "transpose."

It remains to show that every vector bundle over suitable, say, paracompact or compact, spaces admits a metric. This goes as follows. Let the covering $\{U_i\}$ be locally finite and let $\{\psi_i\}$ be a partition of unity with respect to $\{U_i\}$. For each $i \in I$ choose some positive definite form Q_i and define

$$s_j : U_j \to P_n,\quad s_j(x) = \sum_k (\phi_{kj}(x)^T)^{-1}\psi_k(x)Q_k\phi_{kj}(x)^{-1}$$

(Note that the expression on the right hand side as a converse linear combination of positive definite quadratic forms is positive definite). These mappings satisfy the compatibility condition (3.3.20) and hence define a metric.

3.3.31 Subbundles and quotient bundles are direct summands. Let $E \overset{\pi}{\to} X$ be a vector bundle. A *subbundle* is a subset $F \subset E$ such that the restriction of π to F makes F a vector bundle and such that $F \hookrightarrow E$ is a homomorphism of vector bundles. If $F \hookrightarrow E$ is a subbundle we can consider the union $\bigcup_x E_x/F_x$ with the induced topology. There is a natural projection onto X defined by $E_x/F_x \ni v \mapsto x$ and using the obvious quotient vector space structures on E_x/F_x the result is a vector bundle over X which is called a *quotient bundle* and is denoted E/F.

Now let $F \subset E$ be a subbundle. Let s be a metric on E. For each $x \in X$ let $G_x = \{v \in E_x | \langle v, F_x\rangle_x = 0\}$ where $\langle\ ,\ \rangle_x$ denotes the inner product on E_x defined by $s(x)$. Then $\cup G_n$ is a subvector bundle of E and $E = F \oplus G$ so that every subbundle is a direct summand. Analogously if $\alpha : E \to F$ is a homomorphism of vector bundles such that $E_x \to F_x$ is surjective for all x, then there exists a vector bundle homomorphism $\beta : F \to E$ such that $\alpha \cdot \beta = id_F$.

3.3.32 Finite generation of vector bundles. Let $\pi : E \to X$ be an m-dimensional vector bundle over a compact space X. Let $\{U_i | i = (1,\ldots,n)\}$ be a finite open covering of X such that E is trivial over all U_i. For each i let $s_{ij} : U_i \to E|_{U_i}$ be m sections of $\pi^{-1}(U_i) \to U_i$ such that for all $x \in U_i$ the vectors $s_{ij}(x)$, $u = 1,\ldots,m$ form a basis for $\pi^{-1}(x) = E_x$. Now let $\{\phi_i\}$ be a partition of unity with respect to U_i. Then we claim that the maps

$$\phi_i s_{ij}, \quad i \in I, \quad j = 1,\ldots,m$$

(defined by $\phi_i s_{ij}(x) = \phi_i(x) s_{ij}(x)$ if $x \in U_i$, $\phi_i s_{ij}(x) = 0$ if $x \notin U_i$) are continuous sections and are such that for each $x \in X$ the $(\phi_i s_{ij})(x)$ generate E_x. Indeed for each $x \in X$ there is an i_0 such that $\phi_{i_0}(x) \neq 0$ and then the $\phi_{i_0}(x)\phi_{i_0 j}(x)$, $u = 1,\ldots,m$ generate E_x.

3.3.33 Corollary. *Every vector bundle over a compact space is quotient of a finite dimensional trivial bundle.*

Indeed we have seen above that if $E \to X$ is a vector bundle, there exists a finite number of sections $s_1,\ldots,s_r$ such that the $s_1(x),\ldots,s_r(x)$ generate E_x for all x. Now define $\alpha : X \times \mathbb{R}^r \to E$ by $\alpha(x_1(a_1,\ldots,a_r)) = \Sigma a_i s_i(x)$. Then $\alpha : X \times \mathbb{R}^r \to E$ is a homomorphism of vector bundles (exercise) and surjective making E a quotient of $X \times \mathbb{R}^r$ (and hence by 3.3.21) also a direct summand.

3.3.34 Differentiable bundles. A differentiable vector bundle is a vector bundle $\pi : E \to X$ such that E,X are differentiable manifolds and π is a differentiable mapping. Analytic bundles are defined similarly. An example of a differentiable bundle is the tangent bundle $TM \to M$ of a differentiable manifold.

3.3.35 Vector bundles and projective modules. Let M be a differentiable manifold. Then $S(M)$ denotes the ring of differentiable functions on M (pointwise multiplication and addition). Now let $E \to M$ be a differentiable vector bundle over M. Let $s : M \to E$ be a differentiable section of E and $f \in S(M)$. Then for all $m \in M$, $f(m)s(m) \in E_m$ is well defined and this makes the vector space of all differentiable sections a module over the ring $S(M)$. By 3.3.33 and 3.3.31 (or rather their differentiable analogues (which also hold) then modules are direct summand of free modules (the module of sections of $M \times \mathbb{R}^r \to M$ is, of course, $S(M)^r$) and hence then modules are projective modules. Thus giving us a correspondence between differentiable vector bundles over M and finitely generated projective modules over $S(M)$.

Similarly vector bundles over a suitable topological space X correspond to finitely generated projective modules over the ring of continuous functions on X and in algebraic geometry algebriac vector bundles over an affine variety $\mathrm{Spec}(R)$ correspond to finitely generated projective modules over R.

3.3.36 The pullback construction. (Constructing vector bundles 2). Let $\pi : E \to X$ be a vector bundle and let $f : Y \to X$ be a continuous map. Consider

$$E' = \{(e,y) \in Y \times E \mid \pi(e) = f(y)\}$$

There is a natural projection $\pi' : E' \to Y$ defined by $\pi(e,y) = y$. for a fixed $y \in Y$ we have

$$(\pi')^{-1}(y) = \{(e,y) \mid \pi(e) = f(y)\} = E_{f(y)} \times \{y\}$$

which we give the vector space structure of $E_{f(y)}$. Then $\pi' : E' \to Y$ is a vector bundle over Y which is called the pull-back of E along f and which is denoted $f^! E$.

In words $f^! E$ is the vector bundle over Y whose fibre over $y \in Y$ is the fibre of E over $f(y)$.

If E is obtained by patching together local trivial pieces over U_i, $i \in I$ by means of gluing data

$$\phi_{ij} : U_i \cap U_j \to G\ell_m(\mathbb{R})$$

then $f^! E$ is obtained by patching together trivial pieces over the open subsets $f^{-1}(U_i)$, $i \in I$ by means of the gluing data

$$f^{-1}(U_i) \cap f^{-1}(U_j) \xrightarrow{f} U_i \cap U_j \xrightarrow{\phi_{ij}} G\ell_m(\mathbb{R}) \; .$$

From both descriptions it is obvious that if $g : Z \to Y$ is another continuous map then

$$(f \cdot g)^! E = g^! (f^! E) \qquad (3.3.37)$$

3.3.38 Bundle morphisms covering a continuous map. Let $E \to M$ and $F \to N$ be two vector bundles and let $f : M \to N$ be a continuous map. A bundle morphism covering f is a continuous map $\tilde{f} : E \to F$ such that the following diagram is commutative

$$\begin{array}{ccc} E & \xrightarrow{\tilde{f}} & F \\ \downarrow & & \downarrow \\ M & \xrightarrow{f} & N \end{array}$$

and such that the induced maps $\tilde{f}_x : E_x \to F_{f(x)}$ are homomorphisms of vector spaces. There is an obvious 1-1 correspondence between bundle morphisms $E \to F$ covering f and homomorphisms of vector bundles over M from E to $f^! F$. (Exercise)

By now it should be obvious how to describe a bundle morphism covering a continuous map in terms of local pieces and gluing data. (Exercise)

3.3.39 Example (Jacobians). Let M and N be differentiable manifolds of dimension m and n, $f : M \to N$ a differentiable map. Let (U_i, ϕ_i) and (V_i, ϕ_i) be coordinate charts for M and N and suppose that $f(U_i) \subset V_i$. Let $f(i) = \psi_i \cdot f \cdot \phi_i^{-1}$:

$\phi_i(U_i) \to \psi_i(V_i)$. Recall (cf. 3.3.16) that the tangent bundles TM and TN can be obtained by gluing together the $\phi_i(U_i) \times \mathbb{R}^m$ and $\psi_i(V_i) \times \mathbb{R}^n$. Define

$$df(i) : \phi_i(U_i) \times \mathbb{R}^m \to \psi_i(V_i) \times \mathbb{R}^n$$

by the formula

$$(x,v) \to (f(i)(x), J\phi(i)(x)(v))$$

Note that this is compatible with the gluing data for TM and TN so that the $df(i)$ combine to define a differentiable map

$$df : TM \to TN$$

which is (obviously) a morphism of bundles covering f. The induced maps

$$df_x : T_xM \to T_{f(x)}N$$

is called the differential of f at $x \in M$. If M and N are themselves open subsets of $\mathbb{R}^m$ and $\mathbb{R}^n$ then $df_x : \mathbb{R}^m \to \mathbb{R}^n$ is given by the Jacobian matrix of f at x.

3.3.40 Submanifolds (2). Let M,N be differentiable manifolds. A differentiable mapping $f : M \to N$ is *regular* at x if df_x has rank $\max(m,n)$. The manifold M is a submanifold of N if $M \subset N$ set theoretically, $\dim M \le \dim N$ and the inclusion $M \to N$ is a regular differentiable map.

3.4 On Homotopy

3.4.1 Definitions. Two continuous maps $f,g : X \to Y$ are called *homotopic* if there exists a continuous map $F : X \times [0,1] \to Y$ such that $F(x,0) = f(x)$, $F(x,1) = g(x)$ for all $x \in X$.

For $t \in [0,1]$, let $F_t(x) = F(x,t)$. Then the intuitive picture is that f can be continuously deformed into g via the F_t, $0 \le t \le 1$, ($F_0 = f_1$ $F_1 = g$).

3.4.2 Theorem. *Let* $\pi : E \to X$ *be a vector bundle and let* $f,g : Y \to X$ *be two homotopic continuous maps. Then the pullback bundles* $f^!E$ *and* $g^!E$ *are isomorphic over* Y.

3.5 Grassmannians and Classifying Vector Bundles

3.5.1 Grassmann manifolds. Consider the set $G_n(\mathbb{R}^{n+k})$ of n-dimensional subvector spaces of $\mathbb{R}^{n+k}$. Let $\mathbb{R}^{n\times(n+k)}_{reg}$ be the set of all $n\times(n+k)$ matrices of rank n. There is a natural map $\mathbb{R}^{n\times(n+k)}_{reg} \to G_{nl}(\mathbb{R}^{n+k})$ which assigns to an $n\times(n+k)$ matrix A of rank n the n-dimensional subspace of $\mathbb{R}^{n+k}$ spanned by the rows of A. We give $G_n(\mathbb{R}^{n+k})$ the quotient topology. There is a natural differentiable manifold structure on $G_{n,n+k}$ which is described in detail in section 4 of this Introduction (in terms of local pieces and gluing data).

There is a natural embedding $\varepsilon_k : G_n(\mathbb{R}^{n+k}) \hookrightarrow G_n(\mathbb{R}^{n+k+1})$ induced by the map $\mathbb{R}^{n\times(n+k)}_{reg} \to \mathbb{R}^{n\times(n+k+1)}_{reg}$ which adds a column of zeros to an $n\times(n+k)$ matrix A of rank n. We let G_n denote the inductive limit space $\varinjlim_k G_n(\mathbb{R}^{n+k})$. The space G_n can perfectly well be seen as the space of all n-dimensional vector subspaces of $\mathbb{R}^\infty = \{(x_1,x_2,x_3,\dots)\,|\,x_i \in \mathbb{R},$ all but finitely many n_i are zero$\}$.

3.5.2 The "universal" bundle ξ_n. Define

$$\xi_n = \{(x,v) \in G_n(\mathbb{R}^{n+k}) \times \mathbb{R}^{n+k}\,|\,v \in x\} \qquad (3.5.3)$$

There is a natural projection $\xi_n \to G_n(\mathbb{R}^{n+k})$ defined by $(x,v) \to x$ and it is easily seen that this makes ξ_n into a vector bundle whose fibre over $x \in G_n(\mathbb{R}^{n+k})$ "is" the vector space x. A description of this vector bundle in terms of local pieces and gluing data can be found in section 3.4.5 of Professor Hazewinkel's lectures in this volume.

3.5.4 Exercise (easy). Let $\varepsilon_k : G_n(\mathbb{R}^{n+k}) \to G_n(\mathbb{R}^{n+k+1})$ be the embedding described above in 3.5.1 and let ξ_n and ξ'_n be the universal bundles as described above in 3.5.2 over $G_n(\mathbb{R}^{n+k})$ and $G_n(\mathbb{R}^{n+k+1})$ respectively. Then $\varepsilon_k^!\xi'_n = \xi_n$. (This also justifies the notation used).

3.5.5 Classifying vector bundles. Let $\alpha : X \to G_n(\mathbb{R}^{n+k})$ be a continuous map. Then this gives a vector bundle $\alpha^!\xi^n$ over X and homotopic maps give rise to isomorphic vector bundles.

Moreover if k is big enough (and X compact) all vector bundles over X are (up to isomorphism) obtained in this way. The construction which assigns a map into some Grassmann manifold to a bundle over X goes as follows. Let $E \to X$ be a vector bundle. Then there is an $r \in \mathbb{N}$ and a surjective homomorphism of vector bundles $\phi : X \times \mathbb{R}^r \to E$ (cf. 3.3.33 above). Now define $f(x)$ to be the n-dimensional subspace of $\mathbb{R}^r$ consisting of all vectors which are orthogonal to the kernel of $\phi_x : \mathbb{R}^r \to E_x$.

These remarks form the bare bones of the classifying theorem for vector bundles which states that over suitable spaces X

$$[X, G_n] = B_n(X) \tag{3.5.6}$$

where $[X, G_n]$ is the set of homotopy classes of continuous maps $X \to G_n = G_n(\mathbb{R}^\infty)$ and where $B_n(X)$ is the set of isomorphism classes of n-dimensional vector bundles over X. Roughly one can say that if one knows the n-dimensional universal vector bundle ξ_n over $G_n(\mathbb{R}^{n+k})$, k large, that one knows all n-dimensional vector bundles.

4. VARIETIES, VECTOR BUNDLES, GRASSMANNIANS AND INTERSECTION THEORY

In this chapter we will define some of the basic ideas and objects needed for the application of algebraic geometry in systems theory. The material parallels the development of differential topology developed in section 3. We will describe the concepts of affine space, affine varieties, projective spaces and projective varieties. The Grassmannian manifolds will be developed with some care and the various representations that have proven so useful in linear systems theory will be given.

4.1 Affine Spaces and Affine Algebraic Varieties

Let k be an algebraically closed field (for our purposes we can almost always assume that k is the field of complex numbers $\mathbb{C}$). Let k^n denote the point set of n-tuples. We say that a subset X of k^n is *closed* if there are finitely many polynomials $g_1(x_1,\ldots,x_n),\ldots g_m(x_1,\ldots,x_m)$ such that $X = \{x \in k^n : g_1(x) = \ldots = g_m(x) = 0\}$. The set of all closed sets defines a topology on k^n, called the Zariski topology. (The fact that this is a topology is nontrivial--a consequence of the Hilbert Basis Theorem). A closed subset $X \subset k^n$ is given the induced topology and is called an affine algebraic set.

Let $g(x_1,\ldots,x_n)$ be a polynomial and define a set $X_g = \{x \in X : g(x) \neq 0\}$. Clearly the sets X_g are open and form a basis for the topology of X. A regular function on X_g is a function f with domain X_g and range k such that f can be written as $h(x)/g^m(x)$ for some polynomial h and all x. So f is represented by a rational function having no poles on X_g. Let U be an arbitrary open set in X and f a map from U to k. Since U is open, it's the union of X_g's and we say that f is regular if the restriction of f to each X_g is regular. This sequence appears over and over in geometry. We define something simple, then build an object from the simple things and extend the definition.

A closed algebraic set $X \subset k^n$ along with its regular functions on open sets is an affine algebraic variety. An open subset of X together with the ring of regular functions is called a quasi-affine algebraic variety. In the special case that $X = k^n$ the affine variety is denoted by $\mathbb{A}^n$--the affine space of dimension n.

Let $U \subseteq X \subseteq \mathbb{A}^n$ and $V \subseteq Y \subseteq \mathbb{A}^m$ be open subsets of affine varieties X and Y. A map g from U to V is a morphism from U to V if there exist m regular functions $g_1,\ldots,g_m$ defined on U such that $g(x) = (g(x_1),\ldots,g(x_n))$.

In particular, the "coordinate ring" R_X of functions regular on all of X may be thought of in the following seemingly coordinate-dependent way. If $X \subset \mathbb{A}^N$ is an affine algebraic set, then R_X consists of the ring of functions which are restrictions to X of polynomials on $\mathbb{A}^n$. The point is that the ring R_X is intrinsic, i.e., independent of the particular presentation $X \subset \mathbb{A}^N$. Thus, R_X contains not only X as an abstract object (Hilbert Nullstellensaty) but also all possible embeddings of X in affine space. For $X = \mathbb{A}^n$, $R_X = k[x_1,\ldots,x_n]$ which is Noetherian, since k is, by the Hilbert Basis Theorem (2.2.2). More generally, $X \subset \mathbb{A}^n$ gives rise to an algebra homomorphism, restriction,

$$\rho_X : k[x_1,\ldots,x_n] \to R_X \tag{4.1.1}$$

which exhibits R_X as a quotient of $k[x_1,\ldots,x_n]$. Therefore, by lemma 2.1.5, R_X is Noetherian. In this light, it is interesting to examine the geometric content of the ascending chain condition. For affine space $\mathbb{A}^n$, any subvariety $X \subset \mathbb{A}^n$, gives rise to an ideal I_X, viz. the kernel of ρ_X

$$\ker \rho_X = \{f \in k[x_1,\ldots,x_n] \mid f|_X = 0\} \tag{4.1.2}$$

By the Hilbert Basis Theorem, $I_X = (f_1,\ldots,f_m)$ and one sees that X is in fact defined by the equations

$$f_1(x) = \ldots = f_m(x) = 0 \tag{4.1.3}$$

Moreover, this correspondence reverses inclusion; that is to say, if $X \subset Y$ then $I_Y \subset I_X$. Therefore, the ascending chain condition on ideals implies the descending chain condition on subvarieties of $\mathbb{A}^n$. This is true, by similar reasoning, for any affine variety Z.

4.1.4 <u>Theorem</u>. *If* Z *is an affine algebraic variety, then every descending chain* $Z_1 \supset Z_2 \supset \ldots \supset Z_m \supset \ldots$ *of subvarieties of* Z *terminates.*

In one considers the special case in which Z_i is obtained from Z_{i-1} by imposing an additional algebraic constraint

$$f_i(z) = 0 ,$$

then (4.1.4) asserts that no Z can satisfy infinitely many independent constraints. The key to formalizing this notion of independence lies in the concept of dimension.

4.2 Projective space, projective varieties, and quasi-projective varieties

Again let k be an algebraically closed field. Define an equivalence relation on $k^{n+1}\setminus\{(0,\ldots,0)\}$ defining $x \sim y$ iff there is a $\lambda \in k$ such that $\lambda x = y$. Denote the point set of equivalence classes by $\mathbb{P}^n(k)$. Recall that a polynomial g is homogeneous if there is an integer m such that $g(\lambda x) = \lambda^m g(x)$ for all x. We say that a subset X of $\mathbb{P}^n(k)$ is closed if there is a finite set of homogeneous polynomials $g_1,\ldots,g_m$ such that $X = \{[x] \in \mathbb{P}^n(k): g_1(x) = \ldots = g_m(x) = 0\}$. Note that because of homogeneity $g(x) = 0$ implies $g(\lambda x) = 0$ and hence the definition is well founded. The set of closed sets defines a topology on X and this topology is also referred to as the Zariski Topology.

The projective spaces can be developed more prosaically, if k is $\mathbb{C}$, as a compact differentiable manifold. Let V be $\mathbb{C}^{n+1}$ considered as a vector space over $\mathbb{C}$. Let $\mathbb{P}^n(\mathbb{C})$ denote the set of one dimensional subspaces of V. We define open sets in $\mathbb{P}^n(\mathbb{C})$ as follows. Let W be a subspace of V of dimension n and let $U = \{\ell \in \mathbb{P}^n(\mathbb{C}): \ell \cap W = \{0\}\}$. We say that U is open in $\mathbb{P}^n(\mathbb{C})$ and we let $\mathbb{P}^n(\mathbb{C})$ have the topology generated by the U's. This definition coincides with the previous definition for if W is of dimension n then W is the kernel of a non-zero linear functional and hence is the zero set of a homogeneous polynomial. The other direction is more difficult.

We will see later when we discuss the Grassmannian manifolds that the U's can be identified with the affine spaces $\mathbb{C}^n$ exhibiting $\mathbb{P}^n(\mathbb{C})$ as a complex manifold. $\mathbb{P}^1(\mathbb{C})$ can be identified with Riemann sphere or with the real sphere S^2. In a later section we will develop the Grassmannians with more detail. We will also show that $\mathbb{P}^n(\mathbb{C})$ is compact as a manifold in the manifold topology. Note that the Zariski topology is a subtopology of the manifold topology.

Closed subsets of $\mathbb{P}^n(k)$ are called projective varieties and if V is an open subset of a projective variety X then we call V a quasi-projective variety.

We need to extend the definition of regular functions to projective varieties. Note that we have $n+1$ "canonical" embeddings of $\mathbb{A}^n$ into $\mathbb{P}^n(k)$. Define $j_i(x_1,\dots,x_n) = [(x_1,\dots,x_{i-1},1,x_i,\dots,x_n 0)]$. The map j_i is continuous with respect to the Zariski topology and the image of j_i is an open subset of $\mathbb{P}^n(k)$ (it coincides with the open sets defined in the Grassmannian setup).

Let X be a projective variety contained in $\mathbb{P}^n(k)$ and let U be an open subset of X. The set $j_i^{-1}(X) \subseteq \mathbb{A}^n$ is closed for each i and $j_i^{-1}(U)$ is an open subset of $j_i^{-1}(X)$. A regular function f from U to k is defined to be a map such that the composite map from $j_i^{-1}(U) \xrightarrow{j_i} U \to k$ is a regular function on $j_i^{-1}(U)$ for all i.

Morphisms between quasi-projective varieties are defined similarly. First let U be a quasi-projective variety such that $U \subseteq X \subseteq \mathbb{P}^n(k)$ and let V be a quasi-affine variety defined by $V \subseteq Y \subseteq \mathbb{A}^m$. A morphism f from U to V is a map such that there are regular functions $f_1,\dots,f_m$ from U to k such that $f(x) = (f_1(x),\dots,f_m(x)$ for all $x \in U$. Now let W be quasi-projective and defined by $W \subseteq Z \subseteq \mathbb{P}^m(k)$. A morphism f from U to W is a map from U to W with the following properties. Define W_i by $W_i = W \cap j_i(\mathbb{A}^m)$. Let $U_i = f^{-1}(W_i)$. The map f is a morphism iff for each i the induced map from $U_i \to W_i \to j_i^{-1}(W_i) \subseteq \mathbb{A}^m$ is a morphism into the quasi-affine variety $j_i^{-1}(W_i)$. One can easily show that the identity maps are morphisms and that the composition of morphisms is a morphism. Thus we have defined a category whose objects are quasi-projective varieties and whose morphisms are regular maps. Denote the category by qpSch(k).

Let $U \subseteq X \subseteq \mathbb{A}^n$ be a quasi-affine variety and let X be specified by the polynomials $g_i(x_1,\dots,x_n)$ $i = 1,\dots,m$. We must embed X as a closed subset of some $\mathbb{P}^k$. To do this we introduce homogeneous coordinates and let $\hat{g}_i(\lambda, x_1, \ldots, x_n)$ be the corresponding homogeneous polynomial. Let $\hat{X}$ be the zero set of the $\hat{g}_i$ $i = 1, \ldots, m$ in $\mathbb{P}^n$. Then j_0 embeds X as an open subset of $\hat{X}$ and hence j_0 embeds U as an open subset of $\hat{X}$. Thus each quasi-affine variety can be identified with a quasi-projective variety.

Let $U \subset X \subset \mathbb{A}^n$ and $V \subset Y \subset \mathbb{A}^m$ be quasi-affine varieties. The set $\bar{X} \times \bar{Y}$ is easily seen to be an algebraic subset of $\mathbb{A}^{n+m}$ by identifying it with the set $\{(x_1, \ldots, x_n, y_1, \ldots, y_m): g_i(y) = 0 \;\; i = 1, \ldots, m, \;\; f_i(x) = 0, \;\; i = 1, \ldots, n\}$. The reader should convince himself that the topology of $X \times Y$ is not the product topology. (Examine, for example, the Zariski closed sets in A^2 as compared to closed sets in the product topology of $A^1 \times A^1$.) The product set $U \times Y$ is open in $X \times Y$ and hence $U \times V$ is also quasi-affine.

In order to show that the product of quasi-projective varieties admits a quasi-projective structure we must work a bit harder. Let $\mathbb{P}^n(k)$ and $\mathbb{P}^m(k)$ be given and consider the point set $\mathbb{P}^n \times \mathbb{P}^m$. Let $N = (n+1)(m+1) - 1$. Let the coordinates in $\mathbb{P}^N$ be given by w_{ij}, $i = 1, \ldots, n+1$ $j = 1, \ldots, m+1$. Define $\phi: \mathbb{P}^n \times \mathbb{P}^m \to \mathbb{P}^N$ by $\phi(x,y) = (\ldots, w_{ij}, \ldots)$ where $w_{ij} = x_i y_j$. The image of ϕ is a closed subset of $\mathbb{P}^N$ given by $w_{ij} w_{k\ell} = w_{i\ell} w_{kj}$ $i,k = 1, \ldots, n+1$ $j, \ell = 1, \ldots, m+1$ and ϕ is one to one. We give $\mathbb{P}^n \times \mathbb{P}^m$ the projective variety structure of its image in $\mathbb{P}^N$. Now let $U \subset X$ and $V \subset Y$ be quasi-projective then $U \times V$ is given the structure of $\overline{\phi(U \times V)}$. A useful result about products and morphisms is the Closed Graph Theorem for algebraic varieties.

4.2.1 <u>Theorem</u>. *A function* f *from an algebraic variety* X *to an algebraic variety* Y *is a morphism iff the graph of* f *is closed in* $X \times Y$.

A topological space X is called *reducible* if X can be written as a union of closed subsets $X_1 \cup X_2$ with $X_1 \neq X$ and $X_2 \neq X$. The space X is called *irreducible* if X is not reducible. If $U \subset X$ is an open subset of the topological space X and $\bar{U} = X$ (where the bar denotes topological closure) then X

is irreducible if and only if U is irreducible (Elementary). A quasi-projective variety X is said to be irreducible if the underlying space is irreducible. Let X be an affine variety and $[X]$ the k-algebra of regular functions (pointwise addition and multiplication) on X, then one easily checks that X is irreducible iff $A(X)$ has no zero divisors. (If $f,g \in A[X]$ are not identically zero on X and $f(x)g(x) = 0$ for all $x \in X$, $X_f = \{x \in X | f(x) = 0\}$, $X_g = \{x \in X | g(x) = 0\}$ are closed subsets of X satisfying $X = X_f \cup X_g$, $X_f \neq X$, $X_g \neq X$.) Using this we see that the affine spaces $\mathbb{A}^n$ are irreducible. Then by the remarks made above we see that open subsets of $\mathbb{A}^n$ are irreducible and that $P^n(k)$ and its open subsets are irreducible.

If X is an irreducible variety and $U \subset X$ is open, then $\bar{U} = X$. (If $\bar{U}$ were not equal to X then $\bar{U} \cup (X-U) = X$ would show X to be reducible.) For irreducible varieties we have Weyl's irrelevancy principle. Let U be an open subset of an irreducible (quasi) affine variety $X \subset \mathbb{A}^n$, and suppose that $f(x_1,\ldots,x_n)$ is a polynomial over k such that $f(x) = 0$ for all $x \in U$, then $f(x) = 0$ for all $x \in X$. Indeed $f(x) = 0$ defines a closed subset Y of $\mathbb{A}^n$ and we have by hypothesis $U \subset Y$, hence $X \subset \bar{U} \subset \bar{Y} = Y$. Similarly if U is an open subset of an irreducible (quasi) projective variety $X \subset \mathbf{P}^n(k)$ and $g(x)$ is a homogeneous polynomial in $x_0,\ldots,x_n$ such that $g(x) = 0$ for all $x \in U$, then $g(x) = 0$ for all $x \in X$.

Let X be a variety and suppose X is the union of sets $\cup S_i$. We say the union is irredundant iff $S_i \subset S_j$ implies $S_i = S_j$. We have the following theorem

4.2.2 Theorem. *Every algebraic variety is the finite irredundant union of irreducible closed varieites. The decomposition is unique up to permutation.*

The proof of Theorem 4.2.2 follows the following line. Suppose $V = W_1 \cup W_2$ where W_1 and W_2 are closed varieties. If the assertion is false for V then it is false for W_1 or W_2. Applying the theorem again we produce a sequence $W_1 \supset W_3 \supset W_4 \ldots$. The sequence is infinite decreasing and hence corresponds to an infinite increasing sequence of ideals in the coordinate ring. Since the coordinate ring is Noetherian we have a contradiction and hence V can be written as a finite union of irreducible closed subvarieties. The uniqueness of the decomposition can be shown as follows: Suppose $V = \cup W_i$ and $V = \cup V_i$ then $V_j = \cup(W_i \cap V_j)$ and since V_j is irreducible $V_j = W_i \cap V_j$ for some j. On the other hand, $W_i = W_i \cap V_k$ for some k and hence $V_j = V_k = W_i$ thus there is a one to one correspondence between the W_i's and V_i's.

Let V be irreducible. Then the coordinate ring is an integral domain and we can define its field of fractions K_X. Now K_X is a vector space over k and hence has a dimension n. The number $n = \dim K_X$ is the transcendence degree of K_X. We define the *degree of* F to be the number $\dim K_X$. In section 2.3 this is discussed further. We comment here that the dimension of the tangent space at a *nonsingular point* **x** is the same as the degree of V. This can be discovered by considering the ring of derivations of the coordinate ring and considering the derivations as vector fields as in section 3.3.21.

4.3 Algebraic Vector Bundles

In 4.3.1 and 4.3.3 we review some of the material developed in 3.3 in the algebraic geometric setting. In the remaining sections we study the relationship between subvarieties of a variety X and vector bundles on X.

4.3.1 Definition (algebraic vector bundle). An algebraic vector bundle of dimension n over a (quasi-projective) variety X consists of a surjective morphism of varieties $\pi : E \to X$ and an n-dimensional k-vector space structure on each $\pi^{-1}(x) \subset E$, $x \in X$ such that for every $x \in X$ there exists an open neighborhood $x \in U \subset X$ and an isomorphism (of varieties) $\phi: \pi^{-1}(U) \xrightarrow{\sim} U \times \mathbb{A}^n$ which satisfies

(i) $p_U\phi = \pi|U$, where $\pi|\pi^{-1}(U)$ is the restriction of $\pi : E \to X$ to $\pi^{-1}(U)$.

(ii) for every $y \in U$, $\phi : \pi^{-1}(y) \to y \times \mathbb{A}^n$ is a linear isomorphism of k-vector spaces where $y \times \mathbb{A}^n$ is given the obvious k-vector space structure.

We shall often write E_x for $\pi^{-1}(x)$; E_x is called the fibre of E at x.

Let $E_1 \xrightarrow{\pi_1} X$, $E_2 \xrightarrow{\pi_2} X$ be two algebraic vector bundles over the variety X. A homomorphism $\phi : E_1 \to E_2$ of vector bundles over X is a morphism $\phi : E_1 \to E_2$ such that $\pi_2\phi = \pi_1$ and such that the induced maps $\phi_x : E_{1x} \to E_{2x}$ are k-linear homomorphisms of the k-vector spaces E_{1x} into the k-vector spaces E_{2x}. A homomorphism of vector bundles $\phi : E_1 \to E_2$ is an isomorphism of vector bundles if there is a homomorphism $\psi : E_2 \to E_1$ such that $\psi\phi = 1_{E_1}$, $\phi\psi = 1_{E_2}$.

4.3.2 Definition (algebraic sections). A section of the algebraic vector bundle $E \xrightarrow{\pi} X$ is a morphism $s : X \to E$ such that $\pi s = 1_X$. Giving a section of $E \to X$ is equivalent to giving a homomorphism of the trivial one dimensional vector bundle $X \times \mathbb{A}^1 \to X$ into $E \xrightarrow{\pi} X$. The correspondence is as follows: Let $s_1 : X \to X \times \mathbb{A}^1$ be the section $x \mapsto (x,1)$ then $\phi \to \phi s_1$ establishes a one-one onto correspondence between homomorphisms $\phi : X \times \mathbb{A}^1 \to E$ and sections $X \to E$.

4.3.3 Patching data description of bundles and bundle homomorphisms

The definition of vector bundle in 4.4.1 says that every algebraic vector bundle over a variety X can be described (e.g. obtained) by the following data

(i) a (finite) covering $\{U_\alpha\}$ of X by open sets $U_\alpha \subset X$

(ii) for every α a trivial bundle $U_\alpha \times \mathbb{A}^n$ over U_α

(iii) for every α and β an isomorphism of trivial vector bundles

$$\phi_{\alpha\beta} : (U_\alpha \cap U_\beta) \times \mathbb{A}^n \to (U_\beta \cap U_\alpha) \times \mathbb{A}^n$$

where the isomorphisms $\phi_{\alpha\beta}$ are required to satisfy the conditions

(iv) $\phi_{\alpha\beta}\phi_{\beta\alpha} = 1$

(v) $\phi_{\beta\gamma}(x)\phi_{\alpha\beta}(x) = \phi_{\alpha\gamma}(x)$ for every $x \in U_\alpha \cap U_\beta \cap U_\gamma$ where $\phi_{\alpha\beta}(x)$ is the isomorphism $x \times \mathbb{A}^n \to x \times \mathbb{A}^n$ induced by $\phi_{\alpha\beta}$.

We note that giving an isomorphism $\phi_{\alpha\beta} : (U_\alpha \cap U_\beta)\times \mathbb{A}^n$ $(U_\beta \cap U_\alpha) \times \mathbb{A}^n$ is equivalent to giving a morphism $U_\alpha \cap U_\beta \to$ $G\ell_n$ where $G\ell_n$ is the quasiaffine algebraic variety over k with nonzero determinant.

Let E_1 and E_2 be two algebraic vector bundles over the variety X obtained by gluing together trivial bundles $U_\alpha \times \mathbb{A}^n$, resp. $U_\alpha \times \mathbb{A}^m$, where $\{U_\alpha\}$ is an open covering of X (We can take the same covering for E_1 and E_2 by taking if necessary to common refinement of two open coverings).

Let $\phi^1_{\alpha\beta}$ and $\phi^2_{\alpha\beta}$ be the gluing isomorphisms for E_1 and E_2 respectively. A homomorphism $\psi : E_1 \to E_2$ can now be described as follows: ψ consists of homomorphisms $\psi_\alpha : U_\alpha \times \mathbb{A}^n \to U_\alpha \times \mathbb{A}^m$ of trivial bundles such that for every α and β we have

$$\phi^2_{\alpha\beta}(x)\psi_\alpha(x) = \psi_\beta(x)\phi^1_{\alpha\beta}(x) \qquad \text{for all } x \in U_\alpha \cap U_\beta$$

Note that giving a homomorphism $\psi_\alpha : U_\alpha \times \mathbb{A}^n \to U_\alpha \times \mathbb{A}^m$ is equivalent to giving a morphism $U_\alpha \to M(m,n)$, where $M(m,n)$ is the affine algebraic variety of all $m \times n$ matrices with coefficients in k.

Let $E \xrightarrow{\pi} X$ be an algebraic vector bundle over the variety X and let $f : Y \to X$ be a morphism of varieties. We are going to construct a vector bundle $f^!E$ over ϕ. The so-called pull-back (along f) of E. Suppose E is given by patching data $\phi_{\alpha\beta} : U_\alpha \cap U_\beta \to G\ell_n$, then $f^!E$ over ϕ is given by the patching data $f^{-1}(U_\alpha) \cap f^{-1}(U_\beta) \xrightarrow{f} U_\alpha \cap U_\beta \to G\ell_n$.

Similarly if $\psi : E_1 \to E_2$ is a homomorphism of vector bundles given by the local homomorphisms determined by morphisms $\psi_\alpha : U_\alpha \to M(m,n)$, then we define $f^!\psi : f^!E_1 \to f^!E_2$ by means of morphisms $(f^!\psi)_\alpha : f^{-1}(U_\alpha) \xrightarrow{f} U_\alpha \to M(m,n)$.

4.3.4 Subvarieties of X and algebraic vector bundles on X.

Suppose $Z \subset X$ is an irreducible subvariety of an irreducible quasi-projective variety X and for simplicity assume

$$\operatorname{codim}(Z) = \dim X - \dim Z = 1 .$$

If X is smooth, e.g. if X is an open subspace of an algebraic submanifold of $\mathbb{P}^n(\mathbb{C})$, then Z may be locally defined as the zeroes of a single analytic function. More formally, we may cover the manifold X by charts such that on each U_α

$$Z \cap U_\alpha = f_\alpha^{-1}(0) \qquad (4.3.5)$$

for $f_\alpha : U_\alpha \to \mathbb{C}$ an analytic function.

A central question in the classification of subvarieties of a given variety X is whether each codimension 1 subvariety Z may be defined as the locus of a single algebraic or analytic function f. Now, the description (4.3.5) of Z leads to the data

$$\{U_\alpha\} \text{ a cover of } X, \quad g_{\alpha\beta} = f_\alpha/f_\beta : U_\alpha \cap U_\beta \to \mathbb{C} - \{0\} \qquad (4.3.6)$$

But since $g_{\alpha\beta}\, g_{\beta\gamma} = g_{\alpha\gamma}$, (4.3.6) itself constitutes the *local pieces and gluing data* (see 3.3.8) for an analytic rank 1 vector bundle, or preferably an analytic line bundle

$$\pi : L \to X \ . \qquad (4.3.7)$$

Moreover, the description (4.3.5) also yields an analytic section s of the line bundle L, viz. s is given on each U_α by

$$s_\alpha : U_\alpha \to U_\alpha \times \mathbb{C}$$
$$s_\alpha(p) = (p, f_\alpha(p)) \ . \qquad (4.3.8)$$

By (4.3.5), Z arises as the zeroes of the section s. In particular, Z arises as the zeroes of a globally defined analytic function f if, and only if, L is trivial. [We remark that, with more work (see [4], Chap. III) one may show that in (4.3.5) the U_α may be taken to be Zariski open and the f_α to be regular algebraic functions.] As an example, it is fairly easy to show that an algebraic line bundle $L \to \mathbb{A}^n$ an affine space is (algebraically) trivial.

Now, more generally, consider a subvariety Z of X with

$$\operatorname{codim}(Z) = \dim X - \dim Z = r \geq 1.$$

Again, one may cover X by $\{U_\alpha\}$ for which there exist suitable functions $f_\alpha^1, \ldots, f_\alpha^r$ such that

$$Z \cap U_\alpha = \{z | f_\alpha^1(z) = \ldots = f_\alpha^r(z) = 0\} . \qquad (4.3.5)'$$

And, on each intersection $U_\alpha \cap U_\beta$ one has

$$f_\beta^i = \sum_j g_{ij} f_\alpha^j$$

(provided we choose the f_α^j generation for the ideal of analytic functions on U_α vanishing on $Z \cap U_\alpha$), leading to the data

$$\{U_\alpha\} \text{ a cover of } X, \quad g_{\alpha\beta} = (g_{ij}) : U_\alpha \cap U_\beta \to G\ell(r,\mathbb{C}) \qquad (4.3.6)'$$

Now, (4.3.6)' gives the local pieces and gluing data for an analytic rank r vector bundle

$$V \to X,$$

which is trivial if and only if V is definable as the common zeroes of r globally defined analytic functions on X. In a less restrictive setting, if Z is the (complete) intersection of r hypersurfaces Z_i in X,

$$Z = \bigcap_{i=1}^{r} Z_i$$

then

$$V \simeq \bigoplus_{i=1}^{r} L_i \quad .$$

In particular, one is naturally led to the study of algebraic and geometric invariants of vector bundles on X from quite simple considerations involving subvarieties and their intersections or from studying the solution set to a system of simultaneous algebraic equations.

In the next section we will consider the Grassmann variety of p-planes in n-space, developing the algebraic analogues of sections (3.5). In (4.5) some of the basic tools for intersection theory on manifolds will be briefly reviewed.

4.4 Grassmann Manifolds

Let V be a finite dimensional vector space of dimension n over the complex numbers and let $G_p(V)$ be the set of all p-dimensional subspaces of V. The set $G_p(V)$ admits a manifold structure with the following charts. Write $V = U \oplus W$ with $U \in G_p(V)$ and $W \in G_{n-p}(V)$. For $A \in L(U,W)$ define $U_A = \{u + Au : u \in U\}$. The map $A \mapsto U_A$ is a one-to-one map from $L(U,W)$ into $G_p(V)$. It is not onto for we can describe the set of U_A's as exactly those elements of $G_p(V)$ that have zero intersection with W. Let $S_W = \{U_A : A \in L(U,W)\}$. If a basis for U and W is chosen so that A has a matrix representation then S_W along with the map that takes U_A onto the matrix A is a suitable chart. As W ranges over all complements of U the sets S_W form a cover for $G_p(V)$. If ϕ_W is the map from S_W to $L(U,W)$ an easy calculation shows that

$$\phi W_1 \cdot \phi \bar{W}_2^1(A) = A_2(I + A_1)^{-1}$$

where A_1 and A_2 are the unique matrices such that $Au = A_1u + A_2u$ with $A_1u \in U$ and $A_2u \in W_1$. The mapping is defined whenever $A \in \phi_{W_2}(S_{W_1} \cap S_{W_2})$ and being rational it is differentiable. The sets S_W with the maps ϕ_W form an atlas for the manifold.

An important fact about these charts is that S_W is an open dense subset of $G_p(V)$. In fact even more is true because of the fact that the complement of S_W is the subspaces that intersect W. This implies that the complement is algebraic and hence that S_W is Zariski open. The mapping that send A to U_A is thus an embedding of $L(U,W)$ into $G_p(V)$ as an open dense subset.

Let $G\ell(V)$ be the group of all linear automorphisms of V. Any group of automorphisms of V acts naturally on $G_p(V)$ by linear transformation of subspaces. Let α be any element of $G\ell(V)$ and partition α as

$$\begin{bmatrix} \alpha_{11} & \alpha_{12} \\ \alpha_{21} & \alpha_{22} \end{bmatrix}$$

where $\alpha_{11} \in L(U,U)$, $\alpha_{12} \in L(W,U)$, $\alpha_{21} \in L(W,W)$. Then α maps the subspace U_A to the subspace

$$\alpha(U_A) = \{(\alpha_{11} + \alpha_{12}A)u + (\alpha_{21} + \alpha_{22}A)u : u \in U\} .$$

The space $\alpha(U_A)$ is in S_W iff $(\alpha_{11} + \alpha_{12}A)^{-1}$ exists and in that case

$$\alpha(U_A) = U_{(\alpha_{21} + \alpha_{22}A)(\alpha_{11} + \alpha_{12}A)^{-1}} .$$

The $G\ell(V)$ action thus acts locally as a generalized linear fractional transformation. The local behavior of the action is very familiar.

On the other hand, given any two p dimensional subspaces U and W there is a linear automorphism that maps U onto W. Thus the action is transitive and we have that $G_p(V)$ is the homogeneous space $G\ell(V)/H$ for some H. Let U be a fixed element of $G_p(V)$ and W an arbitrary complement. The isotropy subgroup of U is just those transformations with $\alpha_{21} = 0$. Thus we can count dimensions either by the homogeneous space or by the chart.

If we select on V a positive definite bilinear form we choose in each subspace U an orthonormal basis and extend it to basis of V by the Gram-Schmidt process. This shows that the group of orthonormal matrices acts transitively on $G_p(V)$ and thus $G_p(V)$ is compact since $\mathcal{O}(n)$ is compact. This also implies that $G_p(V)$ is projective variety.

Let $U \in G_p(V)$ then each basis of U determines an $n \times p$ matrix B of rank n. Furthermore if B_1 and B_2 are such matrices there is an $p \times p$ matrix P invertible matrix such that $B_1 = B_2P$. Conversely if B_1 and B_2 are $n \times p$ matrices of rank p and there exists a P such $B_1 = B_2P$ the column space of B_1 is the column space of B_2. We have established that their one-to-one correspondence between the orbits of $G\ell(p)$ acting on $n \times p$ matrices of rank p and the $G_p(V)$. The Plucker coordinates of a matrix B is the $\binom{n}{p}$-tuple of determinants of $p \times p$ submatrices of B. It is easy to see that if $B_1 = B_2 P$ then Plucker coordinates of B_1 is scalar multiple of the Plucker coordinates of B_2. Thus we can associate with each point in $G_p(V)$ a line in $\mathbb{C}^{\binom{n}{p}}$. It can be shown, of course, that distinct points map onto distinct lines and that the embedding satisfies a homogeneous algebraic equation and hence $G_p(V)$ is an algebraic subset of

$$\mathbb{P}^{\binom{n}{p}-1} .$$

Thus, $G_p(V)$ is a projective algebraic variety.

The Grassmannian manifolds carry a natural algebraic vector bundle that can be described as follows. Let

$$\eta = \{(x,v) : (x,u) \in G_p(V) \times V \text{ and } v \in x\}.$$

η is a subvariety of $G_p(V) \times V$ and can be shown by the methods of 4.3.1 to be an algebraic vector bundle where the projection $\pi : \eta \to G_p(V)$ is onto the first coordinate. It can be shown that this bundle possesses no sections, but there is no particularly enlightening proof available.

However, if we construct the dual bundle η^* whose fibres are the spaces dual to the fibres of η. Then η^* has a full complement of sections. For let V have a basis $e_1,\ldots,e_n$ and an algebraic innerproduct. Define a section s_i of η^* by $s_i(x)(y) = \langle y,e_i\rangle$ where $x \in G_p(V)$ and y X. The s_i's are linearly independent as sections for consider

$$(\sum \alpha_i s_i)(x)(y) = \langle y, \sum \alpha_i e_i\rangle = 0$$

implies that $\sum \alpha_i e_i = 0$ and hence that the s_i's are independent. Every holomorphic section can be written as a linear combination of the s_i's.

The question whether η or η^* is the natural bundle on $G_p(V)$ depends somewhat on one's background. Traditionally differential geometers consider η to be natural and algebraic geometers prefer η^*.

4.5 Intersections of Subvarieties and Submanifolds

Consider 2 subvarieties X_1, X_2 of $\mathbb{P}^2(\mathbb{C})$ defined by homogeneous functions

$$f_1(x,y,z) = 0 , \qquad f_2(x,y,z) = 0$$

of degress d_1 and d_2, respectively. Bézout's Theorem (1.1.12) asserts that, unless f_1, f_2 have a common factor, the number of points in $X_1 \cap X_2$ counted with multiplicity is given by

$$\#(X_1 \cap X_2) = \deg X_1 \cdot \deg X_2 = d_1 d_2 . \qquad (4.5.1)$$

(1.1.12) was proved in the special case $d_1 = 1$; that is, where X_1 is a line in $\mathbb{P}^2$. We offer a second proof in this case which relies on the "principle of conservation of number."

Now, if f_2 is the product

$$f_2(x,y,z) = \prod_{i=1}^{d_2} \phi_i(x,y,z) \qquad (4.5.2)$$

of pairwise independent linear functionals of$_2$ (x,y,z), then X_1 is the union of d_1 distinct lines in $\mathbb{P}^2$; i.e. X_2 is reducible as

$$X_2 = \bigcup_{i=1}^{d_2} X_2^i \qquad (4.5.2)'$$

However, if X_1 and X_2 contain no common irreducible factors, then

$$\#(X_1 \cap X_2) = \sum_{i=1}^{d_2} \#(X_1 \cap X_2^i)$$

But,

$$\#(X_1 \cap X_2^i) = 1$$

since each pair of distinct lines in $\mathbb{P}^2$ intersect in a unique point.

Consider the case where f_2 is not a product as in (4.5.2). The space $V_{(d_2)}$ of homogeneous polynomials of degree d_2 in (x,y,z) is a finite dimensional vector space. In particular,

f_2 may be joined to a polynomial $\tilde{f}_2$ satisfying (4.5.2) by a path not passing through the 0 polynomial. Indeed, consider the path

$$tf_2 + (1-t)\tilde{f}_2 \subset V_{d_2} . \tag{4.5.3}$$

This deformation from f_2 to $\tilde{f}_2$ also gives rise to a deformation of X_2 to a union of d_2 lines:

$$X_2(t) : X_2(0) = X_2 , \quad X_2(1) = \bigcup_{i=1}^{d_2} X_2^i .$$

The principle of conservation of number asserts that

$$\#(X_1 \cap X_2(t))$$

is independent of t (provided it remains finite), and therefore

$$\#(X_1 \cap X_2) = \sum_{j=1}^{d_2} \#(X_1 \cap X_2^j) = d_2 \tag{4.5.4}$$

proving Bézout's Theorem for X_1 a line. If $\deg X_1 = d_1 > 1$, one may reiterate the above argument deforming X_1 to a union of d_1 lines, say $\tilde{X}_1 = \bigcup_{j=1}^{d_1} X_i^j$, and appealing to the basic principle, i.e.

$$\#(X_1 \cap X_2) = \#(\tilde{X}_1 \cap X_2) = \sum_{i=1}^{d} \#(X_1^j \cap X_2)$$

$$= \sum_{j=1}^{d_1} \sum_{i=1}^{d_2} \#(X_i^j \cap X_2^i) = d_1 d_2 .$$

Now, the successful application of the principle of conservation of number reposes on the introduction of an equivalence relation on submanifolds or subvarieties (of a fixed dimension) so that appropriate deformations of a submanifold do not change the equivalence class of the submanifold and so that intersection numbers, etc. depend only on the equivalence class. In the proof of Bézout's Theorem offered above, such deformations were affected by a continuous change in the coefficients of defining equations and the basic principle amounts to the continuous dependence of the roots on the coefficients on a defining equation. Such a program may be carried out in principle for general varieties, but is far beyond the scope of these notes. The more elementary topological approach employs the equivalence relations defined by homology and homotopy and we list some of the basic results below. For M a smooth manifold of dimension n, and for each r, $0 \leq r \leq n$, one introduces the r-th homology group of M, with integer coefficients in $\mathbb{Z}$ (or $\mathbb{Z}_2$), denoted by $H_r(M;\mathbb{Z})$ or $H_r(M;\mathbb{Z}_2)$. Each submanifold $N \subset M$ determines a $[N] \in H_r(M;\mathbb{Z})$; for example, for $M = \mathbb{P}^2(\mathbb{C})$ it is known that the only nonzero homology groups are

$$H_0(M;\mathbb{Z}) \simeq \mathbb{Z} = ([P])$$

$$H_2(M;\mathbb{Z}) \simeq \mathbb{Z} = ([X_1]) \;, \quad \deg X = 1$$

$$H_4(M;\mathbb{Z}) \simeq \mathbb{Z} = ([\mathbb{P}^2]) \;.$$

In this context, intersection of 2 submanifolds $X_{d_1}, X_{d_2} \subset \mathbb{P}^2(\mathbb{C})$ of dimension 2 (over $\mathbb{R}$) is determined by $[X_{d_1}], [X_{d_2}]$ in a bilinear manner. Thus the intersection theory in Bézout's Theorem amount to the evaluation of the bilinear form

$$i([X_{d_1}],[X_{d_2}]) = d_1 d_2 \;.$$

In general, let M be a orientable connected compact manifold. For each integer n, let

$$H^n(M,\mathbb{R}), \quad H_n(M,\mathbb{R})$$

denote the cohomology and homology vector spaces (with the real numbers $\mathbb{R}$ as coefficients.

For each pair (j,k) of integers, there is a bilinear mapping

$$H^j(M,\mathbb{R}) \times H^k(M,\mathbb{R}) \to H^{j+k}(M,\mathbb{R}) \tag{4.5.5}$$

called the *cup product*. If $\omega_1 \in H^j(M,\mathbb{R})$, $\omega_2 \in H^k(M,\mathbb{R})$, the $\omega_1 \cup \omega_2$. In particular for $k = m-j$ it maps

$$H^j(M,\mathbb{R}) \times H^{m-j}(M,\mathbb{R}) \to H^m(M,\mathbb{R}) = \mathbb{R} \;. \tag{4.5.6}$$

4.5.7 Poincaré Duality Theorem. *The bilinear mapping* (4.5.6) *is nondegenerate. In particular, it identifies* $H^{m-j}(M,\mathbb{R})$ *with the dual vector space of* $H^j(M,\mathbb{R})$, *and identifies* $H^{m-j}(M,\mathbb{R})$ *with* $H_j(M,\mathbb{R})$.

The cup-product (4.5.5) on cohomology then transforms (under this Poincaré duality isomorphism between homology and cohomology) into an algebraic operation on homology--the *intersection* pairing. If

$$j + k = m .$$

and $H_0(M)$ is identified with $\mathbb{R}$, the *intersection* operation defines a bilinear map

$$H_j(M,\mathbb{R}) \times H_k(M,\mathbb{R}) \rightarrow \mathbb{R} .$$

For $\alpha \in H_j(M,\mathbb{R})$, $\beta \in H_k(M,\mathbb{R})$, the real number

$$i(\alpha,\beta)$$

assigned to (α,β) be the operation is called the *intersection number* of the two homology classes α,β.

The above definition of "intersection number" is conceptually very simple, once one understands basic homology theory. To be useful, it must be supplemented by a method of computing it in more familiar geometric terms, for a suitably "generic" situation. Differentiable manifold theory offers such a possibility.

Let N,N' be compact orientable manifolds, with fixed orientation such that

$$\dim M = \dim N + \dim N' .$$

The spaces $H_n(N,\mathbb{R})$, $H_{n'}(N',\mathbb{R})$ have canonical generators ($n = \dim N$, $n' = \dim N'$), which are called the *fundamental homology classes* of the manifolds, denoted by h_N, h_N. Let

$$\phi : N \rightarrow M, \qquad \phi' : N' \rightarrow M$$

be two continuous maps, and let

$$\phi_*(h_N) \in H_n(M,\mathbb{R}) , \qquad \phi_*(h_{N'}) \in H_{n'}(M,\mathbb{R}) ,$$

be the image of these fundamental cycles in the homology of M. The intersection

$$i[\phi_*(h_N),\phi_*(h_{N'})]$$

is called the *intersection number* of the maps ϕ,ϕ', denoted by

$$i(\phi,\phi').$$

Now suppose that ϕ,ϕ' are C^∞ maps. Let $p \in N'$, $p' \in N'$ be two points such that

$$\phi(p) = \phi'(p') .$$

The maps are said to *intersect in general position* at this point if

$$M_{\phi(p)} = d\phi(N_p) \oplus d\phi'(N'_p) . \qquad (4.5.8)$$

(M_q denotes the tangent vector space to M at q; $d\phi$ denotes the induced linear maps on tangent vectors.)

Now, fixing an orientation for N means that it makes sense when a basis for each tangent space is "positively" or "negatively" oriented. Let us say that $\phi(N)$ and $\phi'(N')$ meet at $\phi(p)$ in a positive way if 4.5.8 is satisfied, and if putting together a positively oriented basis for N_p and $N'_{p'}$ provides a positively oriented basis for $M_{\phi(p)}$. Otherwise (and if they meet in general position) they are said to meet at $\phi(p)$ in a *negative* way.

Suppose that $\phi(N)$ and $\phi'(N')$ meet in general position at each point of intersection. Then

4.5.9 Theorem

$$i(\phi,\phi') = \sum_{p\in\phi(N)\cap\phi'(N')} \pm 1 . \qquad (4.5.10)$$

Here, the sign + or - is chosen according to whether the submanifolds meet in a positive or negative way.

Determining the orientations of the intersections is often an obstacle to determining the intersection number using formula (4.5.10). Working in the categories of *complex analytic* instead of *real* manifold removes this obstacle. The manifold M has a *complex manifold structure* if a set of coordinate charts is given, setting up coordinates in $\mathcal{C}^m$, with the transition maps between the charts given by complex analytic functions. A map $\phi : N \to M$ between complex manifolds is complex if it is given, in terms of complex charts, by complex analytic functions. A submanifold $\phi : N \to M$ is said to be complex if the map is complex.

Such a complex structure on manifold M determines an orientation for the manifold M. In terms of this orientation, two complex submanifolds always *meet with positive orientation.* Thus, the sum on the right-hand side of (4.5.10) *only involves plus signs*. In particular, $i(\phi,\phi')$ *is equal to the number of intersections of the submanifolds* $\phi(N)$, $\phi'(M')$, provided they meet in general position.

Here is the situation of greatest importance in algebraic geometry.

$$M = \mathbb{P}_n(\mathbb{C})$$

the complex projective space, of *real* dimension $2n$. It is the quotient of $\mathbb{C}^{n+1} \setminus (0)$ under the dilation group. $\phi(N)$, $\phi(N')$ are subsets determined by *nonsingular*, irreducible algebraic subsets of M. $\mathbb{P}_n(\mathbb{C})$ is a complex manifold, and the algebraic subsets are complex submanifolds. For $n = 2$, this, of course, is just Bézout's Theorem which we proved by purely algebraic methods at the beginning of this section.

REFERENCES

[1] Griffiths, P. A., and Adams, J. Q.: 1974, *Topics in Algebraic and Analytic Geometry*, Princeton University Press Mathematical Notes, Vol. 13, Princeton, N.J.

[2] Griffith, P. A., and Harris, J.: 1978, *Principles of Algebraic Geometry*, New York: Wiley.

[3] Mumford, D.: 1976, *Algebraic Geometry I: Complex Projective Varieties*, Springer-Verlag, New York.

[4] Shafarevitch, I. R.: 1974, *Basic Algebraic Geometry*, Springer-Verlag, New York.

5. LINEAR ALGEBRA OVER RINGS

The solution of linear equations, $AX = B$, and more generally the structure of R-linear transformations on R-modules requires us, in the end, to introduce and study quite a few auxiliary objects which one encounters in only a simplified form over fields. We begin with criterion for surjectivity and injectivity of an R-linear transformation

$$T : M \to N$$

of finitely-generated R-modules. These are always important properties to study, but particular use of these may be made in studying questions of reachability and observability.

5.1 Surjectivity of Linear Transformations, Nakayama's Lemma

We consider an R-linear map

$$T : M \to N \tag{5.1.1}$$

and would like to reduce our questions to a similar question over a field. However, as Example 2.4.8 shows, even when R is a PID, passing to the fraction field K gives us only some of the information we desire, viz. T_K is surjective if, and only if, the cokernel $N/T(M)$ is a torsion module.

Set $\max(R) = \{m \mid m \subset R$ is a maximal ideal of $R\}$, so that $m \in \max(R)$ if, and only if, R/m is a field. If T in (5.1.1) is surjective, then so is

$$\bar{T} : M/mM \to N/mN \tag{5.1.2}$$

5.1.3 Theorem. T *in* (5.1.1) *is surjective if, and only if,* $\bar{T}$ *in* (5.1.2) *is surjective for all* $m \in \max(R)$.

For example, $T : \mathbb{Z} \to \mathbb{Z}$ mapping z to $2z$ gives rise to the map,

$$0 = \bar{T} : \mathbb{Z}/2\mathbb{Z} \to \mathbb{Z}/2\mathbb{Z} ,$$

which fails to be surjective. Similarly

$$T : \mathbb{R}[x,y] \to \mathbb{R}[x,y]$$

$$Tf = (x^2 + y^2)f$$

fails to be surjective, since T "vanishes at the origin." That is, if $m_0 = \{f \mid f(0,0 = 0\}$, then T induces the 0 map

$$0 = \overline{T} : \mathbb{R}[x,y]/m_0 \to \mathbb{R}[x,y]/m_0 \quad .$$

Proof of 5.1.3. The examples above hint at an important special case, let $g \in \mathbb{R}$ and define $T_g : R \to R$ by $T_g(f) = gf$. Then T_g is surjective if, and only if, g is a unit in R. That is, g is a unit if, and only if, g is a unit in R/m for all m. For, g is a unit if, and only if, $g \notin m$ for any $m \in \max(R)$. Consider, on the other hand, those $g \in \bigcap_{m \in \max(R)} m$ $= \text{Jac}(R)$ --the Jacobson radical of R.

$$g \in \text{Jac}(R) \text{ if, and only if, } 1 - gf \text{ is a unit for all } f \in R \;.$$

If $g \in \text{Jac}(R)$, then $1\text{-}fg \equiv 1 \bmod(m)$, for all m, and is therefore a unit of R. Suppose that $1 - fg$ is always a unit, but that $g \notin m$, for some m; i.e., that $(g) + m = R$. Then, for some $f \in R$, $h \in m$, we have the equation

$$fg + h = 1 \;, \quad \text{or} \quad h = 1 - fg \;,$$

implying $m = R$.

Next consider $T : M \to M$ and suppose there exists an ideal I of R such that

$$TM \subset IM \;,$$

then there exists a relation

$$T^n + \sum_{i=1}^{n} r_i T^{n-i} = 0 \;, \quad \text{with } r_i \in I.$$

For, if $\{x_1,\ldots,x_n\}$ generates M, consider

$$Tx_i = \sum_{j=1}^{n} a_{ij} x_j \;, \qquad a_{ij} \in I \;.$$

Equivalently,

$$\sum_{j=1}^{n} (\delta_{ij} T(x_j) - a_{ij} x_j) = 0 \;, \qquad \text{or}$$

$$\sum_{j=1}^{n} (\delta_{ij} T - a_{ij}) x_j = 0 \;.$$

By Cramer's Rule, $\det(\delta_{ij}T - a_{ij})$ annihilates all $x \in M$ and is therefore the 0 endomorphism.

In particular, if $T = I$ one has, setting $r = \sum a_i$.

If $IM = M$, then there exists $r \in R$ such that

(i) $r \equiv 1 \mod I$

(ii) $rM = 0$. (5.1.4)

If $mM = M$, for all $m \in \max(R)$, then $M = (0)$.

For, suppose $0 \neq x \in M$. Consider the ideal

$$\text{Ann}(x) = \{r \in R \mid rx = 0\} \subset R \quad .$$

Since $x \neq 0$, $\text{Ann}(x) \neq R$ and therefore, $\text{Ann}(x) \subset m$ for some m. By hypothesis, there exists $r \in R$ satisfying

$$r \equiv 1 \bmod m, \quad \text{and} \quad rx = 0 \ .$$

But, the second equation asserts $r \in \text{Ann}(x) \subset m$, contrary to the first.

It is now an easy consequence that T is surjective if, and only if, $\overline{T} : M/mM \to N/mN$ is surjective, for all m, for all of the above applies to the module $N/\text{image } T$.

5.1.5 Corollary. [4] In particular, if one considers the linear system,

$$x(t+1) = Ax(t) + Bu(t) \qquad (5.1.6)$$

defined over R, then (5.1.6) is reachable, in the sense that the columns of $(B, AB, \ldots)$ span the state module, if and only if

$$x(t+1) = \overline{A}x(t) + \overline{B}u(t) \qquad (5.1.6)'$$

is reachable over R/m, for all $m \in \max(R)$.

Along the way, we have also developed enough algebra to prove the "fundamental Theorem of Commutative Algebra,"

5.1.7 Nakayama's Lemma. *If* M *is finitely generated over* R, $I \subset \text{Jac}(R)$ *an ideal of* R *such that* $IM = M$, *then* $M = (0)$.

Proof. From (5.1.4) one has an $r \in R$ such that

$$r \equiv 1 \mod I \quad \text{and} \quad rM = (0) .$$

The first equation asserts that $(1-4) \in Jac(R)$ and, by (3.3), r is a unit. The second equation now asserts that $M = (0)$.

This is especially useful when the ring R in question has only one maximal ideal, say m. (R,m) is said to be a local ring--for example, the ring of formal power series $\mathbb{R}[[x_1,\ldots,x_N]]$ is a local ring with $m = \{f \mid \text{the constant term of } f \text{ is } 0\}$, and the ring of germs at 0 of analytic functions in $\mathbb{R}^N$ is a local ring, contained in $\mathbb{R}[[x_1,\ldots,x_N]]$.

If R is local, then $Jac(R) = m$ and we have

5.1.8 Nakayama's Lemma. *If* M *is finitely generated over* R, *and* $mM = 0$, *then* $M = 0$. *In particular,* $\{x_1,\ldots,x_N\}$ *generates* M *if, and only if,* $\{\bar{x}_1,\ldots,\bar{x}_N\}$ *generates* M/mM.

Local rings will arise rather naturally when we study injectivity of R-linear maps in the next section.

5.2 Injectivity of Linear Transformations, Solvability of TX = Y, Localizations

In order to study injectivity as well as a particular equation $Tx = y$, we introduce a refinement of the idea of "evaluating T" at the point $m \in max(R)$, viz. expanding T locally at m. For $m \in max(R)$, denote the ring of fractions of R, with denominators in $R\backslash m$, by R_m (see [1], p. 36). Thus R_m consists of equivalence classes of pairs (f,g), $f \in R$, $g \in R\backslash m$, thought of as fractions f/g. Two pairs are equivalent if there exists $r \in R\backslash m$ such that

$$(f\tilde{g} - \tilde{f}g)r = 0 ,$$

that is, if the corresponding fractions are equal, and pairs are added and multiplies as fractions. As an exercise, one may check that $[(f,g)]$ is invertible in R_m if, and only if, $f \in R\text{-}m$. Therefore, each ideal I of R_m is contained in $\{[f,g] \mid f \in m\}$.

5.2.1 Lemma. R_m *is a local ring, with unqiue maximal ideal* $\{[f,g] \mid f \in m\}$.

If M is an R-module, then one can form the module of fractions, which is a module over the ring R_m. And, R-linear map

$T : M \to N$ induces an R-linear map $T_m : M_m \to N_m$. This is exactly the set-up we need.

5.2.2 Theorem. *The equation* $Tx = y$ *has a solution* $x \in R^{(n)}$, *for a given* $y \in R^{(\ell)}$, *if and only if, the equation*

$$T_m x = y \tag{5.2.3}$$

has a solution over R_m, *for all* $m \in \max(R)$.

Proof. We need only prove sufficiency, set

$$I = \{r \in R \mid Tx = ry \text{ has a solution over } R\} .$$

If the ideal $I = R$, we're done, and if $I \neq R$ then $I \subset m$, for some maximal ideal m of R. Fix such an m and choose a solution $\tilde{x} \in R_m^{(n)}$ to equation (3.11). By clearing denominators, which lie in R-m, one has $r \in R_m$ such that $\tilde{x} = r^{-1}x$, x defined over R, and an $s \in R$ such that $t = rs \equiv 1 \bmod m$. Therefore,

$$T(tx) = sy$$

and $s \in I \subset m$, contrary to assumption.

Remarks 1. If R_m is Noetherian, then the solubility of (5.2.3) can be further reduced, first to the case of a complete local ring and finally [2] to the case of a local Artinian ring, viz. to the solution of (5.2.3) over R/m^k, for each $k \geq 1$.

2. If we consider the question of surjectivity, then Theorem 5.2.2, together with Nakayama's Lemma, implies Theorem 5.1.3 for free (or even projective) state modules. One need not, however, make such hypothesis on M. Indeed, one can show [1]:

Theorem 3.12. *Let* $T : M \to N$, *then*

(i) T *is surjective* $\Leftrightarrow$ $T_m : M_m \to N_m$ *is surjective, for all* m.

(ii) T *is injective* $\Leftrightarrow$ $T_m : M_m \to N_m$ *is injective, for all* m.

5.3 The Structure of Linear Transformations, The Suslin-Quillen Theorem.

We now turn to the structure of linear transformations

$$T : M \to M \;, \qquad M \simeq R^{(n)} \;.$$

If T is not invertible, is M isomorphic to a direct sum of kernel T with image T? In Example 4.8, image T can never be complemented in $\mathbb{Z}$, so we must refine our question. If image T is complemented in M, i.e., is the image of a projection, can we find a basis for image T and complete this, with a basis for $\ker T$, to find a basis for M? The first condition is satisfied, for example, when T itself is a projection and, again, we are led to the question:

(SQ 1) Is every projection $P : R^{(n)} \to R^{(w)}$ diagonalizable?

Suppose, on the other hand, that T is invertible. What does the first column of T look like? Clear $(2,4)^+$ cannot be the first column of an invertible $T \in M_2(\mathbb{Z})$. Indeed, by the classical expansion of a determinant into a linear combination of cofactors one sees that the existence of $r_i \in R$ such that

$$\sum_{i=1}^{n} a_i r_i = \text{unit of } R$$

is a necessary condition that $(a_1,\ldots,a_n)^t$ be the first column of an invertible matrix. By dividing if necessary, one may assume

$$\sum_{v=1}^{n} a_i r_i = 1 \;,$$

that is, $(a_1,\ldots,a_n)$ is unimodular. If P is a rank 1 projection such that image P is free, then by choosing $(a_1,\ldots,a_n)$ to be a generator of image P one might attempt to follow the standard linear algebra route for constructing a T such TPT^{-1} is diagonal. That is, we construct T by setting $(a_1,\ldots,a_n)^+$ as the first row and complete T to an invertible matrix (by adding the basis vectors for $\ker T$). Thus we are led to ask

(SQ 2) Is every unimodular vector $(a_1,\ldots,a_n)$ the first column of an invertible matrix?

For $n=1,2$, (SQ 2) is trivially answered, in the affirmative, for any commutative ring $\mathbb{R}$.

5.3.1 Example. Consider $R = C(S^2)$ = ring of continuous, real-valued functions on the 2-sphere, and consider the free R module M, of rank 3, of $\mathbb{R}^3$-valued functions on S^2. Let $L \subset M$ be the R-submodule of those functions which point in $\pm$ the normal direction, so that L is spanned by the unimodular vector $v = (x,y,z)^t$, where $x^2 + y^2 + z^2 = 1$. Then v cannot be the first row of a unimodular matrix or, equivalently, if $P: M \to L$ is the projection on L, $\ker P$ does not admit a basis. In fact, to exhibit $w \in \ker P$ such that $w(x,y,z) \neq 0$ is to find a nowhere zero vector field on S^2, which is well-known to be contrary to fact.

Thus, the fact that one cannot "comb the hair on a tennis ball," has considerable impact on the linear algebra over $R = C(S^2)$. We note that (SQ 1) is equivalent to the more familiar form of these questions.

> (SQ 3) Is every finitely-generated projective module over R necessarily free?

The connection between (SQ 3) and "combing the hair on a tennis ball" can be made more precise, since the module $\ker P$ of tangent vector fields S^2 is the (finitely-generated, projective) module of continuous sections of a certain vector bundles on S^2, viz. the tangent bundle.

Set $R = \mathbb{C}[x_1,\ldots,x_N]$, then $R^{(1)}$ as a module over R is simply the module of algebraic, scalar valued functions on $\mathbb{A}^N$--which may be regarded as the module of algebraic section of the trivial line bundle

$$\mathbb{A}^N \times \mathbb{C} \to \mathbb{A}^N$$

On the other hand, if $\pi: V \to \mathbb{A}^N$ is a vector bundle, then the additive group $\Gamma(\mathbb{A}^N;V)$ is an R-module, with multiplication $f \in \mathbb{R}$, $\gamma \in \Gamma(\mathbb{A}^N;V)$ defined pointwise

$$f\gamma(p) = f(p)\gamma(p)$$

in the fiber $\pi^{-1}(p)$. If V is trivial, of rank m, then $\Gamma(\mathbb{A}^N;V) \simeq R^{(m)}$. And, we have already noted the converse, for the case $m = 1$. Moreover, any homomorphism $V \to W$ induces, by composition an R-module map $\Gamma(\mathbb{A}^N;V) \to \Gamma(\mathbb{A}^N;W)$.

Thus, we have a correspondence:

$$\{\text{vector bundles on } \mathbb{A}^N\} \to \{\text{modules over } R\} \tag{5.3.2a}$$

such that

$$\{\text{homomorphisms of vector bundles}\} \to \{\text{homomorphism of modules}\} . \tag{5.3.2b}$$

Moreover, this correspondence gives an equivalence

$$\{\text{trivial vector bundles}\} \leftrightarrow \{\text{free modules over } R\} \tag{5.3.3a}$$

$$\{\text{homomorphisms of trivial vector bundles}\} \leftrightarrow \{\text{homomorphisms of free modules}\}$$

In particular, if a trivial vector bundle V of rank m splits

$$V \simeq W_1 \oplus W_2$$

into 2 subbundles, then the homomorphism

$$P_1 : V \to W_1 \subset V , \qquad \text{satisfying} \quad P_1^2 = P_1 ,$$

corresponds to a projection operator

$$\tilde{P}_1 : R^{(m)} \to R^{(m)} \tag{5.3.4)'$$

with image $\tilde{P}_1 \simeq \Gamma(\mathbb{A}^N;W_1)$, a finitely-generated projective R-module. And, conversely, each finitely generated, projective module gives rise to some subbundle W_1 of a trivial bundle, by definition. Now, it can be shown that every vector bundle W_1 is a direct summand in some trivial bundle V and thus the equivalence (5.3.3) extends to an equivalence

$$\{\text{vector bundles}\} \leftrightarrow \{\text{finitely generated, projective modules}\} \tag{5.3.5a}$$

$$\{\text{homomorphisms of vector bundles}\} \leftrightarrow \{\text{homomorphisms of finitely, generated projective modules}\} \tag{5.3.5b}$$

Thus, triviality of a vector bundle is equivalent to freeness of its module of sections, bringing us to ask, for $R = \mathbb{C}[x_1,\ldots x_N]$

(SQ 4) Is every vector bundle on $\mathbb{A}^N$ trivial?

This question was raised by J-P. Serre and settled, in the affirmative, by A. Suslin and D. Quillen [5], [3].

5.3.6 Theorem (SQ) *For* $R = k[x_1,\ldots,x_N]$, *every finitely generated projective module is free; that is,* (SQ1),...,(SQ4) *hold for* R.

We will find all of these forms of Suslin-Quillen quite useful.

Thus, by extending these ideas we see that there exists projective, but not free, modules defined over $\mathbb{R} = C(S^2)$. By using the line bundle over S^1 derived from the Möbius band, this is also true for $C(S^1)$. These facts lie at the heart of the non-existence of continuous canonical forms for realizations, which is, of course, a question of linear algebra with parameters (see Professor Hazewinkel's lectures).

It is somewhat deeper that (SQ2) fails to hold for $R = H^{\infty}(\mathbb{D})$, this calculation comes from certain topological non-triviality of the space, $\max(H^{\infty}(\mathbb{D}))$, as in (SQ4).

REFERENCES

[1] Atiyah, M. F., and Macdonald, I. G.: 1969, *Introduction to Commutative Algebra*, Addison Wesley, Reading, Massachusetts.

[2] Gustafson, W. H.: 1979, *Roth's theorems over commutative rings*, Linear Alg. & Its Application, Vol. 23, pp. 245-251.

[3] Quillen, D.: 1976, *Projective modules over polynomial rings*, Invent. Math. 36, pp. 164-171.

[4] Sontag, E.: 1976, *Linear systems over commutative rings: A survey*, Ricerche Automatica, 7, pp. 1-34.

[5] Suslin, A. A.: 1976, *Projective modules over a polynomial ring are free*, Dokl. Akad. Nauk., S.S.S.R. 229.

ALGEBRAIC AND GEOMETRIC ASPECTS OF THE ANALYSIS OF FEEDBACK SYSTEMS

Christopher I. Byrnes*

Department of Mathematics and the Division of Applied Sciences
Harvard University
Cambridge, Massachusetts 02138

§1. INTRODUCTIONS, NOTATION, AND THE STATEMENTS OF THE PROBLEMS

This manuscript represents two of the three lectures which I gave at this Advanced Study Institute and, for this reason, I shall give two introductions. (The third lecture is historical and may be found in "Introductory Chapter," this volume.) In the first four sections, I shall discuss recent work in algebraic and geometric system theory which centers around the question, "What can be done using state or output feedback." To fix the ideas, it is at least initially sufficient to consider a system σ as being defined by the <u>state-space</u> equations

$$\dot{x}(t) = Fx(t) + Gu(t) \qquad y(t) = Hx(t) \tag{1.1}$$

*Research partially supported by the NASA-AMES under Grant ENG-79-09459 and by the NSF under Grant NSG-2265.

C. I. Byrnes and C. F. Martin, Geometrical Methods for the Theory of Linear Systems, 85-124.

or by the transfer function (the zero initial state Laplace transform)

$$\hat{y}(x) = T(s)\hat{u}(s), \quad T(s) = H(sI-F)^{-1}G \tag{1.1'}$$

which relates the input vector $u \in U \simeq \underline{k}^m$ to the corresponding output $y \in Y \simeq \underline{k}^p$, without explicit mention of the (internal notions of) state $x \in X \simeq \underline{k}^n$. Thus (1.1)' is an external description of σ, as one might see in Ohm's law, where (1.1) is an internal description (i.e., involving states) of σ, as one might see in the non-autonomous differential equations for an RLC network being driven by an applied current $u(t)$ and generating a voltage $y(t)$.

Now, feedback engineering is perhaps the second or third oldest profession and needs little introduction. Indeed, any list of well-known examples of feedback systems should include the oil lamp of Philon, the water clock of Gaza, Christiaan Huygens' construction of a regulator for clock mechanisms, and the centrifugal governor for steam engines, developed by James Watt, followed by a plethora of more sophisticated modern systems. In each of these example, some output--or function of the state--of the system is used to control the evolution of the state in future time and a rather basic question is to determine how much control over the state one can obtain by feeding back the output as an input. For a vehicle driven by a steam engine, one would like to produce a uniform motion in the vehicle by such a feedback law and this is where the mathematics begins to play a role. In an often cited paper [34], J. C. Maxwell linked the intrinsic deviation, of some feedback systems, from uniform motion to the instability of the corresponding differential equations. Now, a feedback law in the linear context is just a linear map

$$K : Y \rightarrow U$$

and the corresponding closed loop system has dynamics given by

$$\dot{x}(t) = (F - GKH)x(t) + Gu(t)$$

$$y(t) = Hx(t) \ , \tag{1.2}$$

or, in external terms, by

$$T(s)(I + KT(x))^{-1} = N(s)(D(s) + KN(s))^{-1} \tag{1.2'}$$

where $N(s)D(s)^{-1}$ is a coprime factorization of $T(s)$ into polynomial matrices. The instability, or rather stability, question is thus whether

$$\chi_\sigma(K) = \det(sI - F + GKH) \tag{1.3}$$

has its roots in the left-half plane. Naturally, the inverse problem is deeper and more applicable: can one find K so that (1.3) has it roots in the left-half plane? More generally, can one adjust arbitrarily via output feedback, the natural frequencies of (1.1)? Since the eigenvalues of F are the poles of T(s) (provided n is minimal), one refers to this problem as pole-placement. For the sake of completeness, this is stated as:

<u>Problem A</u>. Analyze as explicitly as possible the algebraic map

$$\chi_\sigma : \underline{k}^{mp} \to \underline{k}^n$$

defined by regarding the right hand side of (1.3), via its coefficients in s, as a point in k^n. In particular, is χ_σ surjective (pole placement), what are the topological or geometric properties of χ_σ, or of image χ_σ?

In §2, I shall given an exposition of the infinitesimal analysis of χ_σ, viz. a calculation of the Jacobian $d\chi_\sigma$ on k^{mp} and on a certain submanifold $M \subset k^{mp}$. This uses the fact that χ_σ is a polynomial but, except over algebraically closed ground fields, makes more use of differential calculus than of algebraic geometry. However, one of the new results is a proof and sharpening of Kimura's generic pole-placement theorem [29]. This is a simple, geometric proof (taken from [5]) of an honest output pole-placement theorem under the hypothesis $m + p - 1 \geq n$ used by Kimura. The final topic in this section is a classification, due to Brockett [3], of the Lie algebras $\{F,GH\}_{\mathscr{L}}$ associated to a transfer function T(s) for $m = p = 1$, as well as a multi-input-output generalization, with application to Problem A even in the case of time-varying feedback K(t).

In §3, the geometric foundation for §4 is developed, the starting point being the interpretation of graph T(s) as a curve of m-dimensional subspaces of $\mathbb{C}^{m+p}$; i.e., as an algebraic curve in the Grassmannian variety Grass(m,m+p)--due to Hermann-Martin. This geometric approach is actually very close in spirit to Kimura's original proof of this theorem. In this setting, the degree of the curve so obtained is the intersection of this curve with a hyperplane, as in Bézout's Theorem, and the Theorem of Hermann-Martin asserts that the points of one such intersection are precisely the poles of T(s).

In §4, the output feedback group is brought into play, whereas in §3 only the identity element is considered. From this point of view, placing poles by output feedback is the same as prescribing points of intersection of the curve with a hyperplane. This inverse problem in geometry has a long history, making contact with several basic themes in algebraic geometry, and in this context χ_σ may be regarded as the restriction of a central projection, about which several important facts are known. From this "central projection lemma", much in the way of Problem A can be deduced, containing in particular some rather surprising results--especially in view of negative results previously obtained. For example, although over $\mathbb{C}$ all is well, Willems and Hesselink [45] have shown that over $\mathbb{R}$, for $m = p = 2$ and $n = 4$, for generic σ it is a fact that χ_σ misses an open set. Using the Schubert calculus in the case $mp = n$ Brockett and the author [5] have shown that, for $m = 2$, $p = 2^r - 1$ (a Mersenne number), χ_σ is generically onto (over $\mathbb{R}$), although these may be the only such cases (up to symmetries and excluding the scalar cases).

In the remaining sections, I consider linear systems depending on parameters and the corresponding questions of pole placement and stabilization by state feedback. Such parameter dependence arises quite often, through dependence on physical parameters such as altitude or attitude of an aircraft or as the value of a resistor, etc. In these cases, (F,G,H) have entries in an appropriate ring of functions on the parameter space Λ and conversely linear systems defined over rings can be viewed as linear systems depending on parameters--in a slightly more general sense. Two remarkable examples are: first, the representation of linear delay-differential systems, via convolution with finite measures on $\mathbb{R}$, as linear systems defined over a polynomial ring [27] and second, the representation of half-plane digital filters as linear systems defined over ℓ^1, also due to Kamen [28]. Thus, one may pose the problem of pole-placement over a ring R, commutative with identity, such as a ring of functions.

In section 5, I review some of the known positive results, starting with Morse's theorem for P.I.D.'s, a result of the author's for very special systems defined over polynomial rings (or, more generally, projective free rings), and in §6 turn to the recent counterexamples to the general question for certain rings, linking the arithmetic aspects of R with pole-placement.

In section 7, I turn to the more modest question, which is however sufficient for applications:

Problem B. If $(F(\lambda),G(\lambda))$ is defined over an algebra of functions and is stabilizable for each fixed λ, does there exist $K(\lambda)$ defined over the same algebra, such that the closed loop system (1.2) is stable for each λ?

In that section, I use a lemma of D. Delchamps' on smoothness of solutions to a smooth family of Riccati equations to obtain an affirmative answer to Problem B in case where $(F(\lambda),G(\lambda))$ is C^k, $k \geq 0$, and controllable for each λ.

In closing, I would like to apologize for having omitted, primarily for reasons of time and space, recent work which might belong under such a title. Some of the work by Rosenbrock, Fuhrmann, et al. on dynamic conpensation is reported in their lectures, while related work has recently appeared in the thesis of T. Djaferis [15], Djaferis and Mitter [35], and in Emre [16]. It is my intention to report elsewhere on the work of Postlethwaite-MacFarlane [37], et al., which develops root-locus techniques for square multi-input, multi-output systems with respect to scalar gain $K = \lambda I$. One should also mention recent work by Sastry-Desoer [43], which evaluates the asymptotic values of the unbounded root loci, for generic systems.

§2. KIMURA'S THEOREM, INFINITESIMAL ANALYSIS OF χ_σ, LIE ALGEBRAIC INVARIANTS OF χ_σ.

Now, in order to compute the rank of

$$d\chi_\sigma : T_0(\underline{k}^{mp}) \to T_\chi(\underline{k}^n), \quad \text{where } \chi = \det(sI - F),$$

it is efficient to change coordinates by use of the frequency domain. Thus, if $N(s)D(s)^{-1} = T(s) = H(sI-F)^{-1}G$ is a coprime factorization of the transfer function $T(s)$ and if $-K:Y \to U$ is the output gain, the closed loop transfer function is as given in (1.2)':

$$T^{-K}(s) = T(s)(I - KT(s))^{-1} = N(s)(D(s) - KN(s))^{-1} . \tag{2.0}$$

Thus, to solve $p(s) = \chi_\sigma(K)$, with $K \in M$ a subset of matrices, is to solve for rational functions

$$p(s)/\det D(s) = \det(I - KT(s)), \tag{2.1}$$

with $K \in M$. With this change of coordinates on $\underline{k}^n$, χ_σ takes the form:

$$K \mapsto 1 + \sum_{i=1}^{n} c_i(-KT(s)) , \tag{2.2}$$

where the $c_i(R)$ are the characteristic coefficients of R.

Ignoring the constant term, χ_σ is given to first order as

$$\chi_\sigma(K) = tr(-KT(s)) = \langle -K,T(s)\rangle$$

and, since $T(s)$ is rational, the Jacobian is given by the formula

$$d\chi_\sigma(K) = (\langle -K,HF^iG\rangle)_{i=0}^{n-1} . \tag{2.3}$$

From (2.3) one recovers the calculation

$d\chi_\sigma$ *is surjective whenever the Hankel matrices*

$HG, HFG,\ldots,HF^{n-1}G$ *are independent*,

which (since the Hankel matrices are vectors in $\underline{k}^{mp}$) refines the necessary condition, $mp \geq n$, for surjectivity of χ_σ. Indeed, R. Hermann and C. Martin combined this calculation with the dominant morphism theorem to obtain, for $k = \mathbb{C}$,

<u>Theorem</u> ([23]). *If* $mp \geq n$, *then for almost all* (F,G,H) *the image of* χ_σ *is open and dense.*

Several remarks are in order. First, in any such theorem, the "almost all (F,G,H)" hypothesis is necessary. Above, the affine algebraic set which must be excluded is contained in the variety defined by the vanishing of all minors of order n of the $mp\times n$ matrix $(HG,\ldots,HF^{n-1}G)$. But this is as it should be, for in general such conditions must in particular exclude systems which are equivalent to lower order systems, e.g., rank $G = 1$, where image χ_σ is a line. Second, it is in fact true that, for almost all (F,G,H), χ_σ is closed. And finally, over $\mathbb{R}$, J. Willems and W. Hesselink [45] have proved that, for $m = p = 2$, for almost all (F,G,H), image χ_σ is not dense, which illustrates the absence of the "fundamental openness principle" over $\mathbb{R}$.

There is, however, a similar result over $\mathbb{R}$, under stronger hypothesis, due to H. Kimura.

<u>Theorem</u> ([29]). *If* $m + p - 1 \geq n$, *then for generic* (F,G,H) *image* χ_σ *is open and dense.*

In the latter part of this lecture, I shall turn to Kimura's proof, which is quite long. Here, I shall follow a geometric line of reasoning [5] starting from (2.3). First of all, notice that $m + p - 1 = \dim M$, where $M \subset \mathbb{R}^{mp}$ is the submanifold of rank 1 matrices. Surprisingly, it's enough to restrict χ_σ to

M and we wish to compute $d\chi_\sigma$ acting on $T_K M$. In (2.3), this has the effect of restricting K to be of the form ${}^t(x+\varepsilon)(y+\delta)$, for $x \in \mathbb{R}^m - \{0\}$, $y \in \mathbb{R}^p - \{0\}$, $K = {}^t xy$ rank 1, and we therefore consider the vectors

$$d\chi_\sigma({}^t x_j y_j) = (y_j H\tilde{F}^i G x_j)_{i=0}^{n-1} \tag{2.4}$$

where $\tilde{F} = F + GKH$. As before, one sees that, if $m + p - 1 \geq n$, then generically in (F,G,H) there exists matrices ${}^t x_j y_j$, $j = 1,\ldots,n$ such that the vectors (2.4) are linearly independent. In particular, χ_σ is surjective to first order and hence, by the implicit function theorem, image χ_σ contains an open set. Moreover, since $c_i(KT(s)) = 0$ for $i \geq 2$ whenever K has rank one, χ_σ is equal, along M, to $1 + d\chi_\sigma$ and is therefore surjective! Note that, by combining this observation with (2.1)-(2.2), one can develop an algorithm for the solution of (2.1).

More recently, H. Kimura [30] has improved the bound to $m + p + \kappa_1 - 1 \geq n$, where κ_1 is the largest Kronecker index, subject to the constraint $m \geq \mu_1$ (the largest observability index), $p \geq \kappa_1$. This, too, has an amplification to a pole-placement theorem.

Now, as an example of an invariant of χ_σ, which plays a role in the output feedback problem but which is not captured by our previous calculations, we consider a Lie algebra determined by σ. Explicitly, by choosing a minimal realization (F,G,H) of a scalar transfer function $T(s)$, one may form $\mathcal{L}_\sigma = \{F,GH\}_L$--the Lie subalgebra of $\mathfrak{gl}(n,\mathbb{R})$ generated by F and GH. In this way, one obtains not only $\mathcal{L}_\sigma$ but also a representation, $\rho:\mathcal{L}_\sigma \to \mathfrak{gl}(n,\mathbb{R})$, and by the state-space isomorphism theorem, any other realization (F',G',H') give rise to an equivalent representation ρ'. Of course, $\mathcal{L}_\sigma$ is also invariant under output feedback, since $F + KGH$ is contained in $\mathcal{L}_\sigma$ for any scalar K, and this accounts for its importance in the output feedback problem. And symmetries in the representation $\rho: \mathcal{L}_\sigma \to \mathfrak{gl}(n,\mathbb{R})$ reflect symmetries in the closed-loop characteristic equation. For example, if $T(s) = 1/s^2$, then it's not hard to see that $\rho\mathcal{L}_\sigma = \mathfrak{sl}(2,\mathbb{R}) = \mathfrak{sp}(1,\mathbb{R})$, which reflects the equivalent facts that $\mathrm{tr}(F + KGH) = 0$, for all K, and that the closed-loop characteristic polynomial is always an even function. These ideas were developed by R. Brockett in [3], [6] leading to his classification of those $\mathcal{L}_\sigma$ which can occur:

Theorem ([3]). *The following is a list of the* $\mathcal{L}_\sigma$ *which can occur together with the corresponding symmetry properties:*

$\mathcal{L}_\sigma$	symmetries
1. $sp\left(\frac{n}{2}, \mathbb{R}\right)$	$T(s) = T(-s)$
2. $sp\left(\frac{n}{2}, \mathbb{R}\right) + \mathbb{R}$	$T(s) = T(-s+\alpha)$, for some α
3. $s\ell(n, \mathbb{R})$	$tr(F) = tr(GH) = 0$, and none of the above hold
4. $g\ell(n, \mathbb{R})$	none of the above hold

One should also note that this classification gives the same information for time-varying gains $K(t)$. Now, the multivariable case is handled, in part, by a reduction to the scalar case by a lemma (see [2]) reminiscent of Heymann's Lemma. That is, for (F,G,H) minimal, there exists a gain K and input and output channels g and h such that $(F + GKH,g,h)$ is a minimal triple. And, if one defines $\mathcal{L}_\sigma$ to be the smallest Lie subalgebra of $g\ell(n, \mathbb{R})$ containing $\{F + GKH : K \text{ arbitrary}\}$, this reduction enables one to prove:

Theorem ([4]). *If* $\operatorname{rank} T(s) \geq 2$, *then* $\mathcal{L}_\sigma$ *is either* $s\ell(n, \mathbb{R})$ *or* $g\ell(n, \mathbb{R})$, *depending on the vanishing of* trF *and* $tr(HG)$.

§3. BEZOUT'S THEOREM, THE THEOREM OF HERMANN-MARTIN.

There are important external symmetries too, which arise as subgroups of the (output) feedback group. Now, as far as I am aware, the applications of algebraic geometry to linear system theory arise from Laplace transform techniques, from the existence of algebraic groups actions in the form of symmetry groups, and from the interrelation between these 2 points of view. Indeed, perhaps one of the least understood contributions in the Hermann-Martin series is the recognition of the Laplace transform as an intertwining map between the actions of the state and output feedback groups at the state-space level and the classical actions of these groups as linear fractional transformations. This observation is the starting point for our global analysis of χ_σ.

To fix the ideas, I shall begin with a review of Kimura's proof of his theorem, in the case $m = 1$, $p = 2 = n$. Here, one has

$$T(s) = \begin{pmatrix} \frac{q_1(s)}{p(s)} \\ \frac{q_2(s)}{p(s)} \end{pmatrix}, \quad \text{with} \quad \begin{pmatrix} q_1(s) \\ q_2(s) \\ \hline p(s) \end{pmatrix} = \begin{pmatrix} N(s) \\ \hline D(s) \end{pmatrix} \tag{3.1}$$

a coprime factorization of $T(s)$. If $\lambda_1 \neq \lambda_2$ are complex numbers, then the method of proof is to select a non-zero vector from each of the lines

$$\begin{pmatrix} N(\lambda_i) \\ D(\lambda_i) \end{pmatrix} .$$

Geometrically, one has the set-up in Fig. 3.1 where we denote the line through

$$\begin{pmatrix} N(\lambda) \\ D(\lambda) \end{pmatrix}$$

by $T(\lambda)$. This is as it should be, for a choice of coprime factors is only unique up to multiplication by a non-zero scalar. I claim that if one takes the plan π spanned by $T(\lambda_1)$ and $T(\lambda_2)$, then $\pi = \text{graph}(-K)$, where K is a gain for which the closed-loop poles equals $\{\lambda_1,\lambda_2\}$. Notice that to say $\{\lambda_1,\lambda_2\}$ is the polar set of T is to say λ_1,λ_2 are the roots of p in (3.1). Thus, in this case, the linear $T(\lambda_i)$ lie in the Y-plane (Fig. 3.1) and so $K = 0$. This, however, is even far from explaining the minus sign, which occurs for group-theoretic reasons. Since the geometry of lines in $\mathbb{C}^3$ is at issue, it's more efficient to rephrase the observation made above in terms of projective geometry. That is, the transfer function gives a map,

$$T:\mathbb{C}^* \to \mathbf{P}^2 ,$$

of the extended complex line $(T(\infty) = U)$ to the projective plane. The lines in the Y-plane form the projective line $\mathbb{P}^1$, embedded in $\mathbf{P}^2$, while $T(\mathbb{C}^*)$ is a curve in $\mathbf{P}^2$. Moreover, since p has degree 2, T is a curve of degree 2 and intersects the line $\mathbb{P}^1$ twice, as it should according to Bézout's Theorem (see Fig. 3.2).

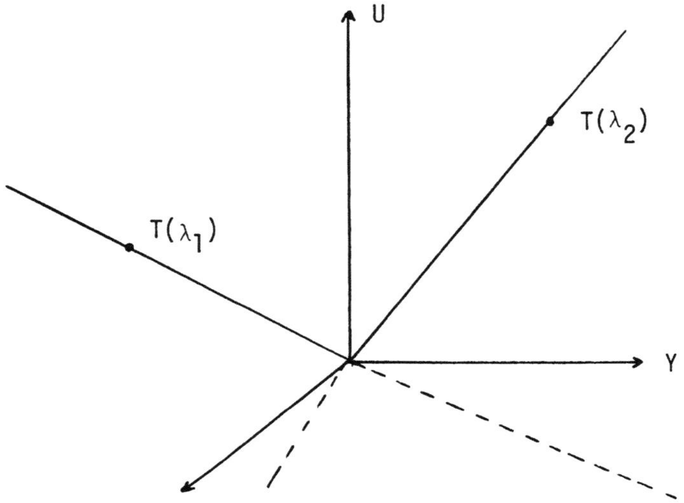

Figure 3.1

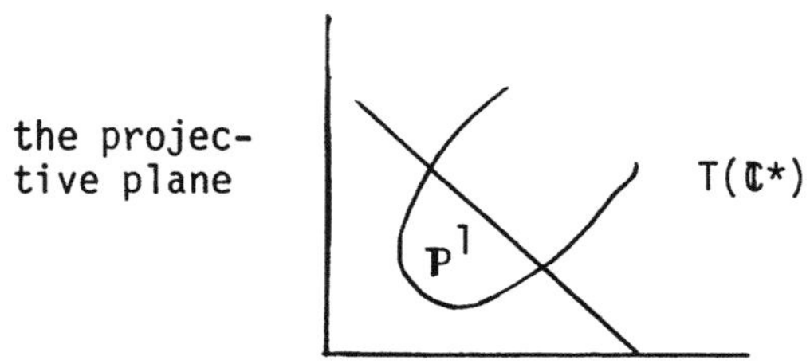

Figure 3.2

In fact, the same reasoning shows that, for any such T with n arbitrary, *the McMillan degree of* T *equals the degree of the curve* T, with

$$T^{-1}(T(\mathbb{C}^*) \cap \mathbb{P}^1) = \text{sing}(T) \quad . \tag{3.2}$$

If we now choose any other plane Y_1 in $\mathbb{C}^3$, complementary to U, then Y_1 determines another line in $\mathbb{P}^2$ and by Bézout's Theorem (or by a little algebra), $T(\mathbb{C}^*)$ intersects this line in n-points. On the other hand, such a plane Y_1 is the graph of a linear map $-K:Y \to U$ (and conversely) and from this point of view, Bézout's Theorem asserts that the McMillan degree is preserved under output feedback, assuming the claim made above. However, the claim is now fairly easy to see. For, any K may be regarded as an element of $GL(U \oplus Y)$ via the representation,

$$\rho:K \to \begin{pmatrix} I_Y & 0 \\ K & I_U \end{pmatrix} \in GL(U \oplus Y) \quad , \tag{3.3}$$

and $GL(U \oplus Y)$ acts on $\mathbf{P}^2$ (the points in $\mathbf{P}^2$ regarded as lines in $U \oplus Y$). One therefore has two possibly distinct actions of K on T: the first is the standard output feedback transformation $T \to T^K$ given in (1.2)', and the second is obtained by composing the map $T:\mathbb{C}^* \to \mathbf{P}^2$ with the classical action $\rho K:\mathbf{P}^2 \to \mathbf{P}^2$. As one can see, ρK leaves the line U fixed and therefore $\rho K \circ T$ is a rational function, vanishing at ∞; i.e., $\rho K \circ T$ is a transfer function. Explicitly, by combining (3.3) with (1.2)' one has

$$\rho K \circ T = \begin{pmatrix} I_Y & 0 \\ K & I_U \end{pmatrix} \begin{pmatrix} N \\ D \end{pmatrix} = \begin{pmatrix} N \\ D+KN \end{pmatrix} = T^K . \tag{3.4}$$

In particular, one may now compute (3.2) acted on by K in two ways:

$$\text{sing}(T^K) = (T^K)^{-1}(T^K(\mathbb{C}^*) \cap \mathbf{P}^1 = T^{-1}(T(\mathbb{C}^*) \cap (-K)\,\mathbf{P}^1) \tag{3.5}$$

where $(-K)\,\mathbf{P}^1$ is the linear $\mathbf{P}^1$(= the plane Y in $\mathbb{C}^3$) acted on by $\rho(-K)$ (= graph of $-K:Y \to U$ in $\mathbb{C}^3$)! Thus, in order to compute the closed loop poles, $\text{sing}(T^K)$, one can alternatively keep the curve T (see Fig. 3.2) fixed and, instead, move $\mathbf{P}^1$ through the inverse "rotation" $-K$.

This proves the claim but in a more general setting, viz. in (3.3) and (3.4) U and Y could just as well be an m-plane and a p-plane, with $GL(U \oplus Y)$ acting as linear transformation on the space of m-planes = $\text{Grass}(m, U \oplus Y)$. In this setting, the generalization of (3.2) is due to R. Hermann and C. Martin, who interpreted the McMillan degree as an intersection number ([24]). The codimension 1 subvariety of $\text{Grass}(m, U + Y)$ which plays the role of the line $\mathbf{P}^1$ in the plane $\mathbf{P}^2$ is the Schubert variety of m-planes:

$$\sigma(Y) = \{W : W \cap Y \supsetneq (0)\} \quad . \tag{3.6}$$

That is, $W \in \sigma(Y)$ if, and only if, W meets Y in at least a line. The beautiful (and useful for our purposes) theorem of R. Hermann-C. Martin is

$$T^{-1}(T(\mathbb{C}^*) \cap \sigma(Y)) = \text{sing}(T) \quad . \tag{3.7}$$

[Alternatively, the extended plane $\mathbb{C}^*$ is the Riemann sphere S^2 (or $\mathbb{P}^1$) and

$$[T(\mathbb{P}^1)] \in \pi_2(\text{Grass}(m, U \oplus Y) \simeq \mathbb{Z}$$

corresponds to the McMillan degree, where the isomorphism is canonical, by virtue of the Hurewicz isomorphism and a choice of complex structure.] As before we can act on (3.6) with K, in two ways, to obtain

$$T^{-1}(T(\mathbb{C}^*) \cap \sigma(-KY)) = \text{sing}(T^K). \tag{3.7}$$

Now, as an illustration of these geometric ideas and in order to return to some of Kimura's algebraic techniques, I shall prove a little pole-placement theorem for state-feedback, i.e., for the case $p = n$. [This combinatorial theorem is a special case of a theorem of Rado ([38]) which also generalizes Ph. Hall's Theorem. Moreover, an elegant application of Rado's Theorem to pole-placement appears, for the first time in Hautus's proof of pole-assignment by state feedback ([19]), published in 1970. I was mistaken in my lectures in ascribing it solely to Kimura.] What I wish to give is a proof of the Wonham-Simon-Mitter-Heymann-Kalman Theorem for distinct poles $\{\lambda_1, \ldots, \lambda_n\}$. The principal lemma in [29] is in fact a celebrated theorem in combinatorics, in disguise. Kimura calls a collection $V_1, \ldots, V_n$, of subspaces of a vector space V, <u>normal</u> just in case one can select vectors $v_i \in V_i$ such that $\{v_1, \ldots, v_n\}$ is independent. The lemma asserts that a collection of subspaces is normal if, and only if, the (general position) condition (*) is satisfied:

$$\text{for each selection } V_{i_1}, \ldots, V_{i_k} \text{ of distinct } V_i\text{'s,}$$

$$\dim\left(V_{i_1} + \ldots + V_{i_k}\right) \geq k \; . \tag{*}$$

Notice, however, that (*) is precisely the diversity condition in Ph. Hall's theorem on distinct representatives, modified to include the set (or subspace) function dim(·)--which, after all, does satisfy a form of the inclusion-exclusion principle.

In order to apply this result to multivariable state-feedback consider, for $\lambda_1,\ldots,\lambda_n$ distinct, the subspaces $T(\lambda_1),\ldots,T(\lambda_n)$ of $U \oplus X$, where $T(s) = (sI-F)^{-1}G$. By Lemma 2 of [29],

$$\dim(T(\lambda_{i_1}) + \ldots + T(\lambda_{i_k})) \geq \ell_k \quad ,$$

where $\ell_k = \dim sp(G,FG,\ldots,F^{k-1}G)$ are the dual Kronecker indices and hence $\ell_k \geq k$. Following Kimura, we may select independent vectors $v_i \in T(\lambda_i)$ and as before define the gain K by the equation

$$gr(K) = sp\{v_1,\ldots,v_n\} \subset U \oplus X \quad .$$

Then,

$$\det(sI - F + GK) = \prod_{i=1}^{n} (s - \lambda_i) \quad .$$

And if $\{\lambda_1,\ldots,\lambda_n\}$ is self-conjugate, $gr(K)$ can be taken to be self-conjugate.

§4. GLOBAL ANALYSIS OF χ_σ, THE CENTRAL PROJECTION LEMMA, POLE PLACEMENT BY OUTPUT FEEDBACK OVER $\mathbb{R}$ AND $\mathbb{C}$.

Theorem. *If* $mp \leq n$, *then generically* χ_σ *is a proper map. In particular, over* $\mathbb{C}$ *(or any algebraically closed field) image* χ_σ *is a subvariety of* $\mathbb{C}^n$. *Over* $\mathbb{R}$, χ_σ *extends to a map* $\bar{\chi}_\sigma : S^{mp} \to S^n$ *of spheres and image* χ_σ *is Euclidean closed in* $\mathbb{R}^n$.

If $mp > n$, then χ_σ is no longer proper--i.e., $C \subset \underline{k}^n$ a compact set implies $\chi_\sigma^{-1}(C) \subset \underline{k}^{mp}$ is compact, although one can still prove that image χ_σ is (generically) closed.

Proof. The proof begins with a study of the map

$$T : \mathbb{C}^* \to Grass(m, U \oplus Y) \quad .$$

Now, $GL(U \oplus Y)$ acts transitively on m-planes in $U \oplus Y$ and so parameterizes $Grass(m,U \oplus Y)$, i.e., there is a map

$$\pi: GL(U \oplus Y) \to Grass(m, U \oplus Y)$$

$$\pi: g \to gU$$

corresponding to the choice of the m-plane U. π, however, is an overparameterization since there are many g's which fix U. In fact, in terms of the decomposition

$$\underline{k}^{m+p} = U \oplus Y \quad ,$$

the subgroup of g's which fix U has the form,

$$g \in \begin{pmatrix} GL(Y) & 0 \\ Hom(Y,U) & GL(U) \end{pmatrix} = \mathscr{F}, \tag{4.1}$$

the output feedback group! By dividing out by $\mathscr{F}$, we get an honest (i.e., 1-1) parameterization of $Grass(m, U \oplus Y)$,

$$Grass(m, U + Y) \simeq GL(U \oplus Y)/\mathscr{F} \quad .$$

This extends the picture in (3.4) quite a bit. In fact, the main idea of the proof is to extend χ_σ by evaluating the left-hand side of (3.7) for all $g \in GL(U \oplus Y)$; that is, we keep T fixed, as a curve in $Grass(m, U \oplus Y)$, and intersect it will all $\sigma(gY)$, for $g \in GL(U \oplus Y)$. Now, when $g \in \mathscr{F}$, gY is complementary to U and is, in fact, the graph of some linear map $K: Y \to U$. In particular, for such a g

$$T^{-1}(T(\mathbb{C}^*) \cap \sigma(gY)) = \text{sing } T^{g^{-1}} = \text{spec}(F - GKH) \tag{4.2}$$

is an unordered set $\{\lambda_1, \ldots, \lambda_n\}$ of points in $\mathbb{C}^* = \mathbb{P}^1$. That is $\{\lambda_1, \ldots, \lambda_n\} \in \mathbb{P}^1 \times \ldots \times \mathbb{P}^1/S_n \simeq \mathbb{P}^n$, the so-called <u>symmetric product</u>. For $g \in \mathscr{F}$, each λ_i is finite, by virtue of (4.2), and

$$\{\lambda_1, \ldots, \lambda_n\} \in \mathbb{C}^n \subset \mathbb{P}^n \quad .$$

Here $\mathbb{C}^n/S_n \simeq \mathbb{C}^n$, where the isomorphism is simply

$$\{\lambda_1, \ldots, \lambda_n\} \to (c_i(\lambda))_{i=1}^n$$

with c_i the elementary <u>symmetric functions</u>. In summary, we have our old picture in this new setting,

$$\chi_\sigma: \sigma(gY) \to T^{-1}(T(\mathbb{C}^*) \cap \sigma(gY)) \in \mathbb{C}^n \subset \mathbb{P}^n \quad ,$$

via

$$g \to \text{sing}(T^{g^{-1}}) = (c_i(F + GKH))_{i=1}^n \ .$$

By conservation of difficulty, one needs more than this restatement of the problem and it is at this point that we consider $\sigma(gY)$ for any $g \in GL(U \oplus Y)$ or, what is the same $\sigma(Y')$ for a p-plane Y' not necessarily complementary to U.

Lemma A. $\sigma(Y')$ *either contains* $T\mathbb{C}^*$ *or intersects it (counting multiplicity) in exactly* n *points. In the latter case, such a point is infinite if, and only if,* Y' *is not complementary to* U.

Proof. The first part of the lemma is an elementary application of value distribution theory and can be found in Chern, "Complex Manifolds with Potential Theory," D. van Nostrand under the topic: "Holomorphic Curves in a Grassmannian." The second part is, in fact, the condition $U \in \sigma(Y')$ and follows from the definition (3.6).

To facilitate the discussion, I shall refer to a p-plane Y' as a generalized feedback (law) while a p-plane Y' complementary to U will be referred to as a classical feedback (law). The idea is therefore to extend χ_σ in (4.3) to all "generalized feedbacks," i.e., to all points in the dual Grassmannian, Grass(p,U ⊕ Y). That is, we wish to define

$$\bar{\chi}_\sigma : \mathrm{Grass}(p, U \oplus Y) \to \mathbb{P}^n \tag{4.4}$$

via

$$Y' \to T^{-1}(T\mathbb{C}^* \cap \sigma(Y')) .$$

Remark. Consider the scalar case and restrict attention to real gains K. Then, the real Grassmannian is the space of lines in $\mathbb{R}^2$, i.e., the circle S^1 and

$$\bar{\chi}_\sigma : S^1 \to \mathbb{P}^n$$

is precisely the root-locus map!

Now the fact that χ_σ is defined at $\infty \in S^1$ is just the fact that $\chi_\sigma(\infty) = \{\text{zeros of } T(s)\}$. For m,p arbitrary the recent formula of Kailath et al. [25], which computes the difference between the number of closed loop poles of $T(s)$ and the number of open-loop zeroes in terms of the left and right Kronecker indices of $T(s)$, shows that there may not be enough open loop zeroes to account for the asymptotic root loci $\bar{\chi}_\sigma(K)$ as $K \to \infty \in S^{mp}$, although this may be the case if $m = p$. In the Grassmannian compactification $K \to \infty$ takes on an entirely new meaning, as ∞ is replaced by the whole subvariety

$\sigma(U) \subset \text{Grass}(p, U \oplus Y)$. This gives much more freedom in the manner in which K "becomes infinite (and this is important for potential applications, allowing for various channels in the gain to grow at various rates) and Lemma B shows that as K "becomes infinite" the root locus $\chi_\sigma(K)$ (still) apporaches an n-tuple of points in the extended complex plane, as in the classical case, for generic systems provided $mp \leq n$. The case $mp \leq n$ is illustrated below

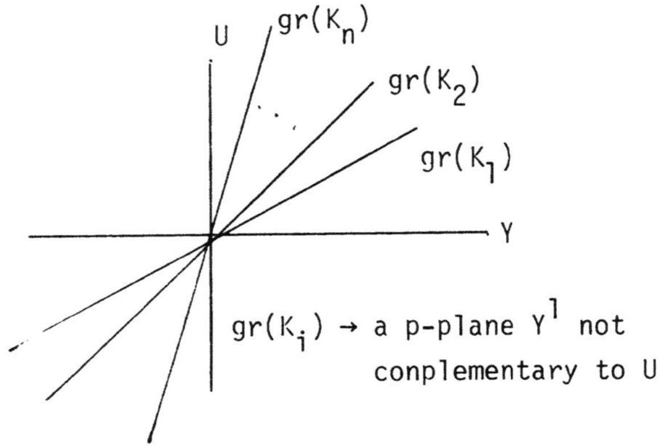

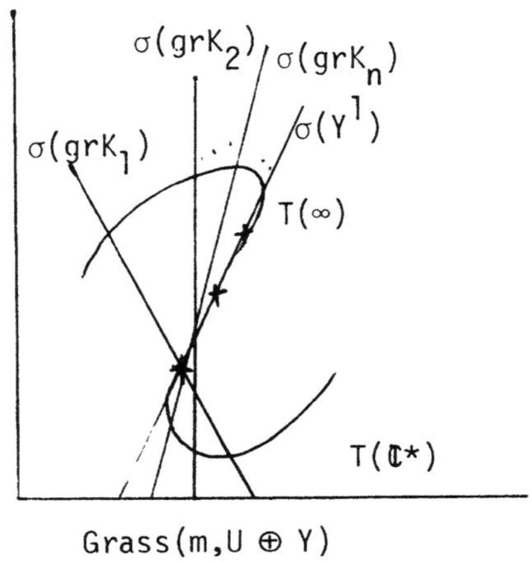

Figure 4.1 Depicting Asymptotic Root Loci as Points of Intersection

The explicit obstruction to this asymptotic extension is in fact explained in the lemma, there can exist p-planes Y' such that $\sigma(Y') \supset T\mathbb{C}^*$. Indeed, if $mp > n$, examples exist in great profusion.

Lemma B. *If* $mp \leq n$, *then for almost all* (F,G,H), $\sigma(Y') \cap T(\mathbb{C}^*)$ *in* n *points, for all* p-*planes* Y'.

Proof. First of all, the set of (F,G,H) for which there exists a Y' with $\sigma(Y') \supset T(\mathbb{C}^*)$ is closed in the variety of all (F,G,H). In fact, if $[(F,G,H)]$ is the corresponding point in the moduli space $\Sigma^n_{m,p}$ then the subset

$$V \subset \Sigma^n_{m,p} \times \mathrm{Grass}(p, U \oplus Y)$$

defined via

$$V = \{([F,G,H],Y') : \sigma(Y') \supset T(\mathbb{C}^*)\}$$

is a subvariety. Since $\mathrm{Grass}(p,U \oplus Y)$ is projective (compact) and $\Sigma^n_{m,p}$ is quasi-projective ([13],[22]) $\pi_1 : \Sigma^n_{m,p} \times \mathrm{Grass}(p,U \oplus Y) \to \Sigma^n_{m,p}$ is closed. In particular, $\pi_1(V)$ is closed in $\Sigma^n_{m,p}$, but $\pi_1(V)$ is precisely the variety we wish to delete in the lemma. To show that, if $mp \leq n$, $\pi_1(V)$ is a proper subvariety, one may appeal to the duality

$$\mathrm{Grass}(m,U \oplus Y) \simeq \mathrm{Grass}(p,U \oplus Y) ,$$

which is related to the duality between inputs and outputs. By Lemma A, to say $\sigma(Y') \supset T(\mathbb{C}^*)$ is to say in particular

$$T(\lambda_1),\dots,T(\lambda_n), \quad T(\infty) \in \sigma(Y') \tag{4.5}$$

for λ_i distinct, finite points. However, to say $T(\lambda) \in \sigma(Y')$ in $\mathrm{Grass}(m,U \oplus Y)$ is to say $Y' \in \sigma(T(\lambda))$ in $\mathrm{Grass}(p,U \oplus Y)$, by the symmetry of the definition (3.6). But $\sigma(T(\lambda))$ is codimension 1 in $\mathrm{Grass}(p,U \oplus Y)$ and hence of dimension $mp-1$. And (4.5) is the assertion

$$Y' \in \bigcap_{i=1}^{n} \sigma(T(\lambda_i)) \cap \sigma(T(\infty)) .$$

Since the Schubert varieties σ are hyperplane sections via the

Plücker imbedding, dim $\cap_{i=1}^{\ell} \sigma(U_i) \geq mp - \ell$ (this is not true for arbitrary varieties, see Section I of this introductory chapter). On the other hand, generically one has dim $\cap_{i=1}^{\ell} \sigma(U_i) \leq mp - \ell$. Therefore Y' must lie, generically, on a subvariety of dimension $mp - (n+1)$, which is impossible unless $mp > n$. ⌑

It is worth remarking that this gives an independent proof that, in case $mp > n$, χ_σ is generically almost onto, assuming the field is algebraically closed. Indeed, using an output feedback invariant version of (2.3) one can give explicitly, the equations defining the generic properly given above.

Returning to the proof of the theorem, by our lemmata, we can generically extend the map χ_σ to the root-locus map

$$\overline{\chi_\sigma}:\mathrm{Grass}(p,U \oplus Y) \to \mathbb{P}^n .$$

In particular, $\overline{\chi_\sigma}$ is a proper map and by the second part of Lemma A, χ_σ is also proper, i.e., since

$$\text{image } \chi_\sigma = \text{image } \overline{\chi_\sigma} \cap \mathbb{C}^n . \tag{4.6}$$

Furthermore, since the real Grassmannian is canonically imbedded in the complex Grassmannian as a compact submanifold, χ_σ remains proper over $\mathbb{R}$. Since the ℓ-sphere is the 1-point compactification of $\mathbb{R}^\ell$, χ_σ extends to a map of spheres. And, by virtue of (4.6), image χ_σ is a subvariety of $\mathbb{C}^n$ and its real points form a closed subspace of $\mathbb{R}^n$. ⌑

In case $mp > n$, one cannot extend χ_σ to the root-locus map as above. However, one can replace the Grassmannian by the closure of the graph of the rational function χ_σ, viz.

$$\overline{\chi_\sigma} = \overline{\text{graph } \chi_\sigma} \subset \mathrm{Grass}(p,U \oplus Y) \times \mathbb{P}^n .$$

And one replaces the map $\overline{\chi_\sigma}$ by the proejction onto the 2nd factor,

$$\pi_2:\overline{\chi_\sigma} \to \mathbb{P}^n .$$

The analog of (4.6) still holds, although one must work a bit harder. In this case, one may still deduce the Theorem, except for the statement that χ_σ is proper. This no longer is valid, as $\chi_\sigma^{-1}(p)$ is a subvariety of positive dimension in affine space and therefore admits unbounded analytic functions.

Remarks. Both cases can be treated in a more unified fashion, but relying on slightly more sophisticated ideas. Denote by P the Plücker imbedding

$$P:\text{Grass}(m,U \oplus Y) \to \mathbb{P}^N$$

of the Grassmannian and suppose for simplicity that, if $\{\lambda_1,\ldots,\lambda_n\} = \text{Sing } T$, for a transfer function $T:\mathbb{P}^1 \to \text{Grass}(m,U \oplus Y)$, then $T(\lambda_i)$ are distinct points in $\text{Grass}(m,U \oplus Y)$ and hence so are the points $V_i = P(T(\lambda_i))$ in $\mathbb{P}^N$. By duality, each point V_i corresponds to linear functional L_i on $\mathbb{C}^{N+1}$ and hence to a hyperplane H_i in $\hat{\mathbb{P}}^N$, the dual projective space. Finally, denote by H_{n+1} (and L_{n+1}) the hyperplane (and functional) corresponding to the point $P(G(\infty))$ and by B, the variety

$$B = \bigcap_{i=1}^{n+1} H_i \quad \text{in} \quad \hat{\mathbb{P}}^N .$$

Following Lemma A, we consider the central projection with base locus B,

$$\phi:\hat{\mathbb{P}}^N - B \to \mathbb{P}^n,$$

defined via

$$\phi(x) = [L_i(x)] , \quad \text{in homogeneous coordinates.}$$

Restricted to $\hat{P}(\text{Grass}(p,U \oplus Y)) - \hat{P}(\text{Grass}(p,U \oplus Y)) \cap B$, one recovers

$$\chi:\text{Grass}(p,U \oplus Y) - B \cap \text{Grass}(p,U \oplus Y) \to \mathbb{P}^n ,$$

as a central projection, with Lemma B asserting that in the correct dimension range χ has no base locus on $\text{Grass}(p,U \oplus Y)$. This admits a particularly nice exploitation of Schubert calculus, especially in the case $mp = n$ (see [5]). For in that case, generically

$$\chi:\text{Grass}(p,U \oplus Y) \to \mathbb{P}^n$$

is globally defined and $\dim \text{Grass}(P,U + Y) = mp = n = \dim \mathbb{P}^n$. The degree of χ is the degree of the subvariety *back provided either* $\min(m,p) = 1$ *or* $\min(m,p) = 2$ *and* $\max(m,p) = 2^r-1$.

We note that Willems and Hesselink ([45]) have shown that, generically in σ, in χ_σ misses an open set in $\mathbb{R}^4$, if $m = p = 2$. This is in harmony with our result but whether the surprisingly combinatorial conditions in this theorem are necessary is at present an open question. On the more positive side, extensions and corollaries of the central projection lemma give rise to sufficient conditions for generic stabilizability in the more general setting $mp \geq n$. These take the form of inequalities

$$C_{m,p} \geq n$$

where $C_{m,p}$ is a function of m and p. These are, however, too complicated to give here.

§5. POLE PLACEMENT OVER RINGS, SOME POSITIVE RESULTS: MORSE'S THEOREM, AND FEEDBACK INVARIANTS

In this section and in the last two, I shall be concerned with the general question of what can be done, particularly in the way of stabilizability and pole-placement, with state feedback where the coefficients of the system lie in a commutative ring R with 1. It is very easy to motivate the study of stabilizability of several classes of systems defined over rings and this has, in turn, motivated the study of general pole-assignability questions over an arbitrary R. This route to stabilizability has the potential advantage (at least over Noetherian R) of providing finite procedures for obtaining a desired gain. However, it is fair to say that, at this point in time, general pole-assignability questions are, in all honesty, primarily mathematical questions about the algebraic structure of dynamical systems, and the reader who wishes to may skip to §7, which deals with the more modest question of stabilizability. On the other hand, pole-assignability questions over a ring are of theoretical interest in their own right and, as recent work has shown, such questions are much harder than anyone had first suspected--even the elementary examples involve non-trivial topological and arithmetic obstructions. We begin with a quick review of the main problems and motivating examples, for the realization theory of such systems we should refer to Professor Rouchaleau's lectures ([40]).

R is a commutative ring with 1, $X \simeq R^{(n)}$ and $U \simeq R^{(m)}$ are free R-modules. It is meaningful to distinguish between two versions of the question: First, the problem of solving the system of equations, for $K \in \mathrm{Hom}_R(X,U)$,

$$\det(sI - F - GK) = p(s) \ , \quad p(s) \in R[s] \text{ monic of deg } n, \tag{5.1}$$

$$\hat{P}(\mathrm{Grass}(p,U+Y)) \subset \hat{\mathbb{P}}^N,$$

since χ is a central projection, and this is well-known ([31]) to be

$$\deg \chi = \frac{1!\ 2! \cdots (p-1)!\ (mp)!}{m!\ (m+1)! \cdots (m+p-1)!} \quad . \tag{4.7}$$

Briefly, what this entails is first the observation that the system of n equations in mp unknowns,

$$\chi(K) = p(s) \ ,$$

can be expressed as an intersection of n hypersurfaces in mp space and these hypersurfaces are well-studied. That is, regarding p, via its roots, as a point $(\lambda_1,\ldots,\lambda_n) \in \mathbb{P}^1)^{(n)}$, one can view p as $\bigcap_{i=1}^{n} H_{\lambda_i}$ of n hyperplanes, H_λ denoting the hyperplane of $(\lambda_1,\ldots,\lambda_n)$ such that $\lambda_i = \lambda$, for some i. In this setting, $\chi^{-1}(p)$ has the form

$$\chi^{-1}\Big(\bigcap_{i=1}^{n} H_{\lambda_i}\Big) = \bigcap_{i=1}^{n} \chi^{-1}(H_{\lambda_i}) = \bigcap_{i=1}^{n} \sigma(T(s_i))$$

provided $\lambda_i = T(s_i)$ lies on the curve $T(\mathbb{P}^1)$. Now, the Schubert calculus enables one to express such intersections in terms of basic, or Schubert, varieties. In particular, if $mp = n$ then a repeated use of one such expression, Pieri's formula, allows us to count the number of points in

$$\bigcap_{i=1}^{n} \sigma(T(s_i)),$$

counting multiplicities, as $\deg \chi$ in (4.7). Thus, the main point of all this is that the output feedback map as a system of equations is actually a well-studied, classical system of equations-- about which much is known (see [31]). As a corollary, one can show that when $\deg \chi$ is odd the map

$$\chi : \mathrm{Grass}_{\mathbb{R}}(p, U \oplus Y) \to \mathbb{RP}^n$$

is surjective, hence we can place poles with real gains! It has been shown ([1]) that $\deg \chi$ is odd if, and only if, either $\min(m,p) = 1$ or $\min(m,p) = 2$ and $\max(m,p) = 2^r - 1$, a Mersenne number.

<u>Theorem</u> ([5]). *Assume* $mp = n$. *It is possible generically in* σ *to place any self-conjugate set of poles by real output feed-*

where $F:X \to X$, $G:U \to X$ are fixed, is referred to as coefficient assignability (over R). Second, if

$$p(s) = \prod_{i=1}^{n} (s-r_i)$$

with $r_i \in R$, then to solve 5.1 for K is to solve a problem of pole, or zero, assignability over R.

In the general situation, one may think of (X,U,F,G) as the data defining the discrete-time system,

$$x(t+1) = Fx(t) + Gu(t) \quad . \tag{5.2}$$

On the other hand, in specific situations, this data may represent continuous-time systems. Explicitly, if one considers a controlled heat equation

$$\frac{\partial u}{\partial t} = \Delta u + f \tag{*}$$

on the n-torus $\mathbb{T}^n$, then it is rather natural (and frequently done) to discretize (*) in the spatial variable relative to a grid in $\mathbb{T}^n$, obtaining a "lumped approximation." If one chooses the grid G_ℓ of points of order ℓ on $\mathbb{T}^n$, then (*) reduces to a linear control system with coefficients in the group algebra $\ell^1(G_\ell)$ of the group $G_\ell \subset \mathbb{T}^n$ (see [7]).

One important class of examples is the class of systems depending on parameters, say in a C^k-fashion,

$$\dot{x}(t) = F(\lambda)x(t) + G(\lambda)u(t), \quad \lambda \in U \subset \mathbb{R}^N \tag{5.2)'$$

where λ is the value of a resistor, or an altitute or attitude. Another class of examples arises in the algebraic theory of delay-differential systems, where a system,

$$\dot{x}(t) = F\star x(t) + G\star u(t) \quad , \tag{5.2)"$$

is regarded as a system defined over a ring of convolution operators ([24],[36]). Explicitly, consider the system

$$\dot{x}(t) = F_0x(t) + F_1x(t-1) + Gu(t) \quad . \tag{5.3}$$

Introducing the convolution operator,

$$(\delta\star x)(t) = x(t-1) \quad ,$$

one may regard (5.3) as

$$\dot{x}(t) = F(\delta)*x(t) + G*u(t) \tag{5.3'}$$

where $F = F_0 + F_1\delta$ is defined over $R = \mathbb{R}[\delta]$, as is G. More generally, a delay-differential system, involving only the commensurate delays by 1,2,... seconds, may be regarded as a system

$$\dot{x}(t) = F(\delta)*x(t) + G(\delta)*u(t) \tag{5.4}$$

defined over $R = \mathbb{R}[\delta]$. The non-commensurate case, of course, leads to a change of scalars $\mathbb{R} \subset \mathbb{R}[\delta_1,\ldots,\delta_N]$.

In both of these latter cases, it is important to know whether there exists a gain K, preferably defined over R, such that the closed loop system $(F+GK,G)$ is asymptotically stable in an appropriate functional analytic sense. In the first case, since ideally K ought to depend on the system (F,G), it is clear that if $(F(\lambda),G(\lambda))$ is C^k for $0 \le k \le \omega$, in $\lambda \in U$ then K ought to be C^k in λ as well; i.e., K ought to be defined over the ring $R = C^k(U)$. Now, in the second case, to ask that K be defined over $R = \mathbb{R}[\delta_1,\ldots,\delta_N]$ is natural from the point of view that K should be constructible from the same components as the system (F,G). And, if one can place the poles of (F,G) over R, then for each $c > 0, \exists K$ defined over R such that $(F+GK,G)$ is exponentially stable (in L^1) with order c ([27], Theorem 8). On the other hand, one should remark that, especially in light of §6, the functional approach has been far more successful ([14],[32]) in obtaining pole-placement results, at the expense of using more general operators K (eg. convolution with continuous measures).

Now, motivated by work of A. S. Morse ([36]) on delay-differential systems, one is led to the

Definition 5.1. *The system* $\sigma = (F,G)$, *defined over* R, *is reachable over* R *just in case the controllability operator*

$$C = (B, AB, \ldots) \quad \bigoplus_{i=1}^{\infty} U \to X \tag{5.5}$$

is surjective.

As observed in [44], if σ is coefficient assignable, then σ is reachable. To see this, setting $\max(R) = M{:}M$ a maximal ideal of $R\}$, recall (see Introductory Chapter) that C is surjective if, and only if,

$$C\colon \bigoplus_{i=1}^{\infty} U \to X \qquad \mathrm{mod}(M)$$

is surjective for all $M \in \max(R)$. For σ defined over R, denote by $\sigma(M)$ the system (F,G) over the field R/M, obtained by reducing (X,U,F,G) modulo $M \in \max(R)$. Now, if σ is coefficient assignable over R, then $\sigma(M)$ is coefficient assignable over R/M and therefore reachable over R/M for any $M \in \max(R)$, as we wished to show. Similarly, zero-assignability implies reachability and the converse, for R a P.I.D., is due to A. S. Morse.

Theorem 5.2 ([36]) *A reachable system* $\sigma = (F,G)$, *defined over* $\mathbb{R}[x]$, *is zero assignable.*

Indeed, Morse's proof applies to reachable systems defined over a P.I.D. and, in this setting, is the best result known at present.

In order to study the coefficient assignability question over R, it is useful to bring in the group of symmetries for state feedback and the Rosenbrock pencil. Explicitly, consider the pencil of equations

$$sx(x) = Fx(s) + Gu(s) \tag{5.6}$$

where $x(s) = \Sigma x_i s^i$, $u(s) = \Sigma u_i s^i$, are polynomial "vectors" with coefficients in the modules X,U, respectively. Thus, the pencil (5.6) is a "formal Laplace transform" of (5.2) and, once again, this transform intertwines the action on systems of the state feedback group with a classical action. It is more precise to regard $x(s)$ as an element of the $R[s]$ module $X[s] \simeq X \otimes \mathbb{Z}[s]$ and $u(s)$ as an element of $U[s]$. Then, the Rosenbrock pencil takes the form

$$\mathbb{R}:(X \oplus U)[s] \to X[s] \quad ,$$

$$(x(s),u(s)) \to (Fx(s) - sIx(s) + Gu(s)) \; , \tag{5.7}$$

and $\mathbb{R}$ is surjective if, and only if, (F,G) is reachable over R. In this case, we are led to the exact sequence,

$$0 \to \ker \mathbb{R} \to (X \oplus U)[s] \to X[s] \to 0 \; , \tag{5.8}$$

where the submodule $\ker \mathbb{R}$ is, at least formally, the Laplace transform of solutions to (5.2) with zero initial data. Now, just as in the case $R = k$, one may show that the strict equivalence of 2 such pencils $\mathbb{R}_1$ and $\mathbb{R}_2$, in the sense of linear algebra, implies the equivalence (under state feedback) of the systems σ_1 and σ_2. That is, to say $\mathbb{R}_1 \sim \mathbb{R}_2$ is to say there exists

$$C:(X \oplus U)[s] \to (X \oplus U)[s]$$

and $D: X[s] \to X[s]$,

invertible maps of $R[s]$-modules, such that

$$DR_1C = \mathbb{R}_2 , \qquad (5.9)$$

and such that C,D are independent of s (i.e. extended from R-module maps). Since $\mathbb{R}_1$ (and $\mathbb{R}_2$) has block diagonal form, it follows from comparing degrees (with respect to s) in (5.9) that C decomposes, with respect to $(X \oplus U)[s] \simeq X[s] \oplus U[s]$, as

$$C = \begin{pmatrix} D^{-1} & 0 \\ KD^{-1} & B \end{pmatrix} \qquad (5.10)$$

where $B \in GL(U)$ and $K \in Hom(X,U)$. In particular, B,D, and K give the desired equivalence of σ_1 and σ_2 modulo the state feedback group $\mathcal{F}_s(R)$. Conversely, this familiar triangular matrix representation of $\mathcal{F}_s(R)$ shows that an equivalence mod $\mathcal{F}_s(R)$ induces a strict equivalence of $\mathbb{R}_1$ with $\mathbb{R}_2$ (see [26]). This is summarized in the classical and well-known proposition:

Theorem 5.3 *The Rosenbrock pencil* $\mathbb{R}$ *of* σ, *up to strict equivalence, is a complete invariant for* σ *modulo the state feedback group* $F_s(R)$.

Now, over a field, Kronecker has given a classification for matrix pencils in terms of the degrees of minimal basis vectors for the submodule $\ker \mathbb{R} \subset (X \oplus U)[s]$. These degrees constitute a partition

$$\sum_{i=1}^{m} n_i = n, \quad n_1 \geq n_2 \geq \ldots \geq n_m \geq 0 \qquad (5.11)$$

of $\dim(X)$, and in this way one obtains a complete set of invariants for σ modulo $\mathcal{F}_s(\underline{k})$, see Professor Rosenbrock's lectures ([39], esp. §5). Now, for general R, one may replace the arithmetic data (5.11) by the isomorphism class of the submodules,

$$0 \to \ker \mathbb{R} \to (X \oplus U)[s], \qquad (5.12)$$

and in some cases this isomorphism class is expressible in a more intrinsic form.

Suppose $R = \underline{k}[\lambda_1,\ldots,\lambda_N]$, with $\underline{k}$ algebraically closed. Since R in (5.7) is surjective R-module map of free R-modules, it is not too hard to see that $\ker \mathbb{R}$ is projective as an R-module and has finite rank, since X and U have finite rank. Recalling the connection between projective modules and vector bundles (see Introductory Chapter, Section III), it is quite plausible to seek a vector bundle characterizing the data (5.12). Indeed, consider the m-vector bundle on $\mathbb{A}^N \times \mathbb{A}^1$ whose fiber over (λ, s) is the vector space

$$\ker[F(\lambda) - sI, G(\lambda)] \subset \underline{k}^n \oplus \underline{k}^m \quad . \tag{5.13}$$

Now, since all m-vector bundles on $\mathbb{A}^N \times \mathbb{A}^1$ are isomorphic (by the Quillen-Suslin Theorem) provided we allow isomorphisms depending on $s \in \mathbb{A}^1$, (5.13) is not fine enough. However, the notion of strict equivalence (i.e., independence of s) suggests an order of growth at $s = \infty$ reminiscent of Lionville's Theorem. That is, by homogenizing all of the above we construct a bundle W_σ on $\mathbb{A}^N \times \mathbf{P}^1$, whose fiber over $(\lambda, [s,t])$ is given by

$$\ker[tF(\lambda) - sI, t\,G(\lambda)] \subset \underline{k}^n \oplus \underline{k}^m. \tag{5.13)'$$

Theorem 5.4 ([10]) *The bundle* W_σ *is a complete invariant for* σ *modulo state feedback.*

Remarks. For $N = 0$, this was studied by R. Hermann and C. Martin, who related the Kronecker invariants (5.11) to the Grothendieck invariants of W_σ, thereby proving Theorem 5.4. For $N > 0$, if one forms the transfer function for the triple $(F(\lambda), G(\lambda), I)$, then just as in the earlier sections one obtains a map

$$T: \mathbb{A}^N \times \mathbf{P}^1 \to \mathrm{Grass}(m, \underline{k}^n \oplus \underline{k}^m)$$

exhibiting (5.13)' as the pullback along the transfer function of the (topological) universal bundle, and Theorem 5.4 follows from Riemann-Roch (see Professor Martin's Lectures, [33]).

Now, if σ is independent of λ then W_σ is independent of λ, i.e. W_σ is a pullback along the second projection $p_2: \mathbb{A}^N \times \mathbf{P}^1 \to \mathbf{P}^1$ of a bundle on $\mathbb{P}^1$. And, Theorem 5.4 asserts that the converse in true. Thus, if W_σ is a pullback, then σ is coefficient assignable--since this result is valid for σ defined over the field $\underline{k}$. On the other hand, one can express this condition more explicitly. For each λ, σ gives rise to a system $\sigma(\lambda)$, by evaluation of λ, over the field $\underline{k}$ and therefore to pointwise Kronecker indices,

$$\sum_{i=1}^{m} n_i(\lambda) = n(\lambda) = n \ .$$

Corollary 5.5 *If the* $n_i(\lambda) = n_i$ *are constant in* λ, *then* σ *is feedback equivalent to a constant system. In particular,* σ *is coefficient assignable.*

Proof. It follows from the main theorem of C. C. Hanna's thesis [18], that constancy in λ of the (Kronecker)-Grothendieck indices implies

$$W_\sigma \simeq \sum_j p_1^*(V_j) \otimes p_2^*(W_j)$$

where V_j is a vector bundle on $\mathbb{A}^N$ and W_j is a bundle on $\mathbb{P}^1$. By the Quillen-Suslin Theorem, V_j is trivial and therefore

$$W_\sigma \simeq \sum_j p_2^*(W_j) \simeq p_2^*(\sum_j W_j) \tag{5.14}$$

Moreover, $\sum_j W_j \simeq \sum_{i=1}^{m} \mathcal{O}(n_i)$. (5.14), however, is enough for our purposes. ⌑

Example 5.6 The use of the Quillen-Suslin Theorem is, in fact, essential. Consider the following reachable pair, defined over $R = C^\infty(S^2)$. Define $U \simeq R^{(3)}$ as the module of smooth sections of a rank 3, trivial vector bundle on S^2--viz., the restriction of the tangent bundle $T(\mathbb{R}^3)$ to S^2. If X is the R-module of smooth sections of $T(S^2)$, i.e. smooth vector fields on S^2, then $X \subset U$. In fact,

$$U \simeq X \oplus R^{(1)}$$

where $R^{(1)}$ is the module of sections of the normal bundle to $S^2 \subset \mathbb{R}^3$. In particular, we are led to the reachable pair $\sigma = (F,G)$ defined by $G = \mathrm{Proj}_1 : U \to X$, and $F = \mathrm{Id} : X \to X$. One easily checks that for $p \in S^2$, the pointwise Kronecker indices of σ are given by

$$(n_1(x), n_2(x)) = (1,1), \quad \text{for } x \in S^2,$$

noting that $\max C^\infty(S^2) \simeq S^2$ in the canonical way. However, the spectrum of (F,G) is not arbitrarily assignable, suppose $FK : X \to U$ such that $\mathrm{spec}(F + GK) = \{0,1\}$. Then $F + GK$ is a projection on X and its kernel and image give rise to a decomposition, $X = M_1 \oplus M_2$, corresponding to a decomposition

$$T(S^2) = L_1 \oplus L_2$$

of the tangent bundle into line subbundles. Now, since $\dim L_i = 1$, each L_i is an integrable distribution and, by Frobenius' theorem, L_i forms a codimension one folition on S^2, as it were, implying that $\chi(S^2) = 0$, contrary to fact.

On the other hand, if one supposes that X and its projective submodules are free, then it becomes harder (see §6) to construct a "counterexample" to pole-assignability for reachable pairs. Moreover, as several authors ([11],[21b],[41]) have noted since Corollary 5.5 appeared, under these conditions on the state module X, a reachable pair (F,G) is coefficient assignable whenever the pointwise Kronecker indices $(n_1(M),\ldots,n_m(M))$ of $\sigma(M)$ are constant in $M \in \max(R)$. More generally, if $\max(R)$ is given the Stone-Jacobson-Zariski topology, where the basic closed sets have the form

$$h(I) = \{M : I \subset M\}, \quad \text{for } I \text{ and ideal of } R,$$

then σ is coefficient assignable whenever the pointwise Kronecker indices are locally constant. Recall that a ring R is said to be "projective-free" just in case each finitely generated projective module over R is free, thus the Quillen-Suslin Theorem asserts that $R = \underline{k}[x_1,\ldots,x_N]$ is projective free.

Proposition 5 *Suppose* R *is projective-free and* $\sigma = (F,G)$ *is a reachable system with free state module and locally constant Kronecker invariants, then* σ *is coefficient assignable.*

Remarks.

1. The basic idea in the proof is to note first that constancy of the pointwise Kronecker indices $(n_1(M),\ldots,n_m(M))$ is equivalent to constancy of the rank

$$r_j(M) = \operatorname{rank}(G(M),F(M)G(M),\ldots,F^{j-1}(M)G(M))$$

(as the $r_j(M)$ form the dual partition to the partition $\Sigma n_i(M) = n$ of n) and that hypothesis on R now implies the freeness of the projective modules

$$(0) \subset \operatorname{span}(G) \subset \operatorname{span}(G,FG) \subset \ldots \subset \operatorname{span}(G,FG,\ldots,F^nG) = X. \quad (5.15)$$

A careful choice of basis now puts $\sigma = (F,G)$ in a standard canonical form, in which form coefficient assignability is immediate. For a very careful proof, see [21b].

2. This is also the route taken in [11]. However, the proposed extension (by working locally and then trying to patch local solutions) of this argument to cover arbitrary R and projective X is incorrect--as Example 5.6 amply demonstrates.

3. I would like to raise the question: Can Proposition 5.7 be improved upon by assuming only that

$$\sum_{i=1}^{m} \mathrm{in}_i(M) = \text{constant} \quad ?$$

This is much more applicable, and I know of no counterexample.

We close this section with a few corollaries to Proposition 5.7.

Corollary 5.8 (Sontag) *If* R *is semilocal, then to say* σ *is reachable is to say* σ *is coefficient assignable.*

Proof ([41]) A semi-local ring has only finitely many maximal ideals, by definition. Thus, max(R) is discrete and every function is locally constant.

Corollary 5.9 (Brockett and Willems) *If* R *is the group algebra of a finite abelian group, then to say* σ *is reachable is to say* σ *is coefficient assignable.*

This follows, although not historically ([17]), from Corollary 5.8 or from somewhat deeper considerations. That is, one measure of the complexity of calculations in R is the structure of set of primes of R,

$$\mathrm{spec}(R) = \{P : P \text{ a prime ideal of } R\} ,$$

and in particular of the Krull dimension of R--i.e., the least upper bound of the lengths of chains of prime ideals of R (see Introductory Chapter, Section VI). Note that any P.I.D. has dimension 1, whereas a field has dimension 0.

Corollary 5.10 *If* R *has Krull dimension* 0, *then to say* σ *is reachable over* R *is to say* σ *is coefficient assignable.*

One example of such a ring, in addition to the group algebras of Corollary 5.9, is furnished by the class of Boolean rings. Indeed, for any ring of dimension 0, max(R) is a Boolean space and based on the general (sheaf-theoretic) structure theory for such rings, we may apply Proposition 5.7.

The present state of affairs is rather intriguing. Morse's Theorem suggests that zero assignability is perhaps related to reachability for rings of Krull dimension 1, while Example 6.1 shows that reachability does not imply zero assignability in dimension 2.

§6. THE COUNTEREXAMPLES OF BUMBY, SONTAG, SUSSMANN AND VASCONCELOS.

In this section we present recent counterexamples ([8]) to zero assignability (over $\mathbb{R}[x,y]$) and to coefficient assignability (over $\mathbb{Z}$, or $\mathbb{R}[x]$). Indeed, all that we do here is for systems σ with rank $X = 2$. Note that, in this case, as an easy consequence ([10], Corollary 4.2) of the results in §5, one knows for R a polynomial ring or the ring $\mathbb{Z}$:

> Let $n = 2$ or 3, and suppose $G(M)$ has constant rank for all $M \in \max(R)$. Then $\sigma = (F,G)$ is reachable if, and only if, σ is coefficient assignable. (*)

<u>Example 6.1</u> Let $R = \mathbb{R}[\lambda_1,\lambda_2]$ and consider $\sigma = (F,G)$:

$$F(\lambda_1,\lambda_2) = \begin{pmatrix} 1 & -1 \\ 1 & 1 \end{pmatrix}, \quad G(\lambda_1,\lambda_2) = \begin{pmatrix} \lambda_1+\lambda_2 & \lambda_1+\lambda_2-1+\lambda_1^2+\lambda_2^2 \\ \lambda_2-\lambda_k & \lambda_2-\lambda_1+1-\lambda_1^2-\lambda_2^2 \end{pmatrix} \tag{6.1}$$

Notice that σ is reachable over $R_{\mathbb{C}} = \mathbb{C}[\lambda_1,\lambda_2]$. Indeed, it is easy to compute the Kronecker indices $(n_1(\lambda),n_2(\lambda))$ of σ. Consider the algebraic sets in $\mathbb{C}^2$,

$$V_1^{\mathbb{C}} = \{(\lambda_1,\lambda_2):\lambda_1^2 + \lambda_2^2 = 1\} ,$$

$$V_2^{\mathbb{C}} = \{(\lambda_1,\lambda_2):\lambda_2 = 0\} .$$

With this notation,

$$(n_1(\lambda),n_2(\lambda)) = \begin{cases} (2,0) & \text{if } \lambda \in V_1 \cup V_2 , \\ (1,1) & \text{otherwise} . \end{cases} \tag{6.2}$$

In either case, $n(\lambda) = n_1(\lambda) + n_2(\lambda) = 2$, so σ is reachable over $R_{\mathbb{C}}$. However, one cannot find $K(\lambda_1,\lambda_2)$ defined over R such that

$$\det(sI - (F+GK)) = s(s+1) .$$

For then, as it were, the submodule

$$\ker(F + GK) \subset R^{(2)}$$

is complemented and hence gives rise to a splitting of R-modules

$$R^{(2)} \simeq M_o \oplus M_{-1} , \tag{6.3}$$

with each M_i free of rank 1. But, to say M_o is free is to say there exists $u \in R^{(2)}$, a unimodular element a posteriori, such that $V = F^{-1}Gu$ is a generator for M_o. Computing along the real points of $V_1^{\mathbb{C}}$, viz. S^1,

$$v(\lambda) = \begin{pmatrix} v_1(\lambda) \\ v_2(\lambda) \end{pmatrix} = \begin{pmatrix} \lambda_2 & \lambda_2 \\ -\lambda_1 & -\lambda_1 \end{pmatrix} \begin{pmatrix} u_1(\lambda_1,\lambda_2) \\ u_2(\lambda_1,\lambda_2) \end{pmatrix} \tag{6.4}$$

is a non-zero tangent vector field to S^1, extendable throughout $\mathbb{R}^2$. By the Poincaré-Bendixson Theorem, $v(\lambda)$ has a zero inside the unit disc, contradicting unimodularity.

Next, consider the question of coefficient assignability for 2×2 systems over a P.I.D. As an example, consider the following system σ defined over $R = \mathbb{Z}$.

Example 6.2

$$F = \begin{pmatrix} 0 & 0 \\ 3 & 1 \end{pmatrix} , \qquad G = \begin{pmatrix} 1 & 0 \\ 0 & 8 \end{pmatrix}$$

For p a prime, the Kronecker indices of $\sigma(p)$ are given by

$$(n_1(p),n_2(p)) = \begin{cases} (2,0) & \text{if } p \text{ is even,} \\ (1,1) & \text{if } p \text{ is odd.} \end{cases}$$

Consider the monic polynomial, $p(s) = s^2 - s - 1$, noting that neither Theorem 5.2 nor Corollary 5.5' apply. In fact, the system of Diophantine equations

$$\mathrm{tr}(F + GK) = 1 \tag{6.4a}$$

$$\det(F + GK) = 1 \tag{6.4b}$$

has no solution

$$K = \begin{pmatrix} x & y \\ w & v \end{pmatrix} \in M_2(\mathbb{Z}) .$$

To see this, substitute a solution of (6.4a) into (6.4b), obtaining

$$100w^2 + 10w - 1 = (-y)(3 + 10v) \tag{6.4$'$}$$

or

$$(100w^2 + 10w - 1) \equiv 0 \bmod(3 + 10v) \tag{6.5}$$

Now (6.5) has a solution if and only if the discriminant $\Delta = 500$ is a square modulo $|3 + 10v|$, if and only if 5 is a square modulo $|3 + 10v|$. It can be shown using Quadratic Reciprocity that this occurs if, and only if, 3 is a square modulo 5, contrary to fact.

One may construct a similar counterexample over $\mathbb{R}[x]$; in contrast, all reachable 2×2 systems over $\mathbb{C}[x]$ are coefficient assignable ([8]). More generally, if R is a P.I.D., then following the matrix operations in [36], (F,G) may be taken (modulo state feedback) in the form

$$F = \begin{pmatrix} 0 & 0 \\ b & 1 \end{pmatrix} \quad , \quad G = \begin{pmatrix} 1 & 0 \\ 0 & c \end{pmatrix}$$

where $(b,c) = 1$. Now following [8], if $f(s) = s^2 - \alpha s + \beta$, then arguing as above leads to the condition: if there exists $K \in M_2(F)$ satisfying

$$\det(sI - F - GK) = f(s), \tag{6.6}$$

then $\alpha^2 - 4\beta$ is a square mod(p), for any irreducible p dividing $b + cv$. That is, the solvability of (6.6) implies

the monic $f(s)$ splits modulo p, for each prime $p|b + cv$, (6.7)

This is in harmony with Morse's Theorem--where $f(s)$ is assumed to split over R. And, if R has the property for $p \in \max(R)$, $\mathrm{char}(R/(p)) \neq 2$, then (6.7) is also sufficient, giving a refinement of Morse's Theorem in the 2×2 case. It appears that the general case lies much deeper. Moreover, the criterion (6.7) involves the unknown quantity v and for this reason is not always easy to apply. As a final remark, we may include $\mathbb{Z}$

but, of course, special care must be taken in including the prime $p = 2$.

§7. STABILIZABILITY OF PARAMETERIZED FAMILIES, DELCHAMP'S LEMMA

Consider a parameterized family of linear systems,

$$\dot{x}(t) = F(\lambda)s(t) + G(\lambda)u(t), \quad x(0,\lambda) = x_0(\lambda), \tag{7.1}$$

real analytic in $\lambda \in \Lambda$, an open subset of $\mathbb{R}^N$ (although we could, of course, take Λ to be a real analytic submanifold of $\mathbb{R}^N$). We seek a real analytic $K(\lambda)$ such that the force-free closed-loop system

$$\dot{x}(t) = (F(\lambda) + G(\lambda)K(\lambda))x(t) \tag{7.2}$$

is asymptotically stable, for all λ. It is natural to find such a $K(\lambda)$ by solving a variational problem, in this case a quadratic optimal control problem leading to an algebraic Riccati equation for $K(\lambda)$. Now, a lemma of D. Delchamps' applies the implicit function theorem--on the manifold of controllable pairs (F,G)--to show that such a $K(\lambda)$ can be chosen real analytic in λ. This also applies to C^k-families, for $k \geq 1$, and by a little global reasoning we extend this to continuous families as well. We begin by giving an exposition of these ideas.

First, suppose $\Lambda = \mathbb{R}^N$ and $\sigma = (F(\lambda),G(\lambda))$ is controllable for all λ. If the Kronecker indices $(n_1(\lambda),\ldots,n_m(\lambda))$ of σ are constant, then we can place the spectrum of $F(\lambda)$ arbitrarily (modulo the constraint that the eigenvalues form a self-conjugate set) and thus, in particular, find a stabilizing $K(\lambda)$ as in (7.2). In order to motivate what follows, we offer another proof of this fact. Set

$$C(n,m) = \{(F,G):(F,G) \text{ is controllable}, \; F^{n\times m}, G^{n\times m}\},$$

and denote the state feedback group of $F_s = F_s(\mathbb{R})$. Thus, one has a real algebraic group action

$$\mathscr{F}_s \times C(n,m) \to C(n,m) \tag{7.3}$$

with finitely many orbits

$$\mathscr{O}_{(F,G)} = \{g(F,G): g \in \mathscr{F}_s\} ,$$

parameterized by the partition

$$\sum_{i=1}^{m} n_i = n, \quad n_1 \geq n_2 \geq \ldots \geq n_m \geq 0$$

of the McMillan degree n into the Kronecker indices. Now, a real analytic family $\sigma = (F(\lambda),G(\lambda))$, for $\lambda \in \Lambda$, is given by a real analytic function,

$$f_\sigma : \Lambda \to C(n,m) \ , \qquad (7.4)$$

and conversely. In this context, to say that the Kronecker indices are constant is to say that the function

$$f_\sigma : \Lambda \to \mathcal{O}_{(F,G)} \subset C(n,m)$$

has its range in a single orbit of the action (7.3). Thus, if

$$\mathcal{H}_{(F,G)} = \{g \in \mathcal{F}_s : g(F,G) = (F,G)\}$$

one has a real analytic map, for (F,G) a point on $\mathcal{O}$,

$$f_\sigma : \Lambda \to \mathcal{F}_s / \mathcal{H}_{(F,G)} \ . \qquad (7.5)$$

A study of the topology of $\mathcal{F}_s / \mathcal{H}_{(F,G)}$ was begun in [4], where (for example) formulae, in terms of the Kronecker indices, for the number of connected components and for the dimension of $\mathcal{F}_s / \mathcal{H}_{(F,G)}$ are given. Here, we need only know that $\mathcal{O}$ is a homogeneous space which is the base of an $\mathcal{H}_{(F,G)}$ fiber bundle $\mathcal{F}_s \to \mathcal{F}_s / \mathcal{H}_{(F,G)}$. In particular, f_σ in (7.5) induces an $\mathcal{H}_{(F,G)}$-bundle on Λ, viz.

$$\mathcal{L}_\sigma^* \mathcal{F}_s \to \Lambda \qquad (7.6)$$

and to say (7.6) is trivial is, of course, to say that by using real analytic state feedback one can put $(F(\lambda),G(\lambda))$ into a canonical form which is independent of λ. For example, by choosing $\mathcal{H}$ to be the isotropy subgroup of the Brunovsky normal form one obtains a global Brunovsky form. In any case, coefficient assignability follows from the result over a field. Assuming $\Lambda = \mathbb{R}^N$, such a bundle is trivial and therefore $(F(\lambda),G(\lambda))$ is coefficient assignable over the ring $R = C^\omega(\mathbb{R}^N)$.

With this notation in mind, consider the variational problem on $C(n,m)$, where $L(\sigma)$ is an arbitrary real analytic, positive definite form in $\sigma \in C(n,m)$: minimize the functional

$$\eta = \int_0^\infty (s'(t)L(\sigma)s(t) + u(t)u'(t))dt$$

along trajectories $(s(t),u(t))$ of the system σ,

$$\frac{dx}{dt}(t) = Fx(t) + Gu(t) \quad ,$$

initialized at some (fixed) real analytic state vector, $x_0 = x_\sigma$.

It is well-known that for a single system $\sigma = (F,G)$ the minimizing control is given by

$$u_\sigma(t) = -G'K(\sigma)\exp\{(F - GG'K(\sigma))t\}\, x_\sigma \quad ,$$

where K_σ is the unique positive definite solution to algebraic Riccati equation

$$F'K(\sigma) + K(\sigma)F - K(\sigma)GG'K(\sigma) + 4(\sigma) = 0 \ . \qquad (7.7)$$

D. Delchamps has shown me a proof that $K(\sigma)$ is real analytic in σ, we only need consider the case $L(\sigma) = I$. In this case, $V = \{$positive definite symmetric forms on $\mathbb{R}^n\}$, we consider the real analytic map π

$$C(n,m) \times V \to C(n,m)$$

restricted to the subvariety $X = \{(\sigma,K)$ satisfying (7.7)$\}$.

<u>Lemma 7.1</u> (Delchamps) X *is a submanifold and* $\pi|_X$ *is a submersion, with* 0-dimensional *fibers. In particular,* π *is a real analytic diffeomorphism with inverse* $K(\sigma) = (\sigma,K_\sigma)$.

Now consider the universal family of systems, $\sigma = (F(\lambda),G(\lambda)) \in C(n,m)$, parameterized by the real analytic manifold $C(n,m)$. Since for each fixed σ, the choice of state feedback $\tilde{K}(\sigma) = G'(\sigma)K(\sigma)$ renders the closed-loop system

$$\dot{x}(t) = (F(\sigma) + G(\sigma)\tilde{K}(\sigma))$$

asymptotically stable Delchamps Lemma implies the existence of a stabilizing gain, for all σ, analytic in σ. In particular, if $\Lambda \subset C(n,m)$ is a submanifold then restricting $\tilde{K}$ one obtains a stabilizing gain for $(F(\lambda),G(\lambda))_{\lambda\in\Lambda}$ analytic in λ. More generally,

Proposition 7.2 *If* $(F(\lambda),G(\lambda))$ *is* C^k *in* Λ, $k \geq 0$, *then there exists a gain* $\tilde{K}(\lambda)$, C^k *in* Λ, *for which the closed-loop system*

$$\dot{x}(t) = (F(\lambda) + G(\lambda)\tilde{K}(\lambda))\, x(t) \qquad (7.8)$$

is asymptotically stable, for all λ. *In fact,* $\tilde{K}$ *is a function of the system* $(F(\lambda),G(\lambda))$.

Proof. $(F(\lambda),G(\lambda))$ defines a C^k-function

$$f:\Lambda \to C(n,m)$$

as in (7.4). By composing the real analytic gain $\tilde{K}(\sigma)$ with f one obtains a C^k-gain, rendering (7.8) stable for all λ.

Remarks.

1. What is surprising here is that the C^0 case comes out so easily, indeed much more is true--for example, similar conclusions hold for Lipschitz continuous functions or L^∞ functions on a finite measure space, by applying the Gel'fand representation to the Banach algebra $L^\infty(X)$ ([12]). In fact, similar questions for ℓ^1 arise in recent work by E. Kamen on half-plane digital filters.

2. D. Delchamps proved a more general form of Lemma 7.1 in order to construct a metric, the Riccati metric, on the state bundle of the moduli space $\{(F,G,H)\}/GL(n,\mathbb{R})$ and to study its properties. Some of his work will appear in the proceedings of this conference, published by the AMS.

3. Constancy of the Kronecker indices is, of course, a very stringent assumption, and it is interesting to study the limiting behavior of the $(n_i(\lambda))$. Thus, if the $(n_i(\lambda))$ are generically constant, then

$$\mathcal{F}_\sigma:\Lambda \to \mathrm{Closure}(\mathcal{O}_1) \subset C(n,m)$$

where $\mathcal{O}_1$ is the orbit corresponding to the generic value of the $(n_i(\lambda))$. As one can see by examining the matrices $(F(\lambda),G(\lambda))$, the partition $(\tilde{n}_i)$ occurs as a limit, or specialization, of the partition (n_i) only if $(\tilde{n}_i) \geq (n_i)$ in the Rosenbrock ordering

$$\tilde{n}_1 \geq n_1,\ \tilde{n}_1 + \tilde{n}_2 \geq n_1 + n_2, \ldots, \qquad (\theta)$$

as one may observe in Example 6.1. From the vector bundle point of view, this illustrates the result of Shatz ([43]) that the Grothendieck decomposition of W_σ rises in the Harder-Narasimhan ordering under specialization ([43]). It has been proven (independently) by Hazewinkel, Kalman, and Martin that $\mathcal{O}_{\tilde{n}}$ Closure ($\mathcal{O}_n$) iff $\tilde{n} \geq n$ in the ubiquitous, or preferably the natural, ordering (*).

REFERENCES

[1] Berstein, I.: 1976, *On the Lusternick-Schnirelmann category of real Grassmannians*, Proc. Cambridge Phil. Soc., 79, pp. 129-139.

[2] Brasch, F. R., and Pearson, J. B.: 1970, *Pole placement using dynamic compensators*, IEEE Trans. Aut. Control, 15, pp. 34-43.

[3] Brockett, R. W.: 1977, *The Lie groups of simple feedback systems*, Proc. 1977 CDC, New Orleans, La.

[4] Brockett, R. W.: 1977, *The geometry of the set of controllable linear systems*, Research Reports of Aut. Control Lab., Nagoya University, 24.

[5] Brockett, R. W., and Byrnes, C. I.: 1979, *On the algebraic geometry of the output feedback pole placement map*, Proc. 1979 CDC, Ft. Lauderdale, Fla.

[6] Brockett, R. W., and Rahimi, A.: 1972, *Lie algebras and linear differential equations*, in Ordinary Differential Equations, edited by Weiss, (Academic Press).

[7] Brockett, R. W., and Willems, J. L.: 1974, *Discretized partial differential equations: examples of control systems defined on modules*, Automatica, 10, pp. 507-515.

[8] Bumby, R., Sontag, E. D., Sussmann, H. J., and Vasconcelos, W.: *Remarks on the pole-shifting problem over rings*, to appear.

[9] Byrnes, C. I.: 1977, *The moduli space for linear dynamical systems*, in Geometric Control Theory, edited by Martin and Hermann, (Math Sci Press).

[10] Byrnes, C. I.: 1978, *On the control of certain infinite-dimensional deterministic systems by algebro-geometric techniques*, Amer. J. of Math., 100, pp. 1333-1381.

[11] Byrnes, C. I.: 1978, *Feedback invariants for linear systems defined over rings*, Proc. 1978 CDC, San Diego, Ca.

[12] Byrnes, C. I.: 1978, *On the stabilization of linear systems depending on parameters*, Proc. 1979 CDC, Ft. Lauderdale, Fla.

[13] Byrnes, C. I., and Hurt, N. E.: 1979, *On the moduli of linear dynamical systems*, Adv. in Math: Studies in Analysis 4, (1978), pp. 83-122 and in Modern Mathematical System Theory, (MIR Press).

[14] Delfour, M., and Mitter, S. K.: 1975, *Hereditary differential systems with constant delays I and II*, J. Diff. Equations, 12 (1972), pp. 213-235, and 18 (1975), pp. 18-28.

[15] Djaferis, T.: 1979, *General pole assignment by output feedback and solution of linear matrix equations from an algebraic point of view*, Ph.D. Thesis, MIT.

[16] Emre, E.: *Dynamic feedback: A system theoretic approach*, to appear.

[17] Fuhrmann, P.: 1980, *Functional models, factorization, and linear systems*, this volume.

[18] Hanna, C. C.: *Decomposing algebraic vector bundles on the projective line*, Proc. Amer. Math. Soc., 61(210), pp. 196-200.

[19] Hautus, M. L. J.: 1970, *Stabilization, controllability, and observability of linear autonomous sytems*, Proc. of Konintel. Nederl. Akademie van Wetenschappen--Amsterdam. Series A, 73, pp. 448-455.

[20] Hazewinkel, M.: 1977, *Moduli and canonical forms for linear dynamical systems III: the algebraic geometric case*, in Geometric Control Theory, edited by Martin and Hermann, (Math Sci Press).

[21a] Hazewinkel, M.: 1980, *(Fine) Moduli (spaces): What are they, and what are they good for?* this volume.

[21b] Hazewinkel, M.: 1979, *A partial survey of the uses of algebraic geometry in systems and control theory*, Severi Centennial Conf., Rome.

[22] Hazewinkel, M., and Kalman, R. E.: 1976, *On invariants, canonical forms and moduli for linear constant finite-dimensional, dynamical systems*, Lecutre Notes in Econo-Math. System Theory, 131, (Springer-Verlag: Berlin).

[23] Hermann, R. and C. F. Martin: 1977, *Applications of algebraic geometry to system theory, part I,* IEEE Trans. Aut. Control, 22, pp. 19-25.

[24] Hermann, R. and C. F. Martin: 1979, *Applications of algebraic geometry to system theory: the McMillan degree and Kronecker indices as topological and holomorphic invariants,* SIAM J. Control, 16, pp. 743-755.

[25] Kailath, T.: 1980, *Linear System Theory,* Englewood Cliffs, N.J.: Prentice Hall.

[26] Kalman, R. E.: 1972, *Kronecker invariants and feedback,* in Ordinary Differential Equations, edited by Weiss, (Academic Press).

[27] Kamen, E. W.: 1978, *An operator theory of linear functional differential equations,* J. Diff. Equs., 27, pp. 274-297.

[28] Kamen, E. W.: *Asymptotic stability of linear shift-invariant two-dimensional digital filters,* to appear.

[29] Kimura, H.: 1975, *Pole assignment by gain output feedback,* IEEE Trans. Aut. Control, 20, pp. 509-516.

[30] Kimura, H.: 1977, *A further result on the problem of pole assignment by output feedback,* IEEE Aut. Control, 22, pp. 458-463.

[31] Kleinman, S. L.: 1976, *Problem 15, rigorous foundations of Schubert's enumerative calculus,* in Proc. Symposia in Pure Mathematics (Mathematical Developments Arising from Hilbert Problems), 28, (Amer. Math. Soc.: Providence).

[32] Manitius, A. and R. Triggiani: 1978, *Function space controllability of linear retarded systems: a derivation from abstract operator conditions,* SIAM J. Control and Optimization.

[33] Martin, C. F.: 1980, *Grassmannian manifolds, Riccati equations and feedback invariants of linear systems,* this volume.

[34] Maxwell, J. C.: 1868, *On governors,* Proc. Royal Soc. London, 16, pp. 270-283.

[35] Mitter, S. K. and T. Djaferis, *The generalized pole-assignment problem,* Proc. 1979 CDC, Ft. Lauderdale, Fla.

[36] Morse, A. S.: 1976, *Ring models for delay differential systems,* Automatica, 12, pp. 529-531.

[37] Postlethwaite, I. and A. G. J. MacFarlane: *A complex variable approach to the analysis of linear multivariable feedback systems*, Lecture Notes in Control and Inf. Sciences, 12, (Springer-Verlag: Berlin).

[38] Rado, R.: 1942, *A theorem on independence relations*, Quart. J. Math (Oxford), 13, pp. 83-89.

[39] Rosenbrock, H.: 1980, *Systems and polynomial matrices*, this volume.

[40] Rouchaleau, Y.: 1980, *Commutative algebra in system theory I, II*, this volume.

[41] Rouchaleau, Y. and E. D. Sontag: 1979, *On the existence of minimal realizations of linear dynamical systems over Noetherian integral domains*, Journal of Computer and System Sciences, Vol. 18, No. 1, pp. 65-75.

[42] Sastry, S. S. and C. A. Desoer: *Asymptotic unbounded root loci by the singular value decomposition*, E. R. L. Memorandum No. UCB/ERL M79/63, Univ. Cal. at Berkeley.

[43] Shatz, S. S.: 1977, *The decomposition and specialization of algebraic families of vector bundles*, Comp. Math., 35, pp. 163-187.

[44] Sontag, E. D.: 1976, *Linear systems over commutative rings: A survey*, Richerche di Automatica, 7, pp. 1-34.

[45] Willems, J. C. and W. H. Hesselink, *Generic properties of the pole placement problem*, Proc. IFAC, (Helsinki).

(FINE) MODULI (SPACES) FOR LINEAR SYSTEMS: WHAT ARE THEY AND WHAT ARE THEY GOOD FOR ?

Michiel Hazewinkel

Erasmus Universiteit Rotterdam
Rotterdam, The Netherlands

ABSTRACT

This tutorial and expository paper considers linear dynamical systems $\dot{x} = Fx + Gu$, $y = Hx$, or, $x(t+1) = Fx(t) + Gu(t)$, $y(t) = Hx(t)$; more precisely it is really concerned with families of such, i.e., roughly speaking, with systems like the above where now the matrices F,G,H depend on some extra parameters σ. After discussing some motivation for studying families (delay systems, systems over rings, n-d systems, perturbed systems, identification, parameter uncertainty) we discuss the classifying of families (fine moduli spaces). This is followed by two straightforward applications: realization with parameters and the nonexistence of global continuous canonical forms. More applications, especially to feedback will be discussed in Chris Byrnes' talks at this conference and similar problems as in these talks for networks will be discussed by Tyrone Duncan. The classifying fine moduli space cannot readily be extended and the concluding sections are devoted to this observation and a few more related results.

1. INTRODUCTION

The basic object of study in these lectures (as in many others at this conference) is a constant linear dynamical system, that is a system of equations

$$\begin{array}{lll} \dot{x} = Fx + Gu & & x(t+1) = Fx(t) + Gu(t) \\ y = Hx & (\Sigma) & y(t) = Hx(t) \\ \text{(a): continuous time} & & \text{(b): discrete time} \end{array} \qquad (1.1)$$

C. I. Byrnes and C. F. Martin, Geometrical Methods for the Theory of Linear Systems, 125-193.

with $x \in k^n$ = state space, $u \in k^m$ = input or control space, $y \in k^p$ = output space, and F,G,H matrices with coefficients in k of the appropriate sizes; that is, there are m inputs and p outputs and the dimension of the state space, also called the dimension of the system Σ and denoted $\dim(\Sigma)$, is n. Here k is an appropriate field (or possibly ring). In the continuous case of course k should be such that differentiation makes sense for (enough) functions $\mathbb{R} \to k$, e.g. $k = \mathbb{R}$ or $\mathbb{C}$. Often one adds a direct feedthrough term Ju, giving $y = Hx + Ju$ in case (a) and $y(t) = Hx(t) + Ju(t)$ in case (b) instead of $y = Hx$ and $y(t) = Hx(t)$ respectively; for the mathematical problems to be discussed below the presence or absence of J is essentially irrelevant.

More precisely what we are really interested in are families of objects (1.1), that is sets of equations (1.1) where now the matrices F,G,H depend on some extra parameters σ. As people have found out by now in virtually all parts of mathematics and its applications, even if one is basically interested only in single objects, it pays and is important to study families of such objects depending on a small parameter ε (deformation and perturbation considerations). This could be already enough motivation to study families, but, as it turns out, in the case of (linear) systems theory there are many more circumstances where families turn up naturally. Some of these can be briefly summed up as delay-differential systems, systems over rings, continuous canonical forms, 2-d and n-d systems, parameter uncertainty, (singularly) perturbed systems. We discuss these in some detail below in section 2.

To return to single systems for the moment. The equations (1.1) define input/output maps $f_\Sigma : u(t) \mapsto y(t)$ given respectively by

$$y(t) = \int_0^t He^{F(t-\tau)}Gu(\tau)d\tau , \quad t \geq 0 \tag{1.2a}$$

$$y(t) = \sum_{i=1}^{t} A_i u(t-1-i), \ A_i = HF^{i-1}G,$$

$$i = 1,2,\ldots, \quad t = 1,2,3,\ldots \tag{1.2b}$$

where we have assumed that the system starts in $x(0) = 0$ at time 0. In both cases the input/output operator is uniquely determined by the sequence of matrices $A_1, A_2, \ldots$. Inversely, realization theory studies when a given sequence $A_1, A_2, \ldots$ is such that there exist F,G,H such that $A_i = HF^{i-1}G$ for all i. Realization with parameters is now the question: given a sequence of matrices $A_1(\sigma)$, $A_2(\sigma)$, $A_3(\sigma)$, ... depending polynomially (resp. continuously, resp. analytically, resp. ...) on parameters

σ when do there exist matrices F,G,H depending polynomially (resp. continuously, resp. analytically, resp. ...) on the parameters σ such that $A_i(\sigma) = H(\sigma)F^{i-1}(\sigma)G(\sigma)$ for all i. And to what extent are such realizations unique? Which brings us to the next group of questions one likes to answer for families.

A single system Σ given by the triple of matrices F,G,H is completely reachable if the matrix R(F,G) consisting of the blocks $G, FG, \ldots, F^nG$

$$R(\Sigma) = R(F,G) = (G \mid FG \mid \ldots \mid F^nG) \tag{1.3}$$

has full rank n. (This means that any state x can be steered to any other state x' by means of a suitable input). Dually the system Σ is said to be completely observable if the matrix Q(F,G) consisting of the blocks $H, HF, \ldots, HF^n$

$$Q(\Sigma) = Q(F,H) = \begin{pmatrix} H \\ HF \\ \vdots \\ HF^n \end{pmatrix} \tag{1.4}$$

has full rank n. (This means that two different states x(t) and x'(t) of the system can be distinguished on the basis of the output $y(\tau)$ for all $\tau \geq t$). As is very well known if $A_1, A_2, \ldots$ can be realized then it can be realized by a co and cr system and any two such realizations are the same up to base change in state space. That is, if $\Sigma = (F,G,H)$ and $\Sigma' = (F',G',H')$ both realize $A_1, A_2, \ldots$ and both are cr and co then $\dim(\Sigma) = \dim(\Sigma') = n$ and there is an invertible $n \times n$ matrix S such that $F' = SFS^{-1}$, $G' = SG$, $H' = HS^{-1}$. (It is obvious that if Σ and Σ' are related in this way then they give the same input/output map). This transformation

$$\Sigma = (F,G,H) \mapsto \Sigma^S = (F,G,H)^S = (SFS^{-1}, SG, HS^{-1})$$

corresponds of course to the base change in state space $x' = Sx$. This argues that at least one good notion of isomorphism of systems is: two systems Σ, Σ' over k are isomorphic iff $\dim(\Sigma) = \dim(\Sigma')$ and there is an $S \in GL_n(k)$, the group of invertible matrices with coefficients in k, such that $\Sigma' = \Sigma^S$. A corresponding notion of homomorphism is: a homomorphism from $\Sigma = (F,G,H)$, $\dim\Sigma = n$, to $\Sigma' = (F',G',H')$, $\dim\Sigma = n'$, is an $n' \times n$ matrix B (with coefficients in k) such that $BG = G'$, $BF = F'B$, $H'B = H$. Or, in other words, it is a linear map from the state space of Σ to the state space of Σ' such that the diagram below commutes.

$$\begin{array}{ccccccc} & & k^n & \xrightarrow{F} & k^n & & \\ & \overset{G}{\nearrow} & \downarrow B & & \downarrow B & \overset{H}{\searrow} & \\ k^m & & & & & & k^p \\ & \underset{G'}{\searrow} & k^{n'} & \xrightarrow{F'} & k^n & \underset{H'}{\nearrow} & \end{array} \qquad (1.6)$$

The obvious corresponding notion of isomorphism for families $\Sigma(\sigma)$, $\Sigma'(\sigma)$ is a family of invertible matrices $S(\sigma)$ such that $\Sigma(\sigma)^{S(\sigma)} = \Sigma'(\sigma)$, where, of course, $S(\sigma)$ should depend polynomially, resp. continuously, resp. analytically, resp. ... on σ if Σ and Σ' are polynomial, resp. continuous, respt. analytical, resp. ... families. One way to look at the results of section 3 below is as a classification result for families, or, even, as the construction of canonical forms for families under the notion of isomorphism just described. As it happens the classification goes in terms of a universal family, that is, a family from which, roughly speaking, all other families (up to isomorphism) can be uniquely obtained via a transformation in the parameters.

Let $L_{m,n,p}(k)$ be the space of all triples of matrices (F,G,H) of dimensions $n \times n$, $n \times m$, $p \times n$, and let $L^{co,cr}_{m,n,p}$ be the subspace of cr and co triples. Then the parameter space for the universal family is the quotient space $L^{co,cr}_{m,n,p}(k)/GL_n(k)$, which turns out to be a very nice space.

The next question we shall take up is the existence or nonexistence of continuous canonical forms. A continuous canonical form on $L^{co,cr}_{m,n,p}$ is a continuous map $(F,G,H) \mapsto c(F,G,H)$ such that $c(F,G,H)$ is isomorphic to (F,G,H) for all $(F,G,H) \in L^{co,cr}_{m,n,p}$ and such that (F,G,H) and (F',G',H') are isomorphic if and only if $c(F,G,H) = c(F',G',H')$ for all (F,G,H), $(F',G',H') \in L^{co,cr}_{m,n,p}$. Obviously if one wants to use canonical forms to get rid of superfluous parameters in an identification problem the canonical form had better be continuous. This does not mean that (discontinuous) canonical forms are not useful. On the contrary, witness e.g. the Jordan canonical form for square matrices under similarity. On the other hand, being discontinuous, it also has very serious drawbacks; cf. e.g. [GWi] for a discussion of some of these. In our case it turns out that there exists a continuous canonical form on all of $L^{co,cr}_{m,n,p}$ if and only if $m = 1$ or $p = 1$.

Now let, again, Σ be a single system. Then there is a canonical subsystem $\Sigma^{(r)}$ which is completely reachable and a canonical quotient system Σ^{co} which is completely observable.

Combining these two constructions one finds a canonical subquotient (or quotient sub) which is both cr and co. The question arises naturally whether (under some obvious necessary conditions) these constructions can be carried out for families as well and also for single time varying systems. This is very much related to the question of whether these constructions are continuous. In the last sections we discuss these questions and related topics like: given two families Σ and Σ' such that $\Sigma(\sigma)$ and $\Sigma'(\sigma)$ are isomorphic for all (resp. almost all) values of the parameters σ; what can be said about the relation between Σ and Σ' as families (resp. about $\Sigma(\sigma)$ and $\Sigma'(\sigma)$ for the remaining values of σ).

2. WHY SHOULD ONE STUDY FAMILIES OF SYSTEMS

For the moment we shall keep to the intuitive first approximation of a family of systems as a family of triples of matrices of fixed size depending in some continuous manner on a parameter σ. This is the definition which we also used in the introduction.

2.1 (Singular) Perturbation, Deformation, Approximation

This bit of motivation for studying families of objects, rather than just the objects themselves, is almost as old as mathematics itself. Certainly (singular) perturbations are a familiar topic in the theory of boundary value problems for ordinary and partial differential equations and more recently also in optimal control, cf. e.g. [OMa]. For instance in [OMa], Chapter VI, O'Malley discusses the singularly perturbed regulator problem which consists of the following set of equations, initial conditions and quadratic cost functional which is to be minimized for a control which drives the state

$$x = \begin{pmatrix} y \\ z \end{pmatrix}$$

to zero at time $t = 1$.

$$\begin{aligned} \dot{y} &= A_1(\varepsilon)y + A_2(\varepsilon)z + B_1(\varepsilon)u \qquad y(0,\varepsilon) = y^0(\varepsilon) \\ \varepsilon\dot{z} &= A_3(\varepsilon)y + A_4(\varepsilon)z + B_2(\varepsilon)u \qquad z(0,\varepsilon) = z^0(\varepsilon) \\ J(\varepsilon) &= x^T(1,\varepsilon)\pi(\varepsilon)x(1,\varepsilon) \\ &\quad + \int_0^1 (x^T(t,\varepsilon)Q(\varepsilon)x(t,\varepsilon) + u^T(t,\varepsilon)R(\varepsilon)u(t,\varepsilon))dt \end{aligned} \tag{2.1.1}$$

with positive definite $R(\varepsilon)$, and $Q(\varepsilon),\pi(\varepsilon)$ positive semidefinite. Here the upper T denotes transposes. The matrices

$A_i(\varepsilon)$, $i = 1,2,3,4$, $B_i(\varepsilon)$, $i = 1,2$, $\pi(\varepsilon)$, $Q(\varepsilon)$, $R(\varepsilon)$ may also depend on t. For fixed small $\varepsilon > 0$ there is a unique optimal solution. Here one is interested, however, in the asymptotic solution of the problem as ε tends to zero, which is, still quoting from [OMa], a problem of considerable practical importance, in particular in view of an example of Hadlock et al. [HJK] where the asymptotic results are far superior to the physically unacceptable results obtained by setting $\varepsilon = 0$ directly.

Another interesting problem arises maybe when we have a system

$$\dot{x} = Fx + G_1u + G_2v, \qquad y = Hx \tag{2.1.2}$$

where v is noise, and there F, G_1, G_2,H depend on a parameter ε. Suppose we can solve the disturbance decoupling problem for $\varepsilon = 0$. I. e., we can find a feedback matrix L such that in the system with state feedback loop L

$$\dot{x} = (F+GL)x + G_1u + G_2v, \quad y = Hx$$

the disturbances v do not show up any more in the output y, (for $\varepsilon = 0$). Is it possible to find a disturbance discoupler $L(\varepsilon)$ by "perturbation" methods, i.e., as a power series in ε which converges (uniformly) for ε small enough, and such that $L(0) = L$.

In this paper we shall not really pay much more attention to singular perturbation phenomena. For some more systems oriented material on singular perturbations cf. [KKU] and also [Haz 4].

2.2 Systems Over Rings

Let R be an arbitrary commutative ring with unit element. A linear system over R is simply a triple of matrices (F,G,H) of sizes $n \times n$, $n \times m$, $p \times n$ respectively with coefficients in R. Such a triple defines a linear machine

$$x(t+1) = Fx(t) + Gu(t), \quad t = 0,1,2,\ldots, \; x \in R^n, \; u \in R^m$$

$$y(t) = Hx(t), \qquad y \in R^p \tag{2.2.1}$$

which transforms input sequences $(u(0),u(1),u(2),\ldots)$ into output sequences $(y(1),y(2),y(3),\ldots)$ according to the convolution formula (1.2.b).

It is now absolutely standard algebraic geometry to consider these data as a family over Spec(R), the space of all prime ideals of R with the Zariski topology. This goes as follows. For each prime ideal p let $i_p : R \to Q(R/p)$ be the canonical map of R into the quotient field $Q(R/p)$ of the integral domain R/p. Let $(F(p), G(p), H(p))$ be the triple of matrices over $Q(R/p)$ obtained by applying i_p to the entries of F,G,H. Then $\Sigma(p) = (F(p),G(p),H(p))$ is a family of systems parametrized by Spec(R).

Let me stress that, mathematically, there is no difference between a system over R as in (2.2.1) and the family $\Sigma(p)$. As far as intuition goes there is quite a bit of difference, and the present author e.g. has found it helpful to think about families of systems over Spec(R) rather than single systems over R. Of course such families over Spec(R) do not quite correspond to families as one intuitively thinks about them. For instance if $R = \mathbb{Z}$ = the integers, then Spec($\mathbb{Z}$) consists of (0) and the prime ideals (p), p a prime number, so that a system over $\mathbb{Z}$ gives rise to a certain collection of systems: one over $\mathbb{Q}$ = rational numbers, and one each over every finite field $\mathbb{F}_p = \mathbb{Z}/(p)$. Still the intuition one gleans from thinking about families as families parametrized continuously by real numbers seems to work well also in these cases.

2.3 Delay-Differential Systems

Consider for example the following delay-differential system

$$\begin{aligned}\dot{x}_1(t) &= x_1(t-2) + x_2(t-\alpha) + u(t-1) + u(t)\\ \dot{x}_2(t) &= x_1(t) + x_2(t-1) + u(t-\alpha) \qquad (2.3.1)\\ y(t) &= x_1(t) + x_2(t-2\alpha)\end{aligned}$$

where α is some real number incommensurable with 1. Introduce the delay operators σ_1, σ_2 by $\sigma_1\beta(t) = \beta(t-1)$, $\sigma_2\beta(t) = \beta(t-\alpha)$. Then we can rewrite (2.3.1) formally as

$$\dot{x}(t) = Fx(t) + Gu(t), \quad y(t) = Hx(t) \qquad (2.3.2)$$

with

$$F = \begin{pmatrix}\sigma_1^2 & \sigma_2\\ 1 & \sigma_1\end{pmatrix}, \quad G = \begin{pmatrix}1+\sigma_1\\ \sigma_2\end{pmatrix}, \quad H = (1 \quad \sigma_2^2) \qquad (2.3.3)$$

and, forgetting so to speak where (2.3.2), (2.3.3) came from, we can view this set of equations as a linear dynamical system over the ring $\mathbb{R}[\sigma_1,\sigma_2]$, and then using 2.2 above also as a family of systems parametrized by the (complex) parameters σ_1, σ_2, a point of view which has proved fruitful, e.g., in [By]. This idea has been around for some time now [ZW,An,Yo,RMY], though originally the tendency was to consider these systems as systems over the fields $\mathbb{R}(\sigma_1,\ldots,\sigma_r)$; the idea to consider them over the rings $\mathbb{R}[\sigma_1,\ldots,\sigma_r]$ instead is of more recent vintage [Mo,Kam].

There are, as far as I know, no relations between the solutions of (2.3.1) and the solutions of the family of systems (2.3.2), (2.3.3). Still many of the interesting properties and constructions for (2.3.1) have their counterpart for (2.3.2), (2.3.3) and vice versa. For example to construct a stabilizing state feedback loop for the family (2.3.2)-(2.3.3) depending polynomially on the parameters σ_1, σ_2 that is finding a stabilizing state feedback loop for the system over $\mathbb{R}[\sigma_1,\sigma_2]$, means finding an $m \times n$ matrix $L(\sigma_1,\sigma_2)$ with entries in $\mathbb{R}[\sigma_1,\sigma_2]$ such that for all complex σ_1,σ_2 $\det(s-(F+GL))$ has its roots in the left half plane. Reinterpreting σ_1 and σ_2 as delays so that $L(\sigma_1,\sigma_2)$ becomes a feedback matrix with delays one finds a stabilizing feedback loop for (the infinite dimensional) system (2.3.1). (cf. [BC], cf. also [Kam], which works out in some detail some of the relations between (2.3.1) and (2.3.2)-(2.3.3) viewed as a system over the ring $\mathbb{R}[\sigma_1,\sigma_2]$).

As another example a natural notion of isomorphism for systems $\Sigma = (F,G,H)$, $\Sigma' = (F',G',H')$ over a ring R is: Σ and Σ' are isomorphic if there exists an $n \times n$ matrix S over R, which is invertible over R, i.e. such that $\det(S)$ is a unit of R, such that $\Sigma' = \Sigma^S$. Taking $R = \mathbb{R}[\sigma_1,\sigma_2]$ and reinterpreting the σ_i as delays we see that the corresponding notion for the delay-differential systems is coordinate transformations with time delays which is precisely the right notion of isomorphism for studying for instance degeneracy phenomena, cf [Kap].

Finally applying the Laplace transform to (2.3.1) we find a transfer function $T(s,e^{-s},e^{-\alpha s})$, which is rational in s,e^{-s} and $e^{-\alpha s}$. It can also be obtained by taking the family of transfer functions

$$T_{\sigma_1,\sigma_2}(s) = H(\sigma_1,\sigma_2)(s-F(\sigma_1,\sigma_2))^{-1}G(\sigma_1,\sigma_2)$$

and then substituting e^{-s} for σ_1 and $e^{-\alpha s}$ for σ_2. Inversely given a transfer function $T(s)$ which is rational in $s,e^{-s},e^{-\alpha s}$ one may ask whether it can be realized as a system with delays which are multiples of 1 and α. Because the functions s, e^{-s}, $e^{-\alpha s}$ are algebraically independent (if α is incommensurable with 1), there is a unique rational function $\tilde{T}(s,\sigma_1,\sigma_2)$ such that $T(s) = \tilde{T}(s,e^{-s},e^{-\alpha s})$ and the realizability of $T(s)$ by means of a delay system, say a system with transmission lines, is now mathematically equivalent with realizing the two parameter family of transfer functions $T(s,\sigma_1,\sigma_2)$ by a family of systems which depends polynomially on σ_1, σ_2.

2.4 2-d and n-d Systems

Consider a linear discrete time system with direct feed-through term

$$x(t+1) = Fx(t) + Gu(t), \quad y(t) = Hx(t) + Ju(t) \qquad (2.4.1)$$

The associated input/output operator is a convolution operator, viz. (cf. (1.2.b))

$$y(t) = \sum_{i=0}^{t} A_i u(t-i), \; A_0 = J, \; A_i = HF^{i-1}G \qquad (2.4.2)$$

$$\text{for} \quad i = 1,2,\ldots$$

Now there is an obvious (north-east causal) more dimensional generalization of the convolution operator (2.4.2), viz.

$$y(h,k) = \sum_{i=0}^{h} \sum_{j=0}^{k} A_{i,j} u(h-i,k-j), \; h,k = 0,1,2,\ldots \qquad (2.4.3)$$

A (Givone-Roesser) realization of such an operators is a "2-d system"

$$\begin{aligned} x_1(h+1,k) &= F_{11}x_1(h,k) + F_{12}x_2(h,k) + G_1u(h,k) \\ x_2(h,k+1) &= F_{21}x_1(h,k) + F_{22}x_2(h,k) + G_2u(h,k) \qquad (2.4.4) \\ y(h,k) &= H_1x_1(h,k) + H_2x_2(h,k) + Ju(h,k) \end{aligned}$$

which yields an input/output operator of the form (2.4.3) with the $A_{i,j}$ determined by the power series development of the 2-d transfer function $T(s_1,s_2)$

$$\sum_{i,j} A_{i,j} s_1^{-i} s_2^{-j} = T(s_1,s_2) = (H_1 \; H_2)\left(\begin{pmatrix} s_1 I_{n_1} & 0 \\ 0 & s_n I_{n_2} \end{pmatrix} - \begin{pmatrix} F_{11} & F_{12} \\ F_{21} & F_{22} \end{pmatrix}\right)^{-1} \begin{pmatrix} G_1 \\ G_2 \end{pmatrix} + J$$

where I_r is the $r \times r$ unit matrix and n_1 and n_2 are the dimensions of the state vectors x_1 and x_2. There are obvious generalizations to n-d systems, $n \geq 3$. The question now arises whether every proper 2-d matrix transfer function can indeed by so realized. (cf. [Eis] or [So2] for a definition of proper.) A way to approach this is to treat one of the s_i as a parameter, giving us a realization with parameters problem.

More precisely let R_g be the ring of all proper rational functions in s_1. In the 2-d case this is a principal ideal domain which simplifies things considerably. Now consider $T(s_1,s_2)$ as a proper rational function in s_2 with coefficients in R_g. This transfer function can be realized giving us a discrete time system over R_g defined by the quadruple of matrices $(F(s_1), G(s_1), H(s_1), J(s_1))$. Each of these matrices is proper as a function of s_1 and hence can be realized by a quadruple of constant matrices. Suppose that

$$(F_F,G_F,H_F,J_F) \quad \text{realizes} \quad F(s_1)$$

$$(F_G,G_G,H_G,J_G) \quad \text{realizes} \quad G(s_1)$$

$$(F_H,G_H,H_H,J_H) \quad \text{realizes} \quad H(s_1)$$

$$(F_J,G_J,H_J,J_J) \quad \text{realizes} \quad J(s_1)$$

Then, as is easily checked, a realization in the sense of (2.4.4) is defined by

$$F = \begin{pmatrix} F_{11} & F_{12} \\ \\ F_{21} & F_{22} \end{pmatrix} = \left(\begin{array}{c|cccc} J_F & H_F & H_G & 0 & 0 \\ \hline G_F & F_F & 0 & 0 & 0 \\ 0 & 0 & F_G & 0 & 0 \\ G_H & 0 & F_H & 0 & 0 \\ 0 & 0 & 0 & 0 & F_J \end{array}\right),$$

$$G = \begin{pmatrix} G_1 \\ \\ G_2 \end{pmatrix} = \left(\begin{array}{c} J_G \\ \hline 0 \\ G_G \\ 0 \\ G_J \end{array}\right)$$

$$H = (H_1 \quad H_2) = (J_F \mid 0 \quad 0 \quad H_H \quad H_J), \quad J = J_J$$

This is the procedure followed in [Eis]; a somewhat different approach, with essentially the same initial step (i.e. realization with parameters, or realization over a ring) is followed in [So2].

2.5 Parameter Uncertainty

Suppose that we have a system $\Sigma = (F,G,H)$ but that we are uncertain about some of its parameters, i.e. we are uncertain about the precise value of some of the entries of F,G or H. That is, what we really have is a family of systems $\Sigma(\beta)$, where β runs through some set B of parameter values, which we assume compact. For simplicity assume that we have a one input-one output system. Let the transfer function of $\Sigma(\beta)$ be $T_\beta(s) = f_\beta(s)/g_\beta(s)$. Now suppose we want to stabilize Σ by a dynamic output feedback loop with transfer function $P(s) = \phi(s)/\psi(s)$, still being uncertain about the value of β. The transfer function of the resulting total system is $T(s)/(1-T(s)P(s))$. So we shall have succeeded if we can find polynomials $\phi(s)$ and $\psi(s)$ such that for all $\beta \in B$ all roots of

$$g_\beta(s)\psi(s) - f_\beta(s)\phi(s)$$

are in the left halfplane, possibly with the extra requirement that $P(s)$ be also stable. The same mathematical question arises from what has been named the blending problem, cf [Tal]. It cannot always be solved. In the special but important case where the uncertainty is just a gain factor, i.e. in the case that B is an interval $[b_1,b_2]$, $b_2 > b_1 > 0$ and $T_\beta(s) = \beta T(s)$, where $T(s)$ is a fixed transferfunction, the problem is solved completely in [Tal].

3. THE CLASSIFICATION OF FAMILIES. FINE MODULI SPACES

3.1 Introductory and Motivational Remarks

(Why classifying families is essentially more difficult than classifying systems and why the set of isomorphism classes of (single) systems should be topologized.)

Obviously the first thing to do when trying to classify families up to isomorphism is to obtain a good description of the set of isomorphism classes of (single) systems over a field k, that is to obtain a good description of the sets $L_{m,n,p}(k)/GL_n(k) = M_{m,n,p}(k)$ and of the quotient map $L_{m,n,p}(k) \to M_{m,n,p}(k)$. This will be done below in section 3.2 for the subset of isomorphism classes (or sets of orbits) of completely reachable systems. This is not particularly difficult (and also well known) nor is it overly complicated to extend this to a description of all of $M_{m,n,p}(k) = L_{m,n,p}(k)/GL_n(k)$, cf [Haz 6]. Though, as we shall see, there are, for the moment, good mathematical reasons, to limit ourselves to cr systems and families of cr systems, or, dually to limit ourselves to co systems.

Now let us consider the classification problem for families of systems. For definiteness sake suppose we are interested (cf. 2.1 and 2.3 above e.g.) in real families of systems $\Sigma(\sigma) = (F(\sigma), G(\sigma), H(\sigma))$ which depend continuously on a real parameter $\sigma \in \mathbb{R}$. The obvious, straightforward and in fact right thing to do is to proceed as follows. For each $\sigma \in \mathbb{R}$ we have a system $\Sigma(\sigma)$, and hence a point $\phi(\sigma) \in M_{m,n,p}(\mathbb{R}) = L_{m,n,p}(\mathbb{R})/GL_n(\mathbb{R})$, the set of isomorphism classes or, equivalently, the set of orbits in $L_{m,n,p}$ under the action $(\Sigma,S) \mapsto \Sigma^S$ of $GL_n(\mathbb{R})$ on $L_{m,n,p}(\mathbb{R})$. This defines a map $\phi(\Sigma):R \to M_{m,n,p}(\mathbb{R})$, and one's first guess would be that two families Σ,Σ' are isomorphic iff their associated maps $\phi(\Sigma)$, $\phi(\Sigma')$ are equal. However, things are not that simple as the following example in $L_{1,2,1}(\mathbb{R})$ shows.

<u>Example</u>

$$\Sigma(\sigma) = \left(\begin{pmatrix} 1 & 0 \\ \sigma^2 & 1 \end{pmatrix}, \begin{pmatrix} 1 \\ 0 \end{pmatrix}, (1,2) \right),$$

$$\Sigma'(\sigma) = \left(\begin{pmatrix} 1 & 0 \\ \sigma & 1 \end{pmatrix}, \begin{pmatrix} 1 \\ 0 \end{pmatrix}, (1,2\sigma) \right) \tag{3.1.1}$$

For each $\sigma \in \mathbb{R}$, $\Sigma(\sigma)$ and $\Sigma'(\sigma)$ are isomorphic via $T(\sigma) = \begin{pmatrix} 1 & 0 \\ 0 & \sigma^{-1} \end{pmatrix}$ if $\sigma \neq 0$ and via $T(\sigma) = \begin{pmatrix} 1 & 2 \\ 0 & 1 \end{pmatrix}$ if $\sigma = 0$. Yet they are not isomorphic as continuous families, meaning that there exists no continuous map $\mathbb{R} \to GL_2(\mathbb{R})$, $\sigma \mapsto T(\sigma)$, such that $\Sigma'(\sigma) = \Sigma(\sigma)^{T(\sigma)}$ for all $\sigma \in \mathbb{R}$. One might guess that part of the problem is topological. Indeed, it is in any case sort of obvious that one should give $M_{m,n,p}(\mathbb{R})$ as much structure as possible. Otherwise the map $\phi(\Sigma): \mathbb{R} \to M_{m,n,p}(\mathbb{R})$ does not tell us whether it could have come from a continuous family. (Of course if $\Sigma(\sigma)$ is a continuous family over $\mathbb{R}$ giving rise to $\phi(\Sigma)$ and $S \in GL_n(\mathbb{R})$ is such that $\Sigma(0)^S \neq \Sigma(0)$ then the discontinuous family $\Sigma'(\sigma)$, $\Sigma'(\sigma) = \Sigma(\sigma)$ for $\sigma \neq 0$, $\Sigma'(0) = \Sigma(0)^S$ given rise to the same map). Similarly we would like to have $\phi(\Sigma)$ analytic if Σ is an analytic family, polynomial if Σ is polynomial, differentiable if Σ is differentiable, ...

One reason to limit oneself to cr systems is now that the natural topology (which is the quotient topology for $\pi: L_{m,n,p}(\mathbb{R}) \to M_{m,n,p}(\mathbb{R})$) will not be Hausdorff unless we limit ourselves to cr systems. (It is clear that one wants to put in at least all co,cr systems).

There are more reasons to topologize $M_{m,n,p}(\mathbb{R})$ and more generally $M_{m,n,p}(k)$, where k is any field. For one thing it would be nice if $M_{m,n,p}(\mathbb{R})$ had a topology such that the isomorphism classes of two systems Σ and Σ' were close together if and only if their associated input/output maps were close together (in some suitable operator topology; say the weak topology); a requirement which is also relevant to the consistency requirement of maximum likelihood identification of systems, cf. [De, DDH, DH, DS, Han]. Yet topologizing $M_{m,n,p}(\mathbb{R})$ does not remove the problem posed by example (3.1.1). Indeed, giving $M_{m,n,p}(\mathbb{R})$ the quotient topology inherited from $L_{m,n,p}(\mathbb{R})$ the maps defined by the families Σ and Σ' of example (3.1.1) are both continuous.

Restricting ourselves to families consisting of cr systems (or dually to families of co systems), however, will solve the problem posed by example (3.1.1). This same restriction will also see to it that the quotient topology is Hausdorff and it will turn out that $M^{cr}_{m,n,p}(\mathbb{R})/GL_n(\mathbb{R})$ is naturally a smooth differentiable manifold. From the algebraic geometric point of view we shall see that the quotient $L^{cr}_{m,n,p}/GL_n$ exists as a smooth scheme defined over $\mathbb{Z}$. It is also pleasant to notice that for pairs of matrices (F,G) the prestable ones (in the sense of [Mu] are precisely the completely reachable ones [Ta2] and they are also the semi-stable points of weight one, [BH].

Ideally it would also be true that every continuous, differentiable, polynomial,... map $\phi : \mathbb{R} \to M^{cr}_{m,n,p}(\mathbb{R})$ comes from a conuous, differentiable, polynomial,... family. This requires assigning to each point of $M^{cr}_{m,n,p}(\mathbb{R})$ a system represented by that point and to do this in an analytic manner. This now really requires a slightly more sophisticated definition of family than we have used up to now, cf. 3.4 below. And indeed to obtain e.g. all continuous map of say the circle into $M_{m,n,p}(\mathbb{R})$ as maps associated to a family one also needs the same more general concept of families of system over the circle.

3.2 Description of the Quotient Set (or Set of Orbits) $L^{cr}_{m,n,p}(k)/GL_n(k)$.

Let k be any field, and fix $n,m,p \in \mathbb{N}$. Let

$$J_{n,m} = \{(0,1),(0,2),\ldots,(0,m);\ (1,1),\ldots,(1,m);\ \ldots\ ;\ (n,1),\ldots,(n,m)\}\ , \qquad (3.2.1)$$

lexicographically ordered (which is the order in which we have written down the $(n+1)m$ elements of $J_{n,m}$). We use $J_{n,m}$ to label the columns of the matrix $R(F,G)$, $F \in k^{n\times n}$, $G \in k^{n\times m}$, cf. 1.3 above, by assigning the label (i,j) to the j-th column of the block F^iG.

A subset $\alpha \subset J_{n,m}$ is called <u>nice</u> if $(i,j) \in \alpha \Rightarrow (i-1,j) \in \alpha$ or $i = 0$ for all i,j. A nice subset with precisely n elements is called a <u>nice selection</u>. Given a nice selection α, a <u>successor index</u> of α is an element $(i,j) \in J_{n,m}\backslash\alpha$ such that $\alpha \cup \{(i,j)\}$ is nice. For every $j_0 \in \{1,\ldots,m\}$ there is precisely one successor index (i,j) of α with $j = j_0$. This successor index will be denoted $s(\alpha,j_0)$.

Pictorially these definitions look as follows. We write down the elements of $J_{n,m}$ in a square as follows $(m=4,\ n=5)$

(0,1)	(1,1)	(2,1)	(3,1)	(4,1)	(5,1)
(0,2)	(1,2)	(2,2)	(3,2)	(4,2)	(5,2)
(0,3)	(1,3)	(2,3)	(3,3)	(4,3)	(5,3)
(0,4)	(1,4)	(2,4)	(3,4)	(4,4)	(5,4)

Using dots to represent elements of $J_{n,m}$ and x's to represent elements of α the following pictures represent respectively a nice subset, a not nice subset and a nice selection.

```
. . . . . .    . x x . . .    . . . . . .
x x x . . .    x . . x . .    x x . . . .
x . . . . .    . x . . . .    . . . . . .
. . . . . .    x x . . . .    x x x . . .
```

The successor indices of the nice selection α of the third picture above are indicated by *'s in the picture below

```
* . . . . .
x x * . . .
* . . . . .        (3.2.2)
x x x * . .
```

We shall use $L_{m,n}(k)$ to denote the set of all pairs of matrices (F,G) over k of sizes $n\times n$ and $n\times m$ respectively;

$L^{cr}_{m,n}(k)$ denotes the subset of completely reachable pairs (cf. 1.3 above). For each subset $\beta \in J_{n,m}$ and each $(F,G) \in L_{m,n}(k)$ we shall use $R(F,G)_\beta$ to denote the matrix obtained from $R(F,G)$ by removing all columns whose index is not in β.

With this terminology and notation we have the following lemma.

3.2.3 <u>Nice Selection Lemma</u>

Let $(F,G) \in L^{cr}_{m,n}(k)$. Then there is a nice selection α such that $\det(R(F,G)_\alpha) \neq 0$.

<u>Proof.</u> Let α be a nice subset of $J_{n,m}$ such that the columns of $R(F,G)_\alpha$ are linearly independent and such that α is maximal with respect to this property. Let

$$\alpha = \{(0,j_1),\dots,(i_1,j_1);$$
$$(0,j_2),\dots,(i_2,j_2); \; \dots \; ; (0,j_s),\dots,(i_s,j_s)\}.$$

By the maximality of α we know that the successor indices $s(\alpha,j)$, $j = 1,\dots,m$ are linearly dependent on the columns of $R(F,G)$. I.e. the columns with indices $(i_1+1,j_1),\dots,(i_s+1,j_s)$ and $(0,t)$, $t \in \{1,\dots,m\}\backslash\{j_1,\dots,j_s\}$ are linearly dependent on the columns of $R(F,G)_\alpha$. Suppose now that with induction we have proved that all columns with indices $(i_r+\ell,j_r)$, $r = 1,\dots,s$ and $(\ell-1,t)$, $t \in \{1,\dots,m\}\backslash\{j_1,\dots,j_s\}$ are linearly dependent on the columns of $R(F,G)_\alpha$, $\ell \geq 1$. This gives us certain relations

$$F^{\ell-1}G_t = \sum_{(i,j)\in\alpha} a(i,j)F^iG_j, \quad F^{i_r+\ell}G_{j_r} = \sum_{(i,j)\in\alpha} b(i,j)F^iG_j$$

(where G_t denotes the t-th column of G). Multiplying on the left with F we find expressions

$$F^{\ell}G_t = \sum_{(i,j)\in\alpha} a(i,j)F^{i+1}G_j,$$

$$F^{i_r+\ell+1}G_{j_r} = \sum_{(i,j)\in\alpha} b(i,j)F^{i+1}G_j$$

expressing $F^{\ell}G_t$ and $F^{i_r+\ell+1}G_{j_r}$ as linear combination of those columns of $R(F,G)$ whose indices are either in α or a successor index of α. The latter are in turn linear combinations of the columns of $R(F,G)_\alpha$, so that we have proved that all columns of $R(F,G)$ are linear combinations of the columns of $R(F,G)_\alpha$. Now (F,G) is cr so that $\text{rank}(R(F,G)) = n$, so that α must have had n elements, proving the lemma.

For each nice selection α we define

$$U_\alpha(k) = \{(F,G,H) \in L_{m,n,p}(k) \mid \det(R(F,G)_\alpha) \neq 0\} \quad (3.2.4)$$

Recall that $GL_n(k)$ acts on $L_{m,n,p}(k)$ by $(F,G,H)^S = (SFS^{-1}, SG, HS^{-1})$.

3.2.5. Lemma. U_α is stable under the action of $GL_n(k)$ on $L_{m,n,p}(k)$. For each $\Sigma \in (F,G,H) \in U_\alpha$ there is precisely one $S \in GL_n(k)$ such that $R(\Sigma^S)_\alpha = R(SFS^{-1},SG)_\alpha = I_n$, the $n \times n$ identity matrix.

Proof. We have

$$R(\Sigma^S) = R(SFS^{-1},SG) = SR(F,G) = S\,R(\Sigma) \quad (3.2.6)$$

It follows that $R(\Sigma^S)_\alpha = SR(\Sigma)_\alpha$, which proves the first statement. It also follows that if we take $S = R(F,G)_\alpha^{-1}$ then $R(\Sigma^S)_\alpha = I_n$ and this is also the only S which does this because in the equation $S\,R(\Sigma)_\alpha = R(\Sigma^S)_\alpha$, $R(\Sigma)_\alpha$ has rank n.

3.2.7. Lemma. Let $x_1,\ldots,x_m$ be an arbitrary m-tuple of n-vectors over k and let α be a nice selection. Then there is precisely one pair $(F,G) \in L^{cr}_{m,n}(k)$ such that $R(F,G)_\alpha = I_n$, $R(F,G)_{s(\alpha,j)} = x_j$, $j = 1,\ldots,m$.

Proof (by sufficiently complicated example). Suppose $m = 4$, $n = 5$ and that α is the nice selection of (3.2.2) above. Then we can simply read off the desired F,G. In fact we find $G_1 = x_1$, $G_2 = e_1$, $G_3 = x_3$, $G_4 = e_2$, $F_1 = e_3$, $F_2 = e_4$, $F_3 = x_2$, $F_4 = e_5$, $F_5 = x_4$. Writing down a fully general proof is a bit tedious and notationally a bit cumbersome and it should now be trivial exercise.

3.2.8. Corollary. The set of orbits $U_\alpha(k)/GL_n(k)$ is in bijective correspondence with $k^{nm} \times k^{pn}$, and $U_\alpha(k) \xrightarrow{\sim} GL_n(k) \times (k^{nm} \times k^{pn})$ (as sets with $GL_n(k)$-action, where $GL_n(k)$ acts on $GL_n(k) \times (k^{nm} \times k^{pn})$ by multiplication on the left on the first factor).

Proof. This follows immediately from lemma 3.2.5 together with lemma 3.2.7. Indeed given $\Sigma = (F,G,H)_\alpha \in U_\alpha$. Take $S = R(F,G)_\alpha^{-1}$ and let $(F',G',H') = \Sigma^S$. Now define $\phi : U_\alpha(k) \to GL_n(k) \times (k^{nm} \times k^{pn})$ by assigning to (F,G,H) the matrix S^{-1}, the m n-vectors $R(\Sigma^S)_{s(\alpha,j)}$, $j = 1,\dots,m$ and the $p \times n$ matrix H'. Inversely given a $T \in GL_n(k)$, m n-vectors x_j, $j = 1,\dots,m$ and a $p \times n$ matrix y. Let $(F',G') \in L^{cr}_{m,n}(k)$ be the unique pair such that $R(F',G')_\alpha = I_n$, $R(F',G')_{s(\alpha,j)} = x_j$, $j = 1,\dots,m$. Take $H' = y$ and define

$$\psi : \ GL_n(k) \times (k^{nm} \times k^{pn}) \to U_\alpha(k)$$

by

$$\psi(T,(x,y)) = (F',G',H')^T \quad .$$

It is trivial to check that $\psi\phi = \text{id}$, $\psi\phi = \text{id}$. It is also easy to check that ϕ commutes with the $GL_n(k)$-actions.

3.2.9. The $c_{\#\alpha}$ (local) canonical forms. For each $\Sigma \in U_\alpha(k)$ we denote with $c_{\#\alpha}(\Sigma)$ the triple:

$$c_{\#\alpha}(\Sigma) = \Sigma^S \quad \text{with} \quad S = R(\Sigma)_\alpha^{-1} \tag{3.2.10}$$

i.e. $c_{\#\alpha}(\Sigma)$ is the unique triple Σ' in the orbit of Σ such that $R(\Sigma')_\alpha = I_n$. Further if $z \in k^{mn} \times k^{np}$, then we let $(F_\alpha(z), G_\alpha(z), H_\alpha(z))$ be the triple $\psi(I_n,z)$; that is if $z = ((x_1,\dots,x_m),y)$ $(F_\alpha(z), G_\alpha(z), H_\alpha(z))$ is the unique triple such that:

$$R(F_\alpha(z), G_\alpha(z))_\alpha = I_n, \quad R(F_\alpha(z), G_\alpha(z))_{\sigma(\alpha,j)} = x_j,$$

$$H_\alpha(z) = y \ , \ z \in ((x_1,\dots,x_m), y) \in k^{nm} \times k^{pn} \tag{3.2.11}$$

3.2.12. Remark. Let $\pi_\alpha : U_\alpha(k) \to k^{nm} \times k^{pn}$ be equal to $\psi : U_\alpha(k) \to GL_n(k) \times (k^{nm} \times k^{pn})$ followed by the projection on the second factor. Then $\tau_\alpha : z \mapsto (F_\alpha(Z), G_\alpha(z), H_\alpha(z))$ is a section of π_α (meaning that $\pi_\alpha\tau_\alpha = \text{id}$), and $c_{\#\alpha}(\tau_\alpha) = \tau_\alpha$. Of course, π_α induces a bijection $U_\alpha(k)/GL_n(k) \to k^{nm} \times k^{pn}$.

3.2.13. Description of the set of orbits. $L^{cr}_{m,n,p}(k)/GL_n(k)$. Order the set of all nice selections from $J_{n,m}$ in some way.

For each $\Sigma \in L^{cr}_{m,n,p}$ let $\alpha(\Sigma)$ be the first nice selection in this ordering such that $R(F,G)_{\alpha(\Sigma)}$ is non-singular. Now assign to Σ the triple $c_{\#\alpha(\Sigma)}(\Sigma)$. This assigned to each $\Sigma \in L^{cr}_{m,n,p}(k)/GL_n(k)$ one particular well defined element in its orbit and this hence gives complete description of the set of orbits

$$L^{cr}_{m,n,p}(k)/GL_n(k).$$

3.3 Topologizing $L^{cr}_{m,n,p}(k)/GL_n(k) = M^{cr}_{m,n,p}(k)$

3.3.1 A more "homogeneous" description of $M^{cr}_{m,n,p}(k)$. The description of the set of orbits of $GL_n(k)$ acting on $L^{cr}_{m,n,p}(k)$ given in 3.2.13 is highly lopsided in the various possible nice selections α. A more symmetric description of $M^{cr}_{m,n,p}(k)$ is obtained as follows. For each nice selection α, let $V_\alpha(k) = k^{nm} \times k^{pn}$ and let for each second nice selection β:

$$V_{\alpha\beta}(k) = \{z \in V_\alpha | \det(R(F_\alpha(z), G_\alpha(z))_\beta) \neq 0\} \qquad (3.3.2)$$

That is, under the section $\tau_\alpha : V_\alpha(k) \to U_\alpha(k)$ of 3.2 above which picks out precisely one element of each orbit in $U_\alpha(k)$ $V_{\alpha\beta}(k)$ corresponds to those orbits which are also in $U_\beta(k)$; or equivalently $V_{\alpha\beta}(k) = \pi_\alpha(U_\alpha(k) \cap U_\beta(k))$. We now glue the $V_\alpha(k)$, α nice, together along the $V_{\alpha\beta}(k)$ by means of the identifications:

$$\phi_{\alpha\beta}: V_{\alpha\beta}(k) \to V_{\beta\alpha}(k),\ \phi_{\alpha\beta}(z) = z' \Leftrightarrow$$

$$(F_\alpha(z), G_\alpha(z), H_\alpha(z))^S = (F_\alpha(z'), G_\alpha(z'), H_\alpha(z'),$$

$$S = R(F_\alpha(z), G_\alpha(z))^{-1}_\beta \qquad (3.3.3)$$

Then, as should be clear from the remarks made just above, $M^{cr}_{m,n,p}(k)$ is the union of the $V_\alpha(k)$ with for each pair of nice selections α, β, $V_{\alpha\beta}(k)$ identified with $V_{\beta\alpha}(k)$ according to (3.3.3).

3.3.4. The analytic varieties $M^{cr}_{m,n,p}(\mathbb{R})$ and $M^{cr}_{m,n,p}(\mathbb{C})$. Now let $k = \mathbb{R}$ or $\mathbb{C}$ and give $V_\alpha(k) = k^{nm} \times k^{pn}$ its usual (real) analytic structure. The subsets $V_{\alpha\beta}(k) \subset V_\alpha(k)$ are then open subsets and the $\phi_{\alpha\beta}(k)$ are analytic diffeomorphisms. It follows that $M^{cr}_{m,n,p}(\mathbb{R})$ and $M^{cr}_{m,n,p}(\mathbb{C})$ will be respectively a real analytic (hence certainly C^∞) manifold and a complex analytic manifold, provided we can show that they are Hausdorff.

First notice that if we give $L_{m,n,p}(\mathbb{R})$ and $L_{m,n,p}(\mathbb{C})$ the topology of $\mathbb{R}^{mn+n^2+np}$ and $\mathbb{C}^{n^2+nm+np}$ respectively and the open subsets $U_\alpha(k)$ and $L^{cr}_{m,n,p}(k)$, $k = \mathbb{R}, \mathbb{C}$ the induced topology, then the quotient topology for $\pi_\alpha: U_\alpha(k) \to V_\alpha(k)$ is precisely the topology resulting from the identification $V_\alpha(k) \simeq k^{nm} \times k^{pn}$. It follows that the topology of $M^{cr}_{m,n,p}(k)$ is the quotient topology of $L^{cr}_{m,n,p}(k) \to L^{cr}_{m,n,p}(k)/GL_n(k) = M^{cr}_{m,n,p}(k)$.

Now let $G_{n,m(n+1)}(k)$ be the Grassmann variety of n-planes in m(n+1)-space. For each (F,G), R(F,G) is an $n \times m(n+1)$ matrix of rank n which hence defines a unique point of $G_{n,m(n+1)}(k)$. Because $R(SFS^{-1},SG) = SR(F,G)$ we have that (F,G) and $(F,G)^S$ define the same point in Grassmann space. It follows that by forgetting H we have defined a map:

$$\bar{R}: M^{cr}_{m,n,p}(k) \to G_{n,m(n+1)}(k),\ (F,G) \mapsto \text{subspace spanned by the rows of } R(F,G). \tag{3.3.5}$$

In addition we let $h: M^{cr}_{m,n,p}(k) \to k^{(n+1)^2mp}$ be the map induced by:

$$\tilde{h}(F,G,H) = \begin{pmatrix} A_1 & A_2 & \cdots & A_{n+1} \\ A_2 & & & \cdot \\ \cdot & & & \cdot \\ \cdot & & & \cdot \\ \cdot & & & \cdot \\ A_{n+1} & & \cdots & A_{2n+1} \end{pmatrix},$$

$$A_i = HF^{i-1}G, \quad i = 1,\dots,2n+1 \tag{3.3.6}$$

It is not particularly difficult to show ([Haz 1-3], cf. also the realization algorithm in 5.2 below) that the combined map $(\bar{R},h): M^{cr}_{m,n,p}(k) \to G_{n,m(n+1)}(k) \times k^{(n+1)^2mp}$ is injective. By the quotient topology remarks above it is then a topological embedding, proving that $M^{cr}_{m,n,p}(k)$ is a Hausdorff topological space. So we have:

3.3.7. Theorem. $M^{cr}_{m,n,p}(\mathbb{R})$ and $M^{cr}_{m,n,p}(\mathbb{C})$ are smooth analytic manifolds. The sets $M^{cr,co}_{m,n,p}(\mathbb{R})$ and $M^{cr,co}_{m,n,p}(\mathbb{C})$ are analytic open sub-manifolds. (These are the sets of orbits of the cr and co systems, or equivalently the images of $L^{cr,co}_{m,n,p}(k)$ under $\pi: L^{cr}_{m,n,p}(k) \to M^{cr}_{m,n,p}(k)$, $k = \mathbb{R},\mathbb{C}$).

3.3.8. Remark. A completely different way of showing that the quotient space $M^{cr}_{m,n,p}(\mathbb{R})$ is a differentiable manifold is due to Martin and Krishnaprasad, [MK]. They show that with respect to a suitable invariant metric of $L^{cr,co}_{m,n,p}(k)$, $GL_n(k)$ acts properly discontinuously.

3.3.9. The algebraic varieties $M^{cr}_{m,n,p}(k)$. Now let k be any algebraically closed field. Giving $L_{m,n,p}(k) = k^{n^2+nm+np}$ the Zariski topology and $U_\alpha(k)$ the induced topology for each nice selection α. Then $U_\alpha(k) \simeq GL_n(k) \times V_\alpha(k)$, $V_\alpha(k) = k^{nm+np}$ also as algebraic varieties. The $V_{\alpha\beta}(k)$ are open subvarieties and the $\phi_{\alpha\beta}(k)$: $V_{\alpha\beta}(k) \to V_{\beta\alpha}(k)$ are isomorphisms of algebraic varieties. The map $(\bar{R},h)$ is still injective and it follows that $M^{cr}_{m,n,p}(k)$ has a natural structure of a smooth algebraic variety, with $M^{cr,co}_{m,n,p}(k)$ an open subvariety.

3.3.10. The scheme $M^{cr}_{m,n,p}$. As a matter of fact, the defining pieces of the algebraic varieties $M^{cr}_{m,n,p}(k)$, that is the $V_\alpha(k)$, and the gluing isomorphisms $\phi_{\alpha\beta}(k)$ are all defined over $\mathbb{Z}$. So there exists a scheme $M^{cr}_{m,n,p}$ over $\mathbb{Z}$ such that for all fields k the rational points over k, $M^{cr}_{m,n,p}(k)$, are precisely the orbits of $GL_n(k)$ acting on $L^{cr}_{m,n,p}(k)$. For details cf. section 4 below.

3.4. A universal family of linear dynamical systems

3.4.1. As has been remarked above it would be nice if we could attach in a continuous way to each point of $M_{m,n,p}(k)$ a system over k representing that point. Also it would be pleasant if every appropriate map from a parameter space V to $M^{cr}_{m,n,p}$ came from a family over V. Recalling from 2.2 above that systems over a ring R can be reinterpreted as families over $\mathrm{Spec}(R)$, this would mean that the isomorphism classes of systems over R would correspond bijectively with the R-rational points $M^{cr}_{m,n,p}(R)$ of the scheme $M^{cr}_{m,n,p}$ over $\mathbb{Z}$, cf. 3.3.10.

Both wishes, if they are to be fulfilled require a slightly more general definition of system than we have used up to now. In the case of systems over a ring R the extra generality means that instead of considering three matrices F,G,H over R, that is three homomorphisms $G: R^m \to R^n$, $F: R^n \to R^n$, $H: R^n \to R^p$ we now generalize to the definition: a projective system over R consists of a projective module X as state module together with three homomorphisms $G: R^m \to X$, $F: X \to X$, $H: X \to R^p$. Thus the extra generality sits in the fact that the state R-module X is not required to be free, but only projective. The geometric counterpart of this is a vectorbundle, cf. below in 3.4.2 for the precise definition of a family and the role the vectorbundle plays.

In some circumstances it appears to be natural, in any case as an intermediate step, to consider even more general families. Thus over a ring R it makes perfect sense to consider arbitrary modules as state modules, and indeed these turn up naturally when doing "canonical" realization theory, cf. [Eil, Ch. XVI], which in terms of families means that one may need to consider more general fibrations by vector spaces than locally trivial ones.

3.4.2. <u>Families of linear dynamical systems (over a topological space)</u>. Let V be a topological space. A continuous family Σ of real linear dynamical systems over V (or parametrized by V) consists of:

(a) a vectorbundle E over V
(b) a vectorbundle endomorphism $F: E \to E$
(c) a vectorbundle morphism $G: V \times \mathbb{R}^m \to E$
(d) a vectorbundle morphism $H: E \to V \times \mathbb{R}^p$

For each $v \in V$ let $E(v)$ be the fibre of E over v. Then we have homomorphisms of vector spaces $G(v):\{v\} \times \mathbb{R}^m \to E(v)$, $F(v): E(v) \to E(v)$, $H(v): E(v) \to \{v\} \times \mathbb{R}^p$. Thus choosing a basis in $E(v)$, and taking the obvious bases in $\{v\} \times \mathbb{R}^m$ and $\{v\} \times \mathbb{R}^p$ we find a triple of matrices $\bar{F}(v)$, $\bar{G}(v)$, $\bar{H}(v)$. Thus the data listed above do define a family over V in the sense that they assign to each $v \in V$ a linear system. Note however that there is no natural basis for $E(v)$ so that the system is really only defined up to base change, i.e. up to the $GL_n(\mathbb{R})$ action, so that what the data (a)-(d) really do is assign a point of $M_{m,n,p}(\mathbb{R})$ to each point $v \in V$.

As E is a vectrobundle we can find for each $v \in V$ an open neighborhood W and n-sections $s_1,\dots,s_n$: $W \to E|_W$ such that $s_1(w),\dots,w_n(w) \in E(w)$ are linearly independent for all $w \in W$. Writing out matrices for $F(w)$, $G(w)$, $H(w)$ with respect to the basis $s_1(w),\dots,s_n(w)$ (and the obvious bases in $\{w\} \times \mathbb{R}^m$ and $\{w\} \times \mathbb{R}^p$), we see that over W the family Σ can indeed be described as a triple of matrices depending continuously on parameters. Inversely if $(\bar{F},\bar{G},\bar{H},)$ is a triple of matrices depending continuously on a parameter $v \in V$, then $E = V \times \mathbb{R}^n$, $F(v,x) = (v,\bar{F}(v)x)$, $F(v,u) = (v,\bar{G}(v)u)$, $H(v,x) = (v,\bar{H}(v)x)$ define a family as described above. Thus locally the new definition agrees (up to isomorphism) with the old intuitive one we have been using up to now; globally it does not.

Here the appropriate notion of isomorphism is of course: two families $\Sigma = (E;\ F,\ G,\ H)$ and $\Sigma' = (E';\ F',\ G',\ H')$ over V are isomorphic if there exists a vectorbundle isomorphism $\phi: E \to E'$ such that $F'\phi = \phi F$, $\phi G = G'$, $H = H'\phi$.

3.4.3. Other kinds of families of systems. The appropriate definitions of other kinds of families are obtained from the one above by means of minor and obvious adjustments. For instance, if V is a differentiable (resp. real analytic) manifold then a differentiable (resp. real analytic) family of systems consists of a differentiable vector bundle E with differentiable morphisms F, G, H (resp. an analytic vectorbundle with analytic morphisms F, G, H). And of course isomorphisms are supposed to be differentiable (resp. analytic).

Similarly if V is a scheme (over k) then an algebraic family consists of an algebraic vectorbundle E over V together with morphisms of algebraic vectorbundles $F: E \to E$, $G: V \times \mathbb{A}^m \to E$, $H: E \to V \times \mathbb{A}^p$, where $\mathbb{A}^r$ is the (vectorspace) scheme $\mathbb{A}^r(R) = R^r$ (with the obvious R-module structure).

Still more variations are possible. E.G. a complex analytic family (or holomorphic family) over a complex analytic space V would consist of a complex analytic vectorbundle E with complex analytic vectorbundle homomorphisms $F: E \to E$, $G: V \times \mathbb{C}^m \to E$, $H: E \to V \times \mathbb{C}^p$.

3.4.4. Convention. From now on whenever we speak about a family of systems it will be a family in the sense of (3.4.2) and (3.4.3) above.

3.4.5. The canonical bundle over $G_{n,r}(k)$. Let $G_{n,r}(k)$ be the Grassman manifold of n-planes in r-space $(r > n)$. Let $E(k) \to G_{n,r}(k)$ be the fibre bundle whose fibre over $x \in G_{n,r}(k)$ is the n-plane in k^r represented by the point x. If $k = \mathbb{R}$ or $\mathbb{C}$ this is an analytic vector bundle over $G_{n,r}(k)$. More generally ally this defines an algebraic vectorbundle E over the scheme $G_{n,r}$.

this defines an algebraic vectorbundle E over the scheme $G_{n,r}$.

In terms of trivial pieces and gluing data this bundle can be described as follows. Let $M^{n\times r}_{reg}(k)$ be the space of all $n\times r$ matrices of rank n and let $\pi: M^{n\times r}_{reg}(k) \to G_{n,r}(k)$ be the map which associates to each $n\times r$ matrix of rank n the n-space in k^{nr} spanned by its row vectors. Then the fibre over $E(x)$ of E over $x \in G_{n,r}(k)$ is precisely the vector space of all linear combinations of any element in $\pi^{-1}(x)$. From this there results the following local pieces the gluing data description of $G_{n,r}(k)$ and $E(k)$. For each subset α of size n of $\{1,2,\ldots,r\}$ let $U'_\alpha(k)$ be the set of all $n\times r$ matrices A such that A_α is invertible, let $V'_\alpha(k) = k^{n(r-n)}$ and for each $z \in V'_\alpha(k)$, $z = (z_1,\ldots,z_{r-n})$, $z_i \in k^n$, let $A_\alpha(z)$ be the unique $n\times r$ matrix such that $(A_\alpha(z))_\alpha = I_n$ and $A_\alpha(z)_{t(j)} = z_j$ where $t(j)$ runs through the elements of $\{1,2,\ldots,r\}-\alpha$ in the natural order, $j = 1,\ldots,r-n$. Then $G_{n,r}(k)$ consists of the $V'_\alpha(k)$ glued together along the $V'_{\alpha\beta}(k) = \{z \in V'_\alpha(k) | A_\alpha(z)_\beta$ is invertible$\}$ by means of the isomorphisms:

$$\phi'_{\alpha\beta}(k): V'_{\alpha\beta}(k) \xrightarrow{\sim} V'_{\beta\alpha}(k),\ z \mapsto z' \Leftrightarrow (A_\alpha(z)_\beta)^{-1}A_\alpha(z) = A_\beta(z') \tag{3.4.7}$$

(Note how very similar this is to the pieces and patching data description of $M^{cr}_{m,n,p}(k)$ given in 3.3.1 above; the reason is understandable if one observes that the map $R: L^{cr}_{m,n,p}(k) \to M^{n\times(n+1)m}_{reg}(k)$, induces a map $\bar{R}: M^{cr}_{m,n,p}(k) \to G_{n,(n+1)m}(k)$, which is compatible with the local pieces and patching data for the two spaces).

The bundle $E(k)$ over $G_{n,r}(k)$ can now be described as follows. Over each $V'_\alpha(k) \subset G_{n,r}(k)$ we can trivialize $E(k)$ as follows:

$$V'_\alpha(k) \times k^n \xrightarrow{\sim} E(k)|_{V'_\alpha(k)},\ (z,x) \mapsto x^T A_\alpha(z)\ . \tag{3.4.8}$$

It follows that the bundle $E(k)$ over $G_{n,r}(k)$ admits the following local pieces and patching data description which is compatible with the local pieces and patching data description given above for $G_{n,r}(k)$. The bundle $E(k)$ consists of the local pieces $E_\alpha(k) = V'_\alpha(k) \times k^n$ glued together along the $E_{\alpha\beta}(k) = V'_{\alpha\beta}(k) \times k^n$ by means of the isomorphisms:

$$\tilde{\phi}'_{\alpha\beta} : V'_{\alpha\beta}(k) \times k^n \xrightarrow{\sim} V'_{\beta\alpha}(k) \times k^n$$
$$(z,x) \mapsto (\phi'_{\alpha\beta}(z),\ (A_\alpha(z)_\beta)^T x) \tag{3.4.9}$$

The bundle which is really of interest to us is the dual bundle E^d to E described by the local pieces $E^d_\alpha(k) = V'_\alpha(k) \times k^n$ glued together by the patching data:

$$\tilde{\phi}^d_{\alpha\beta}\colon\ V'_{\alpha\beta}(k) \times k^n \xrightarrow{\sim} V'_{\beta\alpha}(k) \times k^n$$

$$(z,x) \mapsto (\phi'_{\alpha\beta}(z),\ (A_\alpha(z)_\beta)^{-1}x) \tag{3.4.10}$$

(note that the gluing isomorphisms $\tilde{\phi}^d_{\alpha\beta}$ are compatible with the projections $E^d_\alpha(k) \to V'_\alpha(k)$ and the gluing isomorphisms $\phi'_{\alpha\beta}$ for $G_{n,r}(k)$; note also that all three sets of gluing data $\phi'_{\alpha\beta}$, $\tilde{\phi}'_{\alpha\beta}$, $\tilde{\phi}^d_{\alpha\beta}$ are transitive in the sense that $\tilde{\phi}^d_{\beta\gamma} \circ \tilde{\phi}^d_{\alpha\beta} = \tilde{\phi}^d_{\alpha\gamma}$ are similarly for the $\tilde{\phi}'$ and ϕ').

3.4.11. <u>The underlying vector bundle of the universal family over</u> $M^{cr}_{m,n,p}(k)$. The map $R\colon L^{cr}_{m,n,p}(k) \to M^{n\times(n+1)m}_{reg}(k)$, $(F,G,H) \mapsto R(F,G)$ induces a map.

$$\bar{R}\colon M^{cr}_{m,n,p}(k) \to G_{n,(n+1)m}(k) \tag{3.4.12}$$

(because $R(\Sigma^S) = SR(\Sigma)$, $S \in GL_n(k)$).

If $k = \mathbb{R}$ or $\mathbb{C}$, (3.4.12) is a morphism between analytic manifolds. In general (3.4.12) defines a morphism between the schemes $M^{cr}_{m,n,p}$ and $G_{n,(n+1)m}$. Now let $E^u = \bar{R}^! E^d$, the pull-back by means of $\bar{R}$ of the "canonical" bundle E^d described above in (3.4.5).

Now recall that $M^{cr}_{m,n,p}(k)$ was obtained by gluing the various pieces $V_\alpha(k) = k^{nm} \times k^{pn}$ together, where α runs through all nice selections from $J_{n,m}$. In terms of this description $E^u(k)$ can be described as follows: $E^u(k)$ consists of pieces $E^u_\alpha(k) = V_\alpha(k) \times k^n = k^{nm} \times k^{pn} \times k^n$, one for each nice selection α. For each pair of nice selections $E^u_{\alpha\beta}(k) = V_{\alpha\beta}(k) \times k^n \subset V_\alpha(k) \times k^n$. Now for each pair of nice selections α,β let $\tilde{\phi}_{\alpha\beta}(k)\colon E^u_{\alpha\beta}(k) \to E^u_{\beta\alpha}(k)$ be the isomorphism:

$$\tilde{\phi}_{\alpha\beta}(k)(z,x) = (\phi_{\alpha\beta}(z),\ (R(F_\alpha(z),\ G_\alpha(z))_\beta)^{-1}\ x) \tag{3.4.13}$$

where $\phi_{\alpha\beta}\colon V_{\alpha\beta}(k) \to V_{\beta\alpha}(k)$ is the isomorphism of 3.3 above (which describes how the $V_\alpha(k)$ should be glued together to give

$M_{m,n,p}(k)$, and $V_\alpha(k) \to U_\alpha(k)$, $z \mapsto (F_\alpha(z), G_\alpha(z), H_\alpha(z))$ is the section τ_α described above in (3.2.12). Then $E^u_\alpha(k)$ is obtained by gluing together the $E^u_\alpha(k)$ along the $E^u_{\alpha\beta}(k)$ by means of the isomorphisms (3.4.13).

3.4.14. <u>Construction of a universal family of cr systems</u>. Let $E^u(k)$ over $M^{cr}_{m,n,p}(k)$ be the bundle described above and view it as obtained via the patching data (3.4.13). Recall also that, cf. (3.3.3) above:

$$\begin{aligned} \phi_{\alpha\beta}(z) = z' \Leftrightarrow (F_\alpha(z), G_\alpha(z), H_\alpha(z))^S = \\ (F_\beta(z'), G_\beta(z'), H_\beta(z')) \\ \text{with } S = R(F_\alpha(z), G_\alpha(z))_\beta^{-1} \end{aligned} \tag{3.4.15}$$

For each nice selection α we now define a bundle endomorphism $F^u_\alpha(k)$ of $E^u_\alpha(k) = V_\alpha(k) \times k^n$ and bundle morphisms $G^u_\alpha(k)$: $V_\alpha(k) \times k^m \to E^u_\alpha(k)$, $H^u_\alpha(k)$: $E^u_\alpha(k) \to V_\alpha(k) \times k^p$. These are defined as follows:

$$\begin{aligned} F^u_\alpha(k)\ (z,x) &= (z, F_\alpha(z)x) \\ G^u_\alpha(k)\ (z,u) &= (z, G_\alpha(z)x) \\ H^u_\alpha(k)\ (z,x) &= (z, H_\alpha(z)x) \end{aligned} \tag{3.4.16}$$

We now claim that these bundle morphisms are compatible with the gluing isomorphisms (3.4.13), which means that we must prove the commutativity of the diagram below for each pair of nice selections α, β.

$$\begin{array}{ccccccc} V_{\alpha\beta}\times k^m & \xrightarrow{G^u_\alpha} & E^u_{\alpha\beta} & \xrightarrow{F^u_\alpha} & E^u_{\alpha\beta} & \xrightarrow{H^u_\alpha} & V_{\alpha\beta}\times k^p \\ \downarrow{\scriptstyle \phi_{\alpha\beta}\times id} & & \downarrow{\scriptstyle \tilde{\phi}_{\alpha\beta}} & & \downarrow{\scriptstyle \tilde{\phi}_{\alpha\beta}} & & \downarrow{\scriptstyle \phi_{\alpha\beta}\times id} \\ V_{\beta\alpha}\times k^m & \xrightarrow{G^u_\beta} & E^u_{\beta\alpha} & \xrightarrow{F^u_\beta} & E^u_{\beta\alpha} & \xrightarrow{H^u_\beta} & V_{\beta\alpha}\times k^p \end{array} \tag{3.4.17}$$

where we have abbreviated various notations in obvious ways. Now

$$\tilde{\Phi}_{\alpha\beta}G^u_\alpha(z,u) = \tilde{\Phi}_{\alpha\beta}(z,\ G_\alpha(z)\ u) \qquad \text{by (3.4.16)}$$

$$= (\phi_{\alpha\beta}(z),\ R(F_\alpha(z),\ G_\alpha(z))^{-1}_\beta\ G_\alpha\ (z)\ u) \qquad \text{by (3.4.13)}$$

$$= (\phi_{\alpha\beta}(z),\ G_\beta(z')\ u) \qquad \text{by (3.4.15)}$$

$$= G^u(\phi_{\alpha\beta} \times \mathrm{id}(z,\ u)) \qquad \text{by (3.4.16)}$$

proving the commutativity of the left most square of (3.4.17). Similarly:

$$\tilde{\Phi}_{\alpha\beta}F^u_\alpha(z,x) = \tilde{\Phi}_{\alpha\beta}(z,\ F_\alpha(z)\ x) \qquad \text{by (3.4.16)}$$

$$= (\phi_{\alpha\beta}(z),\ R(F_\alpha(z),\ G_\alpha(z))^{-1}_\beta\ F_\alpha(z)\ x) \qquad \text{by (3.4.13)}$$

$$= (\phi_{\alpha\beta}(z),\ F_\beta(z')\ R(F_\alpha(z),\ G_\alpha(z))^{-1}_\beta x) \qquad \text{by (3.4.15)}$$

$$= F^u_\beta\ \ \tilde{\Phi}_{\alpha\beta}(z,x)$$

proving the commutativity of the middle square of (3.4.17). And finally, and completely analogously:

$$H^u_\beta\tilde{\Phi}_{\alpha\beta}(z,x) = H^u_\beta(\phi_{\alpha\beta}(z),\ R(F_\alpha(z),\ G_\alpha(z))^{-1}_\beta\ x) \quad \text{by (3.4.13)}$$

$$= (\phi_{\alpha\beta}(z),\ H_\beta(z')\ R(F_\alpha(z),\ G_\alpha(z))^{-1}_\beta\ x) \qquad \text{by (3.4.16)}$$

$$= (\phi_{\alpha\beta}(z),\ H_\alpha(z)\ x)$$

$$= (\phi_{\alpha\beta} \times \mathrm{id})\ (H^u_\alpha(z,x))$$

proving the commutativity of the last square of (3.4.17).

Thus the F^u_α, G^u_α, H^u_α combine to define bundle morphisms $F^u(k)\colon E^u(k) \to E^u(k)$, $G^u\colon M^{cr}_{m,n,p}(k) \times k^m \to E^u(k)$, $H^u(k)\colon E^u(k) \to M^{cr}_{m,n,p}(k) \times k^p$.

If $k = \mathbb{R}$ or $\mathbb{C}$, $F^u(k)$, $G^u(k)$, $H^u(k)$ are morphisms of analytic vector bundles. Algebraically speaking the $F^u(k)$, $G^u(k)$, $H^u(k)$ for varying k are parts of morphisms of algebraic vector bundles over the scheme $M^{cr}_{m,n,p}$, which are defined over $\mathbb{Z}$.

3.4.18. The pullback construction. Let V be a topological space and $\phi: V \to M^{cr}_{m,n,p}(\mathbb{R})$ a continuous map. Let $\Sigma^u = (E^u; F^u, G^u, H^u)$ be the universal family of systems constructed above. Then associated to ϕ we have an induced family $\phi^!\Sigma^u$ over V (obtained by pullback). The precise formulas are as follows:

- $\phi^!E^u = \{(v,x) \in V \times E^u | \phi(v) = \pi(x)\}$, where $\pi: E^u \to M^{cr}_{m,n,p}(\mathbb{R})$ is the bundle projection; the bundle projection of $\phi^!E^u$ is defined by $(v,x) \to v$;
- $\phi^!F^u: (v,x) \mapsto (v, F^u x) \in \phi^!E^u$
- $\phi^!G^u: (v,u) \mapsto (v, G^u u) \in \phi^!E^u$
- $\phi^!H^u: (v,x) \mapsto (v, H^u x) \in \phi^!(M^{cr}_{m,n,p}(\mathbb{R}) \times R^p) = V \times \mathbb{R}^p$

Obviously $\phi^!E^u$ is (up to isomorphism) the family of systems over V such that the system over $v \in V$ is (up to isomorphism) the system over $\phi(v)$ in the family Σ^u.

If V and ϕ are differentiable (resp. real analytic) there results a differentiable (resp. real analytic) family over V. If $\phi: V \to M^{cr}_{m,n,p}(\mathbb{C})$ is a morphism of complex analytic manifolds there results a complex analytic family and on the algebraic-geometric side of things if $\phi: V \to M^{cr}_{m,n,p}$ is a morphism of schemes one finds thus an algebraic family over the scheme V.

3.4.19. The topological fine moduli theorem. Let V be a topological space and Σ a continuous family of completely reachable systems over V. Then there exists a unique continuous map $\phi: V \to M^{cr}_{m,n,p}(\mathbb{R})$ such that Σ is isomorphic to $\phi^!\Sigma^u$ (as continuous families; i.e. there is a bijective correspondence between continuous maps $V \to M^{cr}_{m,n,p}(\mathbb{R})$ and isomorphism classes of continuous families over V).

3.4.20. The algebraic-geometric fine moduli theorem. Let V be a scheme and Σ an algebraic family of cr systems over V. Then there exists a unique morphism of schemes $\phi: V \to M^{cr}_{m,n,p}$ such that Σ is isomorphic to $\phi^!\Sigma^u$ over V.

3.4.21. On the proof of these theorems. First consider the topological case. The map ϕ associated to Σ is defined as follows. For each $v \in V$ we have a system $\Sigma(v)$, which uniquely determines an isomorphism class of linear dynamical systems (cf. (3.4.2));

that is, it uniquely defines a point $\phi(v)$ of $M^{cr}_{m,n,p}(\mathbb{R})$ which is the space of all isomorphism classes of cr systems (of the dimensions under consideration). This ϕ is obviously continuous. Now $\Sigma^u(z)$ for all $z \in M^{cr}_{m,n,p}(\mathbb{R})$ represents z. So, by 3.4.18, Σ and $\phi^!\Sigma^u$ are two continuous families of cr systems over V such that for all $v \in V$, $\Sigma(v)$ and $\phi^!\Sigma^u(v)$ are isomorphic. It follows that the families Σ and $\Sigma' = \phi^!\Sigma^u$ are isomorphic as continuous families. The reason is the following rigidity property: if $(F, G, H), (F', G', H') \in L^{cr}_{m,n,p}(\mathbb{R})$ are isomorphic then the isomorphism is unique. Indeed, if S is an isomorphism then we must have $SR(F, G) = R(F', G')$ so that if α is a nice selection such that $R(F,G)_\alpha$ is invertible, then $S = R(F', G')_\alpha (R(F,G)_\alpha)^{-1}$. The statement that Σ and Σ' over V are isomorphic if they are pointwise isomorphic results as follows. For every $v \in V$ there is a $V' \ni v$ such that the bundles E and E' of Σ and Σ' are trivial over V' so that over V' the families Σ and Σ' are simply (up to isomorphism) continuously varying triples of matrices $(F(v'), G(v'), H(v')), (F'(v'), G'(v'), H'(v'))$, $v' \in V'$. Let α be a nice selection such that $R(F(v), G(v))_\alpha$ is invertible. Restricting V' a bit more if necessary we can assume that $R(F(v'), G(v'))_\alpha$ is invertible for all $v' \in V'$. Then $S(v') = R(F'(v'),G'(v'))_\alpha (R(F(v'), G(v'))_\alpha)^{-1}$ is a continuous family of invertible matrices taking $\Sigma(v')$ into $\Sigma'(v')$ for all $v' \in V'$. Thus Σ and Σ' are isomorphic over some small neighborhood of every point of V. The isomorphisms in question must agree on the intersections of these neighborhoods, again by the rigidity property. It follows that these local isomorphisms combine to define a global isomorphism over all of V from Σ to Σ'.

A more formal and also more formula based version of this argument can be found in [Haz1]. The scheme theoretic version (theorem 3.4.20) is based on the same rigidity property, cf. section 4 below for some details.

3.4.22. Remark. In [HK] I claimed that the underlying bundle E^u of the universal family Σ^u was the pullback by means of $\bar{R}$ (cf. 3.3.5)) of the bundle E over $G_{n(n+1)m}$ whose fibre over z was the n-plane represented by z. As we have seen it is not; instead E^u is the pullback of the dual bundle E^d of E. Now the determinant bundle of E^d is a very ample line bundle (rather than the determinant bundle of E) so that the argument in [HK] to prove that $M_{m,n}$ is not quasi affine is correct modulo two errors which cancel each other.

4. THE CLASSIFYING "SPACE" $M^{cr}_{m,n,p}$ IS DEFINED OVER $\mathbb{Z}$ AND CLASSIFIES OVER $\mathbb{Z}$.

Mainly for completeness and tutorial reasons I give in this section the algebraic-geometric details of the remarks 3.3.10 and 3.4.20 that there exists a scheme $M^{cr}_{m,n,p}$ over $\mathbb{Z}$ of which the varieties $M^{cr}_{m,n,p}(k)$, cf. 3.3.9, k an algebraically closed field, are obtained by base change and that this scheme is classifying for algebraic families of cr systems, and thus in particular classifying for cr systems over rings (with possibly a projective module as state module).

Those who are not particularly interested in the algebraic-geometric details can skip this section without consequences for their understanding of the remainder of this paper. There is in any case nothing difficult about what follows below and anyone who has once seen, say, the construction of the Grassmann schemes or projective spaces over $\mathbb{Z}$, will have no difficulties in supplying all details for himself from what has been said in section 3 above. All we are really doing below is rewriting a number of formulas of section 3 above using capital letters instead of small ones. This does take a certain number of pages, though. It seemed desirable to include these, as, judging from the audience's remarks during the oral presentation of these lectures, there is, perhaps rightly so, a distinct unwillingness in accepting without further proof a statement on the part of the lecturer like "the algebraic-geometric version of this theorem is proved similarly."

4.1 <u>Definition of the scheme</u> $M^{cr}_{m,n,p}$. For each nice selection $\alpha \subset J_{n,m}$ let

$$V_\alpha = \mathrm{Spec}(\mathbb{Z}[X^\alpha_{ij}, Y^\alpha_{rs};\ i = 1,\dots,n,\ j = 1,\dots,m,\ r = 1,\dots,p,\ s = 1,\dots,n]) \qquad (4.1.1)$$

Let $H_\alpha(Y)$ be the $p \times n$ matrix (Y^α_{rs}), and let $(F_\alpha(X), G_\alpha(X))$ be the unique pair of matrices over $\mathbb{Z}[X^\alpha_{ij}]$ such that

$$R(F_\alpha(X), G_\alpha(X))_\alpha = I_n,\quad R(F_\alpha(X), G_\alpha(X))_{s(\alpha,j)} = \begin{pmatrix} X^\alpha_{1j} \\ \vdots \\ X^\alpha_{nj} \end{pmatrix},\qquad j = 1,\dots,m \qquad (4.1.2)$$

(where the $s(\alpha,j)$ are the m successor indices of α, cf. 3.2). Finally for each pair of nice selections α,β let $d_{\alpha\beta}(X) \in \mathbb{Z}[X^\alpha_{ij}]$

be the element

$$d_{\alpha\beta}(X) = \det(R(F_\alpha(X),G_\alpha(X))_\beta) \tag{4.1.3}$$

and let $V_{\alpha\beta}$ be the open subscheme of V_α obtained by localizing with respect to $d_{\alpha\beta}(X)$, i.e.

$$V_{\alpha\beta} = \mathrm{Spec}(\mathbb{Z}[X^\alpha_{ij},Y^\alpha_{rs},d_{\alpha\beta}(X)^{-1}]) \tag{4.1.4}$$

Now for each pair of nice selections α,β write down the formulas

$$S_{\alpha\beta}(X)^{-1}F_\alpha(X)S_{\alpha\beta}(X) = F_\beta(X)$$
$$S_{\alpha\beta}(X)^{-1}G_\alpha(X) = G_\beta(X),\ H_\alpha(Y)S_{\alpha\beta}(X) = H_\beta(Y) \tag{4.1.5}$$

where

$$S_{\alpha\beta}(X) = R(F_\alpha(X),G_\alpha(X))_\beta \tag{4.1.6}$$

Because the entries of $F_\beta(X)$ and $G_\beta(X)$ are equal to zero, 1 or X^β_{ij} for some i,j and because the (r,s)-th entry of $H_\beta(Y)$ is Y^β_{rs}, the formulae (4.1.5) provide us with certain expressions for the X^β_{ij} and $Y^\beta_{r,s}$ in terms of the X_{ij}, Y_{rs}, which by (4.1.5) and (4.1.3) (and the usual formula for matrix inversion) can be written as polynomials in X^α_{ij}, Y^α_{rs}, $d_{\alpha\beta}(X)^{-1}$, say

$$X^\beta_{ij} = \phi_{\alpha\beta}(i,j)(X_{ij},d_{\alpha\beta}(X)^{-1}),$$
$$Y^\beta_{rs} = \phi_{\alpha\beta}(r,s)(X^\alpha_{ij},d_{\alpha\beta}(X)^{-1},Y^\alpha_{rs}) \tag{4.1.7}$$

Then

$$\phi^*_{\alpha\beta} : X^\beta_{ij} \mapsto \phi_{\alpha\beta}(i,j)(X^\alpha),\ Y^\beta_{rs} \mapsto \phi_{\alpha\beta}(r,s)(X^\alpha,Y^\alpha) \tag{4.1.8}$$

defines an isomorphism of rings.

$$\mathbb{Z}[X^\beta_{ij},Y^\beta_{rs},d_{\beta\alpha}(X)^{-1}] \simeq \mathbb{Z}[X^\alpha_{ij},Y^\alpha_{rs},d_{\alpha\beta}(X)^{-1}]$$

It follows from 4.1.5 that (with the obvious notations)

$$\phi^*_{\alpha\beta} R(F_\beta(X), G_\beta(X)) = S_{\alpha\beta}(X)^{-1} R(F_\alpha(X), G_\alpha(X))$$
$$\phi^*_{\alpha\beta} H_\beta(Y) = H_\alpha(Y) S_{\alpha\beta}(X) \tag{4.1.9}$$

and these formulae describe $\phi^*_{\alpha\beta}$ completely. It follows that

$$\phi^*_{\alpha\beta} d_{\beta\alpha}(X) = \phi^*_{\alpha\beta} \det(R(F_\beta(X), G_\beta(X))_\alpha) = \det(S_{\alpha\beta}(X))^{-1} = d_{\alpha\beta}(X)^{-1}$$

so that $\phi^*_{\alpha\beta}$ does indeed map $d_{\beta\alpha}(X)^{-1}$ into $\mathbb{Z}[X^\alpha_{ij}, Y^\alpha_{rs}, d_{\alpha\beta}(X)^{-1}]$.

The $\phi^*_{\alpha\beta}$ induce isomorphisms of open subschemes

$$\phi_{\alpha\beta}: V_{\alpha\beta} \to V_{\beta\alpha} \tag{4.1.10}$$

and $M^{cr}_{m,n,p}$ is now the scheme obtained by gluing together the schemes V_α, for all nice selections α, by means of the isomorphisms $\phi_{\alpha\beta}$.

As in section 3 above one can now embed $M^{cr}_{m,n,p}$ into a product of a Grassmannian over $\mathbb{Z}$ and an affine space over $\mathbb{Z}$ to see that $M^{cr}_{m,n,p}$ is a separated scheme.

For each nice selection α let V^{co}_α be the open subscheme of V_α defined by

$$V^{co}_\alpha = \bigcup_\gamma \operatorname{Spec}(\mathbb{Z}[X^\alpha_{ij}, Y^\alpha_{rs}, Q(F_\alpha(X), H_\alpha(Y))^{-1}_\gamma] \tag{4.1.11}$$

where γ runs through all the nice selections of the set of row indices $J_{p,n}$ of $Q(F_\alpha(X), H_\alpha(Y))$. Then the $\phi_{\alpha\beta}$ restrict to give isomorphisms.

$$\phi^{co}_{\alpha\beta}: V^{co}_{\alpha\beta} \to V^{co}_{\beta\alpha} \tag{4.1.12}$$

where $V^{co}_{\alpha\beta} = V^{co}_\alpha \cap V_{\alpha\beta}$. Gluing together the V^{co}_α by means of the $\phi^{co}_{\alpha\beta}$ we obtain the open subscheme $M^{cr,co}_{m,n,p}$ of $M^{cr}_{m,n,p}$.

To see how all these abstract formulas look in concreto consider the case $m = 2$, $n = 2$, $p = 1$. In this case, there are three nice selections $\alpha, \beta, \gamma \subset J_{2,2}$, viz.

$$\alpha = \{(0,1),(0,2)\}, \quad \beta = \{(0,1),(1,1)\},$$
$$\gamma = \{(0,2),(1,2)\} \tag{4.1.13}$$

We have

$$F_\alpha(X) = \begin{pmatrix} X^\alpha_{11} & X^\alpha_{12} \\ X^\alpha_{21} & X^\alpha_{22} \end{pmatrix}, \quad G_\alpha(X) = \begin{pmatrix} 1 & 0 \\ 0 & 1 \end{pmatrix}, \quad H_\alpha(Y) = (Y^\alpha_1, Y^\alpha_2)$$

$$F_\beta(X) = \begin{pmatrix} 0 & X^\beta_{11} \\ 1 & X^\beta_{21} \end{pmatrix}, \quad G_\beta(X) = \begin{pmatrix} 1 & X^\beta_{12} \\ 0 & X^\beta_{22} \end{pmatrix}, \quad H_\beta(Y) = (Y^\beta_1, Y^\beta_2)$$

$$F_\gamma(X) = \begin{pmatrix} 0 & X^\gamma_{12} \\ 1 & X^\gamma_{22} \end{pmatrix}, \quad G_\gamma(X) = \begin{pmatrix} X^\gamma_{11} & 1 \\ X^\gamma_{21} & 0 \end{pmatrix}, \quad H_\gamma(Y) = (Y^\gamma_1, Y^\gamma_2)$$

Thus

$$d_{\alpha\beta}(X) = X^\alpha_{21}, \quad d_{\alpha\gamma}(X) = -X^\alpha_{12}, \quad d_{\beta\gamma}(X) = X^\beta_{12}X^\beta_{12} + X^\beta_{12}X^\beta_{21}X^\beta_{22} - X^\beta_{11}X^\beta_{22}X^\beta_{22}$$

$$d_{\beta\alpha}(X) = X^\beta_{22}, \quad d_{\gamma\alpha}(X) = -X^\gamma_{21}, \quad d_{\gamma\beta}(X) = X^\gamma_{11}X^\gamma_{11} + X^\gamma_{11}X^\gamma_{22}X^\gamma_{21} - X^\gamma_{21}X^\gamma_{21}X^\gamma_{12}$$

$$S_{\alpha\beta}(X) = \begin{pmatrix} 1 & X^\alpha_{11} \\ 0 & X^\alpha_{21} \end{pmatrix}, \quad S_{\beta\alpha}(X) = \begin{pmatrix} 1 & X^\beta_{12} \\ 0 & X^\beta_{22} \end{pmatrix},$$

$$S_{\alpha\gamma}(X) = \begin{pmatrix} 0 & X^\alpha_{21} \\ 1 & X^\alpha_{22} \end{pmatrix}, \quad S_{\gamma\alpha}(X) = \begin{pmatrix} X^\gamma_{11} & 1 \\ X^\gamma_{21} & 0 \end{pmatrix},$$

$$S_{\beta\gamma}(X) = \begin{pmatrix} X^\beta_{12} & X^\beta_{11}X^\beta_{22} \\ X^\beta_{22} & X^\beta_{12} + X^\beta_{21}X^\beta_{22} \end{pmatrix}, \quad S_{\gamma\beta}(X) = \begin{pmatrix} X^\gamma_{11} & X^\gamma_{12}X^\gamma_{21} \\ X^\gamma_{21} & X^\gamma_{11} + X^\gamma_{22}X^\gamma_{21} \end{pmatrix}$$

Thus for example the two isomorphisms $\phi^*_{\alpha\beta}$ and $\phi^*_{\beta\alpha}$ are given by

$$\phi^*_{\alpha\beta} : \mathbb{Z}[X^\beta_{ij}, Y^\beta_r, (X^\beta_{22})^{-1}] \to \mathbb{Z}[X^\alpha_{ij}, Y^\alpha_r, (X^\alpha_{21})^{-1}]$$

$$X^\beta_{12} \mapsto -(X^\alpha_{21})^{-1}X^\alpha_{11},\ X^\beta_{22} \mapsto (X^\alpha_{21})^{-1}$$

$$X^\beta_{11} \mapsto X^\alpha_{12}X^\alpha_{21} - X^\alpha_{11}X^\alpha_{22},\ X^\beta_{21} \mapsto X^\alpha_{11} + X^\alpha_{22}$$

$$Y^\beta_1 \mapsto Y^\alpha_1,\ Y^\beta_2 \mapsto (X^\alpha_{21})^{-1}Y^\alpha_2 - (X^\alpha_{21})^{-1}X^\alpha_{11}Y^\alpha_1$$

$$\phi^*_{\alpha\beta} : \mathbb{Z}[X^\alpha_{ij}, Y^\alpha_r, (X^\alpha_{21})^{-1}] \to \mathbb{Z}[X^\beta_{ij}, Y^\beta_r, (X^\beta_{22})^{-1}]$$

$$X^\alpha_{11} \mapsto -(X^\beta_{22})^{-1}X^\beta_{12},\ X^\alpha_{21} \mapsto (X^\beta_{22})^{-1}$$

$$X^\alpha_{12} \mapsto X^\beta_{11}X^\beta_{22} - X^\beta_{12}X^\beta_{21} - (X^\beta_{22})^{-1}X^\beta_{12}X^\beta_{12}$$

$$X^\alpha_{22} \mapsto (X^\beta_{22})^{-1}X^\beta_{12} + X^\beta_{21}$$

$$Y^\alpha_1 \mapsto Y^\beta_1,\ Y^\alpha_2 \mapsto (X^\beta_{22})^{-1}Y^\beta_2 - (X^\beta_{22})^{-1}Y^\beta_1X^\beta_{12}$$

and one checks without trouble that indeed $d_{\beta\alpha}(X)^{-1} = (X^\beta_{22})^{-1}$ gets mapped into $\mathbb{Z}[X^\alpha_{ij}, Y^\alpha_r, (X^\alpha_{21})^{-1}]$ and $d_{\alpha\beta}(X)^{-1} = (X^\alpha_{21})^{-1}$ into $\mathbb{Z}[X^\beta_{ij}, Y^\beta_r, (X^\beta_{22})^{-1}]$ and that indeed $\phi^*_{\alpha\beta} \circ \phi^*_{\beta\alpha} = \mathrm{id}$, $\phi^*_{\beta\alpha} \circ \phi^*_{\alpha\beta} = \mathrm{id}$. (The formulas are not always so simple; for instance the formulas for $\phi^*_{\beta\gamma}$ and $\phi^*_{\gamma\beta}$ are a good deal more complicated.

4.2. Small Intermezzo: Completely reachable systems over a ring.

A system $\Sigma = (F,G,H)$ over a ring R is said to be completely reachable if $R(F,G)\colon R^r \to R^n$, $r = (n+1)m$ is a surjective map, cf. e.g. [Sol] or [Rou]. This is equivalent to each element of the family $\Sigma(p) = (F(p),G(p),H(p))$, $p \in \mathrm{Spec}(R)$ being completely reachable. Indeed $R(F,G)\colon R^r \to R^m$ is surjective if it is surjective mod every maximal ideal [Bou, Ch. II, §3.3, Prop. 11] and the statement follows.

4.3. The Algebraic Geometric Version of the Nice Selection Lemma.

The next thing to do is to discuss the algebraic-geometric version of the nice selection lemma, 3.2.3. Recall that this lemma says that if the system (F,G,H) over a field k is cr then there is a nice selection α such that $R(F,G)_\alpha$ is invertible. Now let (F,G,H) be a cr system over a ring R, which per definition means that $R(F,G): R^r \to R^n$, $r\ (n+1)m$, is surjective, which in turn is equivalent to condition that the systems $\Sigma(p) = (F(p),G(p),H(p)$ over $k(p)$, the quotient field of R/p, are cr for all prime ideals p. Then of course one does not expect the existence of a nice selection α such that $R(F,G)_\alpha$ is an invertible matrix over R; after all $\Sigma = (F,G,H)$ should be interpreted as a family and not as a single system.

For a continuous topological family $\Sigma(\sigma)$ over a topological space M the nice selection lemma implies that there is a finite covering $M = \cup U_\alpha$ such that for all $\sigma \in U_\alpha$, $R(F(\sigma),G(\sigma))_\alpha$ is invertible. And this property generalizes nicely.

4.3.1. Lemma. Let $\Sigma = (F,G,H)$ be a cr system over a ring R. For each nice selection α let $d_\alpha = \det(R(F,G)_\alpha)$. Then the ideal generated by the d_α is the whole ring R. (This means of course that the $U_\alpha = \mathrm{Spec}(R[d_\alpha^{-1}])$ cover all of $\mathrm{Spec}(R)$).

Proof. Let I be the ideal generated by the d_α, α nice. Suppose that $I \neq R$. Then there is a maximal ideal m such that $I \subset m$. Consider $\Sigma(m) = (F(m),G(m),H(m))$. Then $\det(R(\Sigma(m))_\alpha) = 0$ in R/m for all α, showing that $\Sigma(m)$ is not cr (by the old nice selection lemma 3.2.3 over the field R/m) which contradicts the assumption that Σ was cr.

To state the more global version of this lemma we need a bit of notation. Let Σ be a family of cr systems over a scheme V. For each nice selection α we define

$$U_\alpha = \{v \in V \mid \det(R(\Sigma(v))_\alpha) \neq 0\} \qquad (4.3.2)$$

This definition seems a bit ambiguous at first because $R(\Sigma(v))$ depends on what basis we choose in the state space of $\Sigma(v)$ and hence is only defined up to multiplication on the left by an $n \times n$ invertible matrix with coefficients in $k(v)$. This matrix being invertible, however, means that the whole symbol group $\det(R(\Sigma(v))_\alpha) \neq 0$ makes perfectly good sense so that U_α is well defined. Of course U_α is an open subscheme of V.

4.3.3. Lemma. Let Σ be a family of cr systems over a scheme V. For each nice selection α let U_α be as in (4.3.2). Then $\bigcup_{\alpha \text{ nice}} U_\alpha = V$.

This follows immediately from lemma 4.3.1 because V can be covered with affine schemes $\mathrm{Spec}(R_i)$ (such that moreover the underlying bundle of Σ is trivial over each $\mathrm{Spec}(R_i)$).

4.4. <u>The Universal Bundle</u> E^u <u>over</u> $M^{cr}_{m,n,p}$. The universal bundle E^u over $M^{cr}_{m,n,p}$ is constructed just as in 3.4.11 above. Writing things out in relentness detail one obtains the following algebraic-geometric local pieces and patching data description.

For each nice selection α let

$$E_\alpha = \mathrm{Spec}(\mathbb{Z}[X^\alpha_{ij}, Y^\alpha_{rs}] \otimes \mathbb{Z}[Z^\alpha_1, \dots, Z^\alpha_n]) = V_\alpha \times \mathbb{A}^n \tag{4.4.1}$$

where $\mathbb{Z}[X^\alpha_{ij}, Y^\alpha_{rs}]$ is as in 4.1.1; i.e. $\mathrm{Spec}\ \mathbb{Z}[X^\alpha_{ij}, Y^\alpha_{rs}] = V_\alpha$. Let

$$\pi_\alpha : E_\alpha \to V_\alpha \tag{4.4.2}$$

be the projection induced by the natural inclusion

$$\pi^*_\alpha : \mathbb{Z}[X^\alpha_{ij}, Y^\alpha_{r,s}] \subset \mathbb{Z}[X^\alpha_{ij}, Y^\alpha_{r,s}, Z^\alpha_t] .$$

Define for each pair of nice selections α, β.

$$E_{\alpha\beta} = \mathrm{Spec}\ \mathbb{Z}[X^\alpha_{ij}, Y^\alpha_{r,s}, Z^\alpha_t, d_{\alpha\beta}(X)^{-1}]) = V_{\alpha\beta} \times \mathbb{A}^n \tag{4.4.3}$$

and let

$$\tilde{\phi}_{\alpha\beta} : E_{\alpha\beta} \to E_{\beta\alpha} \tag{4.4.4}$$

be the isomorphism given by the ring isomorphism

$$\tilde{\phi}^*_{\alpha\beta} : \mathbb{Z}[X^\beta_{ij}, Y^\beta_{rs}, Z^\beta_t, d_{\beta\alpha}(X)^{-1}] \to \mathbb{Z}[X^\alpha_{ij}, Y^\alpha_{rs}, Z^\alpha_t, d_{\alpha\beta}(X)^{-1}] \tag{4.4.5}$$

given by

$$X^\beta_{ij} \to \phi_{\alpha\beta}(i,j)(X^\alpha), Y^\beta_{r,s} \to \phi_{\alpha\beta}(r,s)(X\ ,Y\ ),$$

$$Z^\beta_t \to \tilde{\phi}_{\alpha\beta}(t)(X^\alpha, Z^\alpha) \tag{4.4.6}$$

where the $\tilde{\phi}_{\alpha\beta}(t)(X^\alpha, Z^\alpha)$ are defined by the equality

$$\begin{pmatrix} \tilde{\phi}_{\alpha\beta}(1)(X^\alpha, Z^\alpha) \\ \vdots \\ \tilde{\phi}_{\alpha\beta}(n)(X^\alpha, Z^\alpha) \end{pmatrix} = S_{\alpha\beta}(X)^{-1} \begin{pmatrix} Z^\alpha_1 \\ \vdots \\ Z^\alpha_n \end{pmatrix}$$

The $\tilde{\phi}_{\alpha\beta}$ are compatible (by their definition) with the $\phi_{\alpha\beta}$ in that the following diagram commutes for each pair of nice selections α,β.

$$\begin{array}{ccc} E_{\alpha\beta} & \xrightarrow{\tilde{\phi}_{\alpha\beta}} & E_{\beta\alpha} \\ \downarrow{\scriptstyle \pi_\alpha} & & \downarrow{\scriptstyle \pi_\beta} \\ V_{\alpha\beta} & \xrightarrow[\tilde{\phi}_{\alpha\beta}]{} & V_{\beta\alpha} \end{array} \qquad (4.4.8)$$

It follows that by gluing the E_α together by means of the $\tilde{\phi}_{\alpha\beta}$ we obtain a vector bundle E^u.

$$\pi : E^u \to M^{cr}_{m,n,p} \qquad (4.4.9)$$

4.5. <u>The Morphism into $M^{cr}_{m,n,p}$ Associated to an Algebraic Family of cr Systems.</u> We start with the case that the underlying vector bundle E of the family Σ is trivial and that the parametrizing scheme V is affine. Σ is then described by a ring R, $V = \mathrm{Spec}(R)$, $E = \mathrm{Spec}(R[Z_1,\ldots,Z_n]$, $\pi : E \to V$ induced by the natural inclusion $R \to R[Z_1,\ldots,Z_n]$, and vector bundle homomorphisms $F: E \to E$, $G: \mathrm{Spec}(R[U_1,\ldots,U_m]) \to E$, $H: E \to \mathrm{Spec}(R[Y_1,\ldots,Y_m])$. The fact that these morphisms are vector bundle homomorphisms is reflected by the fact that the associated homomorphisms of rings

$$F^* : R[Z_1,\ldots,A_n] \to R[Z_1,\ldots,Z_n], \quad G^* : R[Z_1,\ldots,Z_n] \to R[U_1,\ldots,U_m], \quad H^* : R[Y_1,\ldots,Y] \to R[Z_1,\ldots,Z_n]$$

are firstly R-algebra homomorphisms and further of the form

$$F^*(Z_i) = \sum_{j=1}^{n} f_{ij}Z_j, \quad G^*(Z_i) = \sum_{j=1}^{m} g_{ij}U_j, \quad H^*(Y_i) = \sum_{j=1}^{n} h_{ij}Z_j \qquad (4.5.1)$$

where the f_{ij}, g_{ij}, h_{ij} are elements of R. This defines a triple of matrices $\bar{F} = (f_{ij})$, $\bar{G} = (g_{ij})$, $\bar{H} = (h_{ij})$. For each nice selection α let $S_\alpha = R(\bar{F},\bar{G})_\alpha$, $d_\alpha = \det(S_\alpha) \in R$, let $U_\alpha = \mathrm{Spec}(R[d_\alpha^{-1}])$, and let $V_\alpha = \mathrm{Spec}(\mathbb{Z}[X^\alpha_{ij}, Y^\alpha_{rs}])$ be "the nice-selection-α-piece of $M^{cr}_{m,n,p}$" of 4.1 above. Now define

$$\psi_\alpha : U_\alpha \to V_\alpha \qquad (4.5.2)$$

by the morphism of rings

$$\psi_\alpha^* : \mathbb{Z}[X_{ij}^\alpha, Y_{rs}^\alpha] \to R[d_\alpha^{-1}] \tag{4.5.3}$$

given by

$$\begin{aligned} X_{ij}^\alpha &\mapsto \text{i-th entry of the column vector } S_\alpha^{-1}R(\bar{F},\bar{G})_{s(\alpha,j)} \\ Y_{rs}^\alpha &\mapsto \text{r-th entry of the column } s \text{ of the matrix } \bar{H}S_\alpha \end{aligned} \tag{4.5.4}$$

where $s(\alpha,j)$ is the j-th successor index of the nice selection α, cf. 3.2 above. Or, using the obvious notation, ψ_α^* is defined by

$$\psi_\alpha^*(R(F_\alpha(X),G_\alpha(X)) = S_\alpha^{-1}R(\bar{F},\bar{G}), \quad \psi_\alpha^* H_\alpha(Y) = \bar{H}S_\alpha \tag{4.5.5}$$

Now let β be a second nice selection. We claim that the ψ_α and ψ_β agree on $U_\alpha \cap U_\beta = \mathrm{Spec}(R[d_\alpha^{-1},d_\beta^{-1}])$. In view of how the V_α, V_β are glued together to obtain $M_{m,n,p}^{cr}$ this means that we must prove the commutativity of the diagram

$$\begin{array}{ccc} \mathbb{Z}[X_{ij}^\alpha, Y_{rs}^\alpha, d_{\alpha\beta}(X)^{-1}] & \xrightarrow{\psi_\alpha^*} & \\ \uparrow \phi_{\alpha\beta}^* & & R[d_\alpha^{-1}, d_\beta^{-1}] \\ \mathbb{Z}[X_{ij}^\beta, Y_{rs}^\beta, d_{\beta\alpha}(X)^{-1}] & \xrightarrow{\psi_\beta^*} & \end{array} \tag{4.5.6}$$

Note first that

$$\psi_\alpha^*(S_{\alpha\beta}(X)) = \psi_\alpha^*(R(F_\alpha(X),G_\alpha(X)_\beta) = S_\alpha^{-1}R(\bar{F},\bar{G})_\beta = S_\alpha^{-1}S_\beta \tag{4.5.7}$$

so that ψ_α^* does indeed map $d_{\alpha\beta}(X)^{-1}$ into $R[d_\alpha^{-1},d_\beta^{-1}]$. Now ψ_β^* is described by

$$\psi_\beta^*(R(F_\beta(X),G_\beta(X)) = S_\beta^{-1}R(\bar{F},\bar{G}), \quad \psi_\beta^* H_\beta(Y) = \bar{H}S_\beta \tag{4.5.8}$$

and on the other hand

$$\psi_\alpha^*\phi_{\alpha\beta}^* R(F_\beta(X),G_\beta(X)) = \psi_\alpha^* (S_{\alpha\beta}(X)^{-1}R(F_\alpha(X),G_\alpha(X))) \quad \text{(by (4.1.9))}$$

$$= S_\beta^{-1} S_\alpha S_\alpha^{-1} R(\bar{F},\bar{G})$$

(by (4.5.7) and (4.5.5))

$$= S_\beta^{-1} R(\bar{F},\bar{G})$$

which fits perfectly with (4.5.8). Similarly $\psi_\alpha^* \phi_{\alpha\beta}^* H_\beta(X) = \psi_\alpha^* H_\alpha(Y) S_{\alpha\beta}(X) = \bar{H} S_\alpha S_\alpha^{-1} S_\beta = \bar{H} S_\beta = \psi_\beta^* H_\beta(Y)$, so that (4.5.6) is indeed commutative. Thus the $\psi_\alpha : U_\alpha \to V_\alpha$ are compatible, and because $\bigcup_{\alpha \text{ nice}} U_\alpha = \mathrm{Spec}(R)$ we obtain a morphism of schemes

$$\psi_\Sigma : V = \mathrm{Spec}(R) \to M^{cr}_{m,n,p}$$

4.5.9. Lemma. The morphism ψ_Σ depends only on the isomorphism class of Σ (so in particular ψ_Σ does not depend on how E is trivialized).

Proof. Let Σ' be a second family of cr systems over $V = \mathrm{Spec}(R)$ with trivial underlying vectorbundle $E' = \mathrm{Spec}(R[Z_1',Z_2',\ldots,Z_n'])$. Suppose Σ' is isomorphic to Σ and let the isomorphism be $\mu : E \to E'$. Because μ is a morphism of vectorbundles over $V = \mathrm{Spec}(R)$ its ring homomorphism

$$\mu^* : R[Z_1',Z_2',\ldots,Z_n'] \to R[Z_1,Z_2,\ldots,Z_n]$$

is an R-algebra homomorphism of the form

$$\mu^*(Z_i') = \sum_{j=1}^{n} s_{ij} Z_j, \qquad s_{ij} \in R$$

Let S be the matrix (s_{ij}). Then S is invertible (over R) because μ is an isomorphism. Now because μ defines an isomorphism $\Sigma' \simeq \Sigma$ we have $F'\mu = \mu F$, $\mu G = G'$, $H = H'\mu$ which in terms of the matrices $\bar{F},\bar{G},\bar{H}$ associated to Σ (cf. (4.5.1) above) and the analogous matrices $\bar{F}',\bar{G}',\bar{H}'$ of Σ' means that

$$S\bar{F} = \bar{F}'S,\ S\bar{G} = \bar{G}',\ \bar{H} = \bar{H}'S$$

It follows that if d_α', S_α', U_α' are defined analogously to d_α, S_α, U_α then $S_\alpha' = SS_\alpha$. $d_\alpha' = \det(S)d_\alpha$ so that $U_\alpha' = U_\alpha$ and $\psi_\alpha' = \psi_\alpha$ all because $SR(\bar{F},\bar{G}) = R(\bar{F}',\bar{G}')$, $\bar{H}S^{-1} = \bar{H}'$, which proves the lemma.

4.5.10. Construction of ψ_Σ for families whose underlying bundle is not necessarily trivial.

Now let $\Sigma = (E; F,G,H)$ be a family of cr systems over a scheme V. We can cover V with affine pieces $U_i = \mathrm{Spec}(R_i)$ such that E is trivializable over U_i. By the construction above and lemma 4.5.9 this gives us morphisms (independent of the trivialization chosen)

$$\psi_i : U_i \to M^{cr}_{m,n,p}$$

Now on $U_i \cap U_j$ the ψ_i and ψ_j must agree, because by lemma 4.5.9 again ψ_i and ψ_j agree on all affine pieces $\mathrm{Spec}(R) \subset U_i \cap U_j$. Hence the ψ_i combine to define a morphism

$$\psi_\Sigma : V \to M^{cr}_{m,n,p}$$

which, again by lemma 4.5.9 depends only on the isomorphism class of Σ.

4.6. The universal family Σ^u of cr systems over $M^{cr}_{m,n,p}$. Let E^u be the vectorbundle over $M^{cr}_{m,n,p}$ constructed in 4.4 above. In this section I describe a (universal) family of cr systems over $M^{cr}_{m,n,p}$ whose underlying bundle is E^u. (That this family is indeed universal will be proved in 4.9 below).

Recall that E^u was constructed out of affine pieces $E_\alpha = \mathrm{Spec}(\mathbb{Z}[X^\alpha_{ij}, Y^\alpha_{rs}, Z^\alpha_t])$ glued together by means of certain isomorphisms $\tilde{\phi}_{\alpha\beta}$, cf. 4.4. Let $\mathbb{A}^r = \mathrm{Spec}(\mathbb{Z}[U_1,\dots,U_r])$. To define $\Sigma^u = (E^u; F^u, G^u, H^u)$ it suffices to define vectorbundle homomorphisms

$$F_\alpha : E_\alpha \to E_\alpha,\ G_\alpha : V_\alpha \times \mathbb{A}^m \to E_\alpha,\ H_\alpha : E_\alpha \to V_\alpha \times \mathbb{A}^p \tag{4.6.1}$$

which are compatible with the identifications

$$\tilde{\phi}_{\alpha\beta} : E_{\alpha\beta} \to E_{\beta\alpha},\ \phi_{\alpha\beta} \times \mathrm{id} : V_{\alpha\beta} \times \mathbb{A}^m \to V_{\beta\alpha} \times \mathbb{A}^m,$$

$$\phi_{\alpha\beta} \times \mathrm{id} : V_{\alpha\beta} \times \mathbb{A}^p \to V_{\beta\alpha} \times \mathbb{A}^p$$

in the sense that the following diagram must be commutative

$$\begin{array}{ccccccc} V_{\alpha\beta}\times \mathbb{A}^m & \xrightarrow{G_\alpha} & E_{\alpha\beta} & \xrightarrow{F_\alpha} & E_{\alpha\beta} & \xrightarrow{H_\alpha} & V_{\alpha\beta}\times \mathbb{A}^p \\ \downarrow{\scriptstyle \phi_{\alpha\beta}\times \mathrm{id}} & & \downarrow{\scriptstyle \tilde\phi_{\alpha\beta}} & & \downarrow{\scriptstyle \tilde\phi_{\alpha\beta}} & & \downarrow{\scriptstyle \phi_{\alpha\beta}\times \mathrm{id}} \\ V_{\beta\alpha}\times \mathbb{A}^m & \xrightarrow{G_\beta} & E_{\beta\alpha} & \xrightarrow{F_\beta} & E_{\beta\alpha} & \xrightarrow{H_\beta} & V_{\beta\alpha}\times \mathbb{A}^p \end{array}$$

(4.6.2)

(cf. also (3.4.17)). We now describe $F_\alpha, G_\alpha, H_\alpha$ as those morphisms which on the ring level are given by the $\mathbb{Z}[X^\alpha_{ij}, Y^\alpha_{rs}]$ - algebra homomorphisms

$$F^*_\alpha\colon \mathbb{Z}[X^\alpha_{ij}, Y^\alpha_{rs}, Z^\alpha_t] \to \mathbb{Z}[X^\alpha_{ij}, Y^\alpha_{rs}, Z^\alpha_t], \quad Z^\alpha \mapsto F_\alpha(X)Z^\alpha \tag{4.6.3}$$

$$G^*_\alpha\colon \mathbb{Z}[X^\alpha_{ij}, Y^\alpha_{rs}, Z^\alpha_t] \to \mathbb{Z}[X^\alpha_{ij}, Y^\alpha_{rs}, U_1,\dots,U_m], \quad Z^\alpha \mapsto G_\alpha(X)U \tag{4.6.4}$$

$$H^*_\alpha\colon \mathbb{Z}[X^\alpha_{ij}, Y^\alpha_{rs}, V_1,\dots,V_p] \to \mathbb{Z}[X^\alpha_{ij}, Y^\alpha_{rs}, Z^\alpha_t], \quad V \mapsto H_\alpha(Y)Z^\alpha \tag{4.6.5}$$

where Z^α, U, V are respectively the column vectors $(Z^\alpha_1,\dots,Z^\alpha_n)^T$, $(U_1,\dots,U_m)^T$, $(V_1,\dots,V_p)^T$.

It remains to check that the diagram (4.6.2) is indeed commutative, which is done by checking that the dual diagram of rings homorphisms is commutative.

This comes down to precisely the same calculations as in 3.4.14. As an example we check that the diagram

$$\begin{array}{ccc} \mathbb{Z}[X^\alpha_{ij}, Y^\alpha_{rs}, Z^\alpha_t, d_{\alpha\beta}(X)^{-1}] & \xleftarrow{F^*_\alpha} & \mathbb{Z}[X^\alpha_{ij}, Y^\alpha_{rs}, Z^\alpha_t, d_{\alpha\beta}(X)^{-1}] \\ \uparrow{\scriptstyle \tilde\phi^*_{\alpha\beta}} & & \uparrow{\scriptstyle \tilde\phi^*_{\alpha\beta}} \\ \mathbb{Z}[X^\beta_{ij}, Y^\beta_{rs}, Z^\beta_t, d_{\beta\alpha}(X)^{-1}] & \xleftarrow{F^*_\beta} & \mathbb{Z}[X^\beta_{ij}, Y^\beta_{rs}, Z^\beta_t, d_{\beta\alpha}(X)^{-1}] \end{array}$$

is commutative. Because $\tilde{\phi}^*_{\alpha\beta}$ maps $\mathbb{Z}[X^\beta_{ij},Y^\beta_{rs},d_{\beta\alpha}(X)^{-1}]$ into $\mathbb{Z}[X^\alpha_{ij},Y^\alpha_{rs},d_{\alpha\beta}(X)^{-1}]$ and because F^*_α and F^*_β are respectively $\mathbb{Z}[X^\alpha_{ij},Y^\alpha_{rs}]$-algebra and $\mathbb{Z}[X^\beta_{ij},Y^\beta_{rs}]$-algebra homomorphisms it suffices to check that

$$\tilde{\phi}^*_{\alpha\beta}\, F^*_\beta(Z^\beta_t) = F^*_\alpha\phi^*_{\alpha\beta}(Z^\beta_t), \quad t = 1,\dots,n$$

By the definitions (4.4.7), (4.6.3) and using the definition of $\phi^*_{\alpha\beta}$, cf. 4.1, we have

$$F^*_\alpha\tilde{\phi}^*_{\alpha\beta}(Z^\beta) = F^*_\alpha(S_{\alpha\beta}(X)^{-1}Z^\alpha) = S_{\alpha\beta}(X)^{-1}F_\alpha(X)Z^\alpha$$

$$\phi^*_{\alpha\beta}F^*_\beta(Z^\beta) = \tilde{\phi}^*_{\alpha\beta}(F_\beta(X)Z^\beta) = \phi^*_{\alpha\beta}(F_\beta(X))S_{\alpha\beta}(X)^{-1}Z^\alpha$$

$$= S_{\alpha\beta}(X)^{-1}F_\alpha(X)S_{\alpha\beta}(X)S_{\alpha\beta}(X)^{-1}Z^\alpha$$

$$= S_{\alpha\beta}(X)^{-1}F_\alpha(X)Z^\alpha$$

The remaining two squares of diagram (4.6.2) are similarly shown to be commutative.

4.7. A rigidity lemma.

The key to the proof of theorem 3.4.20 (the algebraic-geometric classifying theorem) is (as was remarked before) a rigidity property which in this context takes the following form.

4.7.1. Proposition. Let Σ, Σ' be two families of cr systems over a scheme V. Suppose that there is a covering by open subschemes (U_i) of V such that the two families Σ and Σ' restricted to U_i are isomorphic for all i. Then Σ and Σ' are isomorphic as algebraic families over V.

We note that no such proposition holds for arbitrary families of systems cf. [HP] for a counterexample.

Proof. We can assume that the underlying vectorbundles E and E' have been obtained by gluing together trivial pieces over affine subschemes of V. Refining the covering (U_i) if necessary (this does not change the validity of the hypothesis of the proposition) we can therefore assume that E and E' have been obtained by gluing together trivial bundles $U_i \times A^n$ over affine schemes U_i.

Our data are then as follows. We have for each i an affine scheme $U_i = \mathrm{Spec}(R_i)$ for each i,j isomorphisms of (trivial) bundles

$$\phi_{ij},\phi'_{ij} : (U_i \cap U_j) \times \mathbb{A}^n \to (U_j \cap U_i) \times \mathbb{A}^n$$

which respectively define the bundles E and E'. The remaining ingredients of the two families of systems Σ and Σ' are then given by vectorbundle homomorphisms

$$F_i, F'_i \colon U_i \times \mathbb{A}^n \to U_i \times \mathbb{A}^n, \quad G_i, G'_i \colon U_i \times \mathbb{A}^m \to U_i \times \mathbb{A}^n$$

$$H_i, H'_i \colon U_i \times \mathbb{A}^n \to U_i \times \mathbb{A}^p \tag{4.7.1}$$

such that the following diagrams are commutative for all i,j (where U_{ij} is short for $U_i \cap U_j$)

$$\begin{array}{ccccccc} & & U_{ij}\times\mathbb{A}^n & \xrightarrow{F_i,F'_i} & U_{ij}\times\mathbb{A}^n & & \\ & \nearrow^{G_i,G'_i} & & & & \searrow^{H_i,H'_i} & \\ U_{ij}\times\mathbb{A}^m & & \downarrow{\scriptstyle \phi_{ij},\phi'_{ij}} & & \downarrow{\scriptstyle \phi_{ij},\phi'_{ij}} & & U_{ij}\times\mathbb{A}^p \\ & \searrow_{G_j,G'_j} & & & & \nearrow_{H_j,H'_j} & \\ & & U_{ij}\times\mathbb{A}^n & \xrightarrow{F_j,F'_j} & U_{ij}\times\mathbb{A}^n & & \end{array} \tag{4.7.2}$$

Finally the fact that Σ and Σ' are isomorphic over each U_i means that there are vectorbundle isomorphisms $\phi_i \colon U_i \times \mathbb{A}^n \to U_i \times \mathbb{A}^n$ such that the following diagram is commutative for all i

$$\begin{array}{ccccccc} & & U_i\times\mathbb{A}^n & \xrightarrow{F_i} & U_i\times\mathbb{A}^n & & \\ & \nearrow^{G_i} & & & & \searrow^{H_i} & \\ U_i\times\mathbb{A}^m & & \downarrow{\scriptstyle \phi_i} & & \downarrow{\scriptstyle \phi_i} & & U_i\times\mathbb{A}^p \\ & \searrow_{G'_i} & & & & \nearrow_{H'_i} & \\ & & U_i\times\mathbb{A}^n & \xrightarrow{F'_i} & U_i\times\mathbb{A}^n & & \end{array} \tag{4.7.3}$$

We now claim that the ϕ_i are compatible and combine to define an isomorphism $\phi \colon E \to E'$ (it then follows, because this is locally true, that $\phi F = F'\phi$, $\phi G = G'$, $H'\phi = H$). To prove this we must show that for each $\mathrm{Spec}(R) = U \subset U_{ij} = U_i \cap U_j$ the following diagram commutes

$$\begin{array}{ccc} U \times \mathbb{A}^n & \xrightarrow{\phi_{ij}} & U \times \mathbb{A}^n \\ \downarrow{\scriptstyle \phi_i} & & \downarrow{\scriptstyle \phi_j} \\ U \times \mathbb{A}^n & \xrightarrow{\phi'_{ij}} & U \times \mathbb{A}^n \end{array} \qquad (4.7.4)$$

Now vectorbundle homomorphisms of trivial vectorbundles over an affine scheme $U = \mathrm{Spec}(R)$ are given by matrices with coefficients in R as we explained *en passant* in the first few paragraphs of 4.5 above. Let $\bar{G}_i$, $\bar{G}'_i$, F_i, F'_i, H_i, H'_i, S_{ij}, S'_{ij}, S_i, S_j be the matrices of the morphisms of vectorbundles G_i, G'_i, F_i, F'_i, H_i, H'_i, ϕ_{ij}, ϕ'_{ij}, ϕ_i, ϕ_j restricted to U. The commutativity relations (4.7.2) and (4.7.3) then imply for these matrices with coefficients in R that

$$S_{ij}\bar{G}_i = \bar{G}_j,\ S_{ij}\bar{G}'_i = \bar{G}'_j,\ S_{ij}\bar{F}_i = \bar{F}_j S_{ij},\ \ S_{ij}\bar{F}'_i = \bar{F}'_j S'_{ij}$$

$$S_{ij}\bar{H}_i = \bar{H}_j,\ S'_{ij}\bar{H}'_i = \bar{H}'_j,\ S_i\bar{G}_i = \bar{G}'_i,\ \ S_i\bar{F}_i = \bar{F}'_i S_i$$

$$\bar{H}'_i S_i = \bar{H}_i, \qquad (4.7.5)$$

$$S_j\bar{G}_j = \bar{G}'_j,\ S_j\bar{F}_j = \bar{F}'_j S_j,\ \bar{H}'_j S_j = \bar{H}_j$$

and the matrices S_i, S_j, S_{ij}, S'_{ij} are all invertible because they come from vectorbundle isomorphisms.

It follows that

$$\begin{aligned} S_j S_{ij} R(\bar{F}_i,\bar{G}_i) &= S_j R(\bar{F}_j,\bar{G}_j) = R(\bar{F}'_j,\bar{G}'_j) \\ &= S'_{ij} R(\bar{F}'_i,\bar{G}'_i) = S'_{ij} S_i R(\bar{F}_i,\bar{G}_i) \end{aligned} \qquad (4.7.6)$$

Now Σ is a family of cr systems and hence so is its restriction to $U = \mathrm{Spec}(R)$. It follows (cf. 4.2 above) that $R(\bar{F}_i,\bar{G}_i): R^r \to R^n$ $r = (n+1)m$, is a surjective map. Hence, (4.7.6) implies that $S_j S_{ij} = S'_{ij} S_i$ proving the commutativity of (4.7.4) and hence the proposition.

4.8. On the pullback construction. Let $\Sigma = (E;\ F,G,H)$ be a family of systems over a scheme M and let $\psi : V' \to M$ be a morphism of schemes. Assume that everything is given in terms of local affine pieces and patching data; i.e. Σ is given by trivial bundles $U_i \times \mathbb{A}^n \to U_i = \mathrm{Spec}(R_i) \subset M$ with vectorbundle isomorphisms $\phi_{ij}: U_{ij} \times \mathbb{A}^n \to U_{ij} \times \mathbb{A}^n$ and vector bundle mormorphisms $F_i: U_i \times \mathbb{A}^n \to U_i \times \mathbb{A}^n$, $G_i: U_i \times \mathbb{A}^m \to U_i \times \mathbb{A}^n$, $H_i: U_i \times \mathbb{A}^n \to U_i \times \mathbb{A}^p$ such the nonprime diagram (4.7.2) is commutative, and ψ is given by affine morphisms $\psi_i: U_i' \to U_i, U_i' = \mathrm{Spec}(R_i')$. Let $\psi_i^*: R_i \to R_i'$ be the ring homomorphism of ψ_i. Let, as before, $\bar{F}_i, \bar{G}_i, \bar{H}_i$ be the matrices of the vectorbundle morphisms F_i, G_i, H_i.

Then the local pieces of the pullback family $\psi^!\Sigma = \Sigma'$ are the trivial bundles $U_i' \times \mathbb{A}^n \to U_i'$ with the vectorbundle homomorphisms $F_i': U_i' \times \mathbb{A}^n \to U_i' \times \mathbb{A}^n$, $G_i': U_i' \times \mathbb{A}^m \to U_i' \times \mathbb{A}^n$, $H_i': U_i' \times \mathbb{A}^n \to U_i' \times \mathbb{A}^p$ given by the matrices $\bar{F}_i' = \psi_i^*\bar{F}_i$, $\bar{G}_i' = \psi_i^*\bar{G}_i$, $\bar{H}_i' = \psi_i^*\bar{H}_i$. The patching data are defined as follows. If $U' = \mathrm{Spec}(R) \subset U_i' \cap U_j'$ maps into $U = \mathrm{Spec}(R) \subset U_i \cap U_i$ under ψ and $\psi_1^*: R \to R'$ is the associated homomorphism of rings, then over Spec(R') the isomorphim $\phi_{ij}': U' \times \mathbb{A}^n \to U' \times \mathbb{A}^n$ is given by the matrix $S_{ij}' = \psi_1^* S_{ij}$ if S_{ij} is the matrix of $\phi_{ij}: U \times \mathbb{A}^n \to U \times \mathbb{A}^n$.

This can be taken as the definition of the pullback family $\psi^!\Sigma$. It agrees of course with the more informal description given in section 3 above.

4.9. The classifying theorem for algebraic families of cr systems over schemes ($M_{m,n,p}^{cr}$ is classifying over $\mathbb{Z}$).

We can now prove the algebraic-geometric classifying theorem for families of cr systems, i.e. theorem 3.4.20. Stated more precisely this theorem says

4.9.1 Theorem. Let Σ be an algebraic family of cr systems over a scheme V. Then there exists a unique morphism of schemes $\psi_\Sigma: V \to M_{m,n,p}^{cr}$ (defined in 4.5 above) such that $\psi_\Sigma^!\Sigma^u \simeq \Sigma$ where Σ^u is the universal family constructed in section 4.6 above. That is the map $\Sigma \mapsto \psi_\Sigma$ and the map $\psi \mapsto \psi^!\Sigma^u$ (of 4.8 above) set up a bijective correspondence between the set of scheme morphisms

$V \to M^{cr}_{m,n,p}$ and isomorphism classes of families of cr systems over V. Moreover this isomorphism is functorial.

Proof. First let $\psi: V \to M^{cr}_{m,n,p}$ be a morphism of schemes; let $\Sigma = \psi^! \Sigma^u$. Then we must show that $\psi_\Sigma = \psi$. To do this it suffices to show that ψ_Σ and ψ agree on all elements of some affine covering (U_i) of V. We can take this covering to be finer than the covering $(\psi^{-1}(V_\alpha), \alpha \text{ nice})$ where $V_\alpha \subset M^{cr}_{m,n,p}$ is the piece belonging to the nice selection α, cf. 4.1. Let therefore $U = \mathrm{Spec}(R)$ be such that $\psi(U) \subset V_\alpha$, and let

$$\psi^* : \mathbb{Z}[X^\alpha_{ij}, Y^\alpha_{rs}] \to R$$

be the associated ring homomorphism. Then according to 4.8 above and the definition of Σ^u, cf. 4.6, the family Σ over U is described by the three matrices

$$\bar{F} = \psi^* F_\alpha(X),\ \bar{G} = \psi^* G_\alpha(X),\ \bar{H} = \psi^* H_\alpha(Y)\ . \qquad (4.9.2)$$

By 4.5 above the morphism $\psi^*_\Sigma : \mathbb{Z}[X^\alpha_{ij}, Y^\alpha_{rs}] \to R$ associated to this family is characterized by

$$\psi^*_\Sigma(R(F_\alpha(X),\ G_\alpha(X)) = S^{-1}_\alpha R(\bar{F},\bar{G}),\ \psi^*_\alpha H_\alpha(Y) = \bar{H} S_\alpha \quad (4.9.3)$$

where $S_\alpha = R(\bar{F},\bar{G})_\alpha$. Because $R(F_\alpha(X),\ G_\alpha(X))_\alpha = I_n$, $S_\alpha = I_n$ in this case (cf. (4.9.2)) so that indeed (comparing (4.9.2) and (4.9.3)) $\psi^*_\Sigma = \psi^*$.

Now let Σ over V be a family of cr systems and let $\psi_\Sigma: V \to M^{cr}_{m,n,p}$ be the associated morphism as defined in 4.5. We have to show that $\psi^!_\Sigma \Sigma^u$ is isomorphic to Σ. By the rigidity result 4.7.1 it suffices to show that $\psi^!_\Sigma \Sigma^u$ and Σ are isomorphic over each element of some affine covering (U_i) of V, which we can take fine enough so that the underlying bundle E of Σ is trivial over each U_i. Let therefore $U = \mathrm{Spec}(R)$ be such that Σ over U is described by the triple of matrices $\bar{F},\bar{G},\bar{H}$. Let $d_\alpha = \det(\bar{R}(\bar{F},\bar{G})_\alpha)$ for each nice selection α. Then U in turn is covered by the $U_\alpha = \mathrm{Spec}(R[d^{-1}_\alpha])$ (by the nice selection lemma). So taking a still finer covering (if necessary) we can assume that $U = \mathrm{Spec}(R)$ is such that for a certain nice selection α we have that $S_\alpha = R(\bar{F},\bar{G})_\alpha$ is invertible over R. Then by 4.5 ψ_Σ is given on U by the ring homomorphism

$$\psi^* : \mathbb{Z}[X^\alpha_{ij}, Y^\alpha_{r,s}] \to R$$

characterized by

$$\psi^* R(F_\alpha(X), G_\alpha(X)) = S_\alpha^{-1} R(\bar{F},\bar{G}),\ \psi^* H_\alpha(Y) = \bar{H}S_\alpha \quad (4.9.4)$$

By 4.8 the family of cr systems $\psi^!_\Sigma \Sigma^u$ is defined by the matrices

$$\bar{F}' = \psi^* F_\alpha(X),\ \bar{G}' = \psi^* G_\alpha(X),\ \bar{H}' = \psi^* H_\alpha(Y) \quad (4.9.5)$$

Comparing (4.9.4) and (4.9.5) we see that over U the families defined by $\bar{F},\bar{G},\bar{H}$ and by $\bar{F}',\bar{G}',\bar{H}'$ are indeed isomorphic with the isomorphism being defined by S_α (which is invertible over R). This concludes the proof of the theorem.

4.10. <u>On cr systems over rings</u>. The classifying theorem 4.9.1 of course also applies to systems over rings R. Such a system (with finitely generated projective state module X) gives rise to a family of cr systems over R iff $R(F,G): R^r \to X$, $r = (n+1)m$, is surjective (cf. 4.2). If R is such that all finitely generated projective modules are free (which happens e.g. if R is a ring of polynomials over a field by the Quillen-Suslin theorem [Qu,Sus], then theorem 4.9.1 says that the R-rational points of $M^{cr}_{m,n,p}$ are precisely the $GL_n(R)$ orbits in $L^{cr}_{m,n,p}(R)$, i.e.

$$M^{cr}_{m,n,p}(R) \simeq L^{cr}_{m,n,p}(R)/GL_n(R) \quad \text{(if R is projective free)}.$$

In general the theorem gives a canonical injection

$$L^{cr}_{m,n,p}(R)/GL_n(R) \to M^{cr}_{m,n,p}(R)$$

with the remaining points of $M^{cr}_{m,n,p}(R)$ corresponding to systems over R whose state module is projective but not free.

4.11. <u>A few final remarks</u>. There is a completely dual theory from the co instead of cr point of view. Also the open subscheme $M^{cr,co}_{m,n,p}$ is of course classifying for families of co and cr systems. This scheme is embeddable (over $\mathbb{Z}$) in an affine scheme $\mathbb{A}^{(n+1)mp}$ as a locally closed subscheme.

5. EXISTENCE AND NONEXISTENCE OF GLOBAL CONTINUOUS CANONICAL FORMS

As a first application of the fine moduli spaces of sections 3 and 4 above we discuss existence and nonexistence of global continuous canonical forms for linear dynamical systems.

5.1. <u>The topological case</u>. Let L' be a $GL_n(\mathbb{R})$-invariant subspace of $L_{m,n,p}(\mathbb{R})$. A canonical form for $GL_n(\mathbb{R})$ acting on L' is a mapping $c: L' \to L'$ such that the following three properties hold

$$c(\Sigma^S) = c(\Sigma) \quad \text{for all } \Sigma \in L',\ S \in GL_n(\mathbb{R}); \tag{5.1.1}$$

$$\text{for all } \Sigma \in L' \text{ there is an } S \in GL_n(\mathbb{R}) \text{ such that } c(\Sigma) = \Sigma^S ; \tag{5.1.2}$$

$$c(\Sigma) = c(\Sigma') \Rightarrow \exists S \in GL_n(\mathbb{R}) \text{ such that } \Sigma' = \Sigma^S . \tag{5.1.3}$$

(Note that (5.1.3) is implied by (5.1.2).)

Thus a canonical form selects precisely one element out of each order of $GL_n(\mathbb{R})$ acting on L'. We speak of a continuous canonical form if c is continuous.

Of course there exist (many) canonical forms. E.G. order the set of all nice selection α in $J_{n,m}$ in some way. For each $\Sigma \in L^{cr}_{m,n,p}(\mathbb{R})$ let $\alpha(\Sigma)$ be the first α such that $R(\Sigma)_\alpha$ is nonsingular. Then

$$\Sigma \mapsto c_{\alpha(\Sigma)}(\Sigma) = \Sigma^S, \quad S = R(\Sigma)^{-1}_{\alpha(\Sigma)} \tag{5.1.4}$$

is a canonical form on $L^{cr}_{m,n,p}(\mathbb{R})$ (Luenberger canonical forms à la Bryson). This mapping is not continuous, however, except when $m = 1$ (in which case there is only one nice selection), which entails a number of drawbacks, e.g. in numerical calculations and in identification procedures, cf [GWi] for a discussion in the similar case of Jordan canonical forms.

5.1.5. <u>Theorem</u>. There is a continuous canonical form on $L^{cr,co}_{m,n,p}(\mathbb{R})$ if and only if $p = 1$ or $m = 1$.

<u>Proof</u>. If $m = 1$ let $\alpha \subset J_{1,n} = \{(0,1),(1,1),\dots,(n,1)\}$ be the unique nice selection $(0,1),\dots,(n-1,1)$. Then

$$c_{\#\alpha}: \Sigma \mapsto c_{\#\alpha}(\Sigma) = \Sigma^S, \quad S = R(\Sigma)_\alpha^{-1} \tag{5.1.6}$$

is a continuous canonical form, because $R(\Sigma)_\alpha$ is always invertible for Σ cr.

Similarly if $p = 1$, let $\beta \subset J_{n,1}$, be the unique nice row selection. Then $\Sigma \mapsto \Sigma$, $S = Q(\Sigma)_\beta$ is a continuous form because $Q(\Sigma)_\beta$ is invertible for all co Σ (if $p = 1$).

It remains to show that there cannot be a continuous canonical form c on all of $L^{cr,co}_{m,n,p}(\mathbb{R})$ if both $m > 1$, $p > 1$.

To do this we construct two families of linear dynamical systems as follows for all $a \in \mathbb{R}$, $b \in \mathbb{R}$ (We assume $n \geq 2$; if $n = 1$ the examples must be modified somewhat).

$$G_1(a) = \left(\begin{array}{cc|ccc} a & 1 & 0 & \cdots & 0 \\ 1 & 1 & 0 & \cdots & 0 \\ \hline 2 & 1 & & & \\ \vdots & \vdots & & B & \\ 2 & 1 & & & \end{array}\right),$$

$$G_2(b) = \left(\begin{array}{cc|ccc} 1 & b & 0 & \cdots & 0 \\ 1 & 1 & 0 & \cdots & 0 \\ \hline 2 & 1 & & & \\ \vdots & \vdots & & B & \\ 2 & 1 & & & \end{array}\right)$$

where B is some (constant) $(n-2) \times (m-2)$ matrix with coefficients in $\mathbb{R}$,

$$F_1(a) = \begin{pmatrix} 1 & 0 & \cdots & 0 \\ 0 & 2 & & \vdots \\ \vdots & & \ddots & 0 \\ 0 & \cdots & 0 & n \end{pmatrix} = F_2(b)$$

$$H_1(a) = \left(\begin{array}{cc|ccc} y_1(a) & 1 & 2 & \dots & 2 \\ y_2(a) & 1 & 1 & \dots & 1 \\ \hline 0 & 0 & & & \\ \vdots & \vdots & & C & \\ 0 & 0 & & & \end{array}\right),$$

$$H_2(b) = \left(\begin{array}{cc|ccc} x_1(b) & 1 & 2 & \dots & 2 \\ x_2(b) & 1 & 1 & \dots & 1 \\ \hline 0 & 0 & & & \\ \vdots & \vdots & & C & \\ 0 & 0 & & & \end{array}\right)$$

where C is some (constant) real $(p-2) \times (n-2)$ matrix. Here the continuous functions $y_1(a)$, $y_2(a)$, $x_1(b)$, $x_2(b)$ are e.g. $y_1(a) = a$ for $|a| \leq 1$, $y_1(a) = a^{-1}$ for $|a| \geq 1$, $y_2(a) = \exp(-a^2)$, $x_1(b) = 1$ for $|b| \leq 1$, $x_1(b) = b^{-2}$ for $|b| \geq 1$, $x_2(b) = b^{-1}\exp(-b^{-2})$ for $b \neq 0$, $x_2(0) = 0$. The precise form of these functions is not important. What is important is that they are continuous, that $x_1(b) = b^{-1}y_1(b^{-1})$, $x_2(b) = b^{-1}y_2(b^{-1})$ for all $b \neq 0$ and that $y_2(a) \neq 0$ for all a and $x_1(b) \neq 0$ for all b.

For all $b \neq 0$ let $T(b)$ be the matrix

$$T(b) = \begin{pmatrix} b & 0 & \dots & 0 \\ 0 & 1 & & \vdots \\ \vdots & & \ddots & 0 \\ 0 & \dots & 0 & 1 \end{pmatrix} . \tag{5.1.7}$$

Let $\Sigma_1(a) = (F_1(a), G_1(a), H_1(a))$, $\Sigma_2(b) = (F_2(b), G_2(b), H_2(b))$. Then one easily checks that

$$ab = 1 \Rightarrow \Sigma_1(a)^{T(b)} = \Sigma_2(b) \quad . \tag{5.1.8}$$

Note also that $\Sigma_1(a)$, $\Sigma_2(b) \in L^{co,cr}_{m,n,p}(\mathbb{R})$ for all $a,b \in \mathbb{R}$; in fact

$$\Sigma_1(a) \in U_\alpha, \quad \alpha = ((0,2),(1,2),\ldots,(n-1,2)) \text{ for all } a \in \mathbb{R} \tag{5.1.9}$$

$$\Sigma_2(b) \in U_\beta, \quad \beta = ((0,1),(1,1),\ldots,(n-1,1)) \text{ for all } b \in \mathbb{R} \tag{5.1.10}$$

which proves the complete reachability. The complete observability is seen similarly.

Now suppose that c is a continuous canonical form on $L^{co,cr}_{m,n,p}(\mathbb{R})$. Let $c(\Sigma_1(a)) = (\bar{F}_1(a),\bar{G}_1(a),\bar{H}_1(a))$, $c(\Sigma_2(b)) = (\bar{F}_2(b),\bar{G}_2(b),\bar{H}_2(b))$. Let $S(a)$ be such that $c(\Sigma_1(a)) = \Sigma_1(a)^{S(a)}$ and let $\bar{S}(b)$ be such that $c(\Sigma_2(b)) = \Sigma_2(b)^{\bar{S}(b)}$.

It follows from (5.1.9) and (5.1.10) that

$$\begin{aligned} S(a) &= R(\bar{F}_1(a),\bar{G}_1(a))_\alpha R(F_1(a),G_1(a))_\alpha^{-1} \\ \bar{S}(b) &= R(F_2(b),G_2(b))_\beta R(F_2(b),G_2(b))_\beta^{-1} \end{aligned} \tag{5.1.11}$$

Consequently $S(a)$ and $S(b)$ are (unique and are) continuous functions of a and b.

Now take $a = b = 1$. Then $ab = 1$ and $T(b) = I_n$ so that (cf. (5.1.7), (5.1.8) and (5.1.11)) $S(1) = \bar{S}(1)$. It follows from this and the continuity of $S(a)$ and $\bar{S}(b)$ that we must have

$$\text{sign}(\det S(a)) = \text{sign}(\det \bar{S}(b)) \text{ for all } a,b \in \mathbb{R} \tag{5.1.12}$$

Now take $a = b = -1$. Then $ab = 1$ and we have, using (5.1.8),

$$\begin{aligned} \Sigma_1(-1)^{(\bar{S}(-1)T(-1))} &= (\Sigma_1(-1)^{T(-1)})^{\bar{S}(-1)} \\ &= \Sigma_2(-1)^{\bar{S}(-1)} = c(\Sigma_2(-1)) \\ &= c(\Sigma_1(-1)) = \Sigma_1(-1)^{S(-1)} \quad . \end{aligned}$$

It follows that $S(-1) = \bar{S}(-1)T(-1)$, and hence by (5.1.7), that

$$\det(S(-1)) = -\det(\bar{S}(-1))$$

which contradicts (5.1.12). This proves that there does not exist a continuous canonical form $L^{co,cr}_{m,n,p}(\mathbb{R})$ if $m \geq 2$, and $p \geq 2$.

5.1.13. Remark. By choosing the matrices B,C in $G_1(a)$, $G_2(b)$, $H_1(a)$, $H_2(b)$ judiciously we can also see to it that rank $G_1(a) = m =$ rank $G_2(b)$, rank $H_1(a) = p =$ rank $H_2(b)$ if $p < n$ and $m < n$. Note also that F in the example above has n distinct real eigenvalues so that a restriction like "F must be semi-simple" also does not help much.

5.1.14. Discussion of the proof of theorem 5.1.5. The proof given above, though definitely a proof, is perhaps not very enlightening. What is behind it is the following. Consider the natural projection

$$\pi : L^{cr,co}_{m,n,p}(\mathbb{R}) \to M^{cr,co}_{m,n,p}(\mathbb{R}) \quad . \qquad (5.1.15)$$

Let c be a continuous canonical form. Because c is constant on all orbits c induces a continuous map $\tau : M^{cr,co}_{m,n,p}(\mathbb{R}) \to L^{cr,co}_{m,n,p}(\mathbb{R})$ which clearly is a section of π, (cf. (5.1.1) - (5.1.3)). Inversely if τ is a continuous section of π then $\tau\cdot\pi : L^{co,cr}_{m,n,p}(\mathbb{R}) \to L^{co,cr}_{m,n,p}(\mathbb{R})$ is a continuous canonical form.

Now (5.1.15) is (fairly easily at this stage, cf [Haz 1]), seen to be a principal $GL_n(\mathbb{R})$ fibre bundle. Such a bundle is trivial iff it admits a continuous section. The mappings

$$a \mapsto \Sigma_1(a), \quad b \mapsto \Sigma_2(b)$$

of the proof above now combine to define a continuous map of $\mathbb{P}^1(\mathbb{R})$ = circle into $M^{cr,co}_{m,n,p}(\mathbb{R})$ such that the pullback of the fibre bundle (5.1.15) is nontrivial. In fact the associated determinant $GL_1(\mathbb{R})$ fibre bundle is the Möbius band (minus zero section) over the circle.

5.2. The algebraic-geometric case. The result corresponding to theorem 5.1.5 in the algebraic-geometric case is the following. For simplicity we state it for varieties (over algebraically closed fields).

5.2.1. Theorem. Let k be an algebraically closed field. Then there exists a canonical form $c: L^{cr,co}_{m,n,p}(k) \to L^{cr,co}_{m,n,p}(k)$ which is a morphism of algebraic varieties if and only if $m = 1$ or $p = 1$.

Here of course a canonical form is defined just as in 5.1 above; simply replace $\mathbb{R}$ with k everywhere in (5.1.1)-(5.1.3) and replace the word "continuous" with "morphism of algebraic varieties," which means that locally c is given by rational expressions in the coordinates.

The proof is rather similar to the one briefly indicated in 5.1.14 above. In this case $L^{cr,co}_{m,n,p} \to M^{cr,co}_{m,n,p}(k)$ is an algebraic prinicpal $GL_n(k)$ bundle and one again shows that it is trivial if and only if $m = 1$ or $p = 1$. The only difference is the example used to prove nontriviality. The map used in 5.1.14 is non-algebraic, nor is there an algebraic injective morphism $\mathbf{P}^1(k) \to M^{co,cr}_{m,n,p}(k)$. Instead one defines a three dimensional manifold much related to the families $\Sigma_1(a)$, $\Sigma_2(b)$ together with an injection into $M^{cr,co}_{m,n,p}(k)$ such that the pullback of this principal bundle is easily seen to be nontrivial. Cf. [Haz 2] for details.

6. REALIZATION WITH PARAMETERS AND REALIZING DELAY-DIFFERENTIAL SYSTEMS

As a second application of the existence of fine moduli spaces for cr systems we discuss realization with parameters (cf. also [By]) and realization of delay-differential systems. A preliminary step for this is the following bit of realization theory.

6.1. <u>Resumé of some realization theory</u>. Let $T(s)$ be a proper rational matrix-valued function of s with the (formal) power series expansion (around $s = \infty$)

$$T(s) = A_1 s^{-1} + A_2 s^{-2} + \ldots, \quad A_i \in k^{p\times m} . \tag{6.1.1}$$

One says that $T(s)$ is realizable by a linear system of dimension $\leq n$, if $T(s)$ is the Laplace transform (resp. z-transform) of a linear differentiable (resp. difference) system $\Sigma = (F,G,H) \in L_{m,n,p}(k)$. This means that

$$T(s) = H(sI_n - F)^{-1}G \tag{6.1.2}$$

or, equivalently

$$A_i = HF^{i-1}G, \quad i = 1,2,3,\ldots \tag{6.1.3}$$

A necessary and sufficient condition that $T(s)$ be realizable by a system of dimension n is that the associated Hankel matrix $h(\mathscr{A})$ of the sequence $\mathscr{A} = (A_1, A_2, A_3, \ldots)$ be of rank $\leq n$. Here $h(\mathscr{A})$ is the block Hankel matrix.

$$h(\mathscr{A}) = \begin{pmatrix} A_1 & A_2 & A_3 & \cdots \\ A_2 & A_3 & & \\ A_3 & & & \\ \vdots & \vdots & & \end{pmatrix} .$$

More precisely we have the partial realization result which says that there eixst $F,G,H \in L^{co,cr}_{m,n,p}(k)$ such that $A_i = HF^{i-1}G$ iff rank $h_n(\mathscr{A})$ = rank $h_{n+1}(\mathscr{A}) = n$, where $h_i(\mathscr{A})$ is the block matrix consisting of the first i block rows and the first i block-columns of $h(\mathscr{A})$.

Now suppose that rank $h(\mathscr{A})$ is precisely n, and let F,G,H realize $\mathscr{A}$.

We have

$$h(\mathscr{A}) = \begin{pmatrix} H \\ HF \\ HF^2 \\ \vdots \end{pmatrix} (G \mid FG \mid F^2G \mid \ldots)$$

and it follows by the Cayley-Hamilton theorem that $R(F,G)$ and $Q(F,H)$ are both of rank n so that $\Sigma = (F,G,H)$ is in this case both cr and co.

Finally we recall that if Σ and Σ' are both cr and co and both realize $\mathscr{A}$, then Σ and Σ' are isomorphic, i.e., there is an $S \in GL_n(k)$ such that $\Sigma' = \Sigma^S$.

For all these facts, cf e.g. [KFA] or [Haz 3].

6.2. A realization algorithm. Now let $\mathscr{A}$ be such that rank $h(\mathscr{A}) = n$. We describe a method for calculating a

$$\Sigma = (F,G,H) \in L^{cr,co}_{m,n,p}(k)$$

which realizes $\mathscr{A}$. By the above we know that there exist a nice selection $\alpha_c \subset J_{m,n}$ the set of column indices of

$$h_{n+1}(\mathscr{A}) = \begin{pmatrix} A_1 & A_2 & \cdots & A_{n+1} \\ A_2 & & & \\ \cdot & & & \\ \cdot & & & \\ \cdot & & & \\ A_{n+1} & & \cdots & A_{2n+1} \end{pmatrix} \qquad (6.2.1)$$

and a nice selection $\alpha_r \subset J_{p,n}$, the set of row indices of $h_{n+1}(\mathscr{A})$, such that the $n \times n$ matrix $h_{n+1}(\mathscr{A})_{\alpha_r,\alpha_c}$ has rank n. Here $h_{n+1}(\mathscr{A})_{\alpha_r,\alpha_c}$ is the matrix obtained from $h_{n+1}(\mathscr{A})$ by removing all rows whose index is not in α_r and all columns whose index is not in α_c. We now describe a method for finding a $\Sigma = (F,G,H) \in L^{cr,co}_{m,n,p}(k)$ such that Σ realizes $\mathscr{A}$ and such that $R(F,G)_{\alpha_c} = I_n$. (Such a Σ is unique).

Let γ_r be the subset of $J_{p,n}$ of the first p row indices, so that $h_{n+1}(\mathscr{A})_{\gamma_r}$ consists of the first row of blocks in (6.2.1). Now let

$$H = h_{n+1}(\mathscr{A})_{\alpha_r,\alpha_c} \quad . \qquad (6.2.2)$$

Now let

$$S = h_{n+1}(\mathscr{A})_{\alpha_r,\alpha_c} \qquad (6.2.3)$$

and define $R' = S^{-1}(h_{n+1}(\mathcal{A})_{\alpha_r})$. Then $(R')_{\alpha_c} = I_n$ and we let F,G be the unique $n \times n$ and $n \times m$ matrices such that

$$R(F,G) = R' \tag{6.2.4}$$

Recall, cf 3.2.7 above, that the columns of F and G can be simply read from the columns of R', being equal to either a standard basis vector or equal to a column of R'.

For every field k and each pair of nice selections $\alpha_c \subset J_{m,n}$, $\alpha_r \subset J_{p,n}$ let $W(\alpha_r,\alpha_c)(k)$ be the space of all sequences of $p \times m$ matrices $\mathcal{A} = (A_1,\ldots,A_{2n+1})$ such that $\mathrm{rank}(h_{n+1}(\mathcal{A})) = n$ and $\mathrm{rank}(h_{n+1}(\mathcal{A})_{\alpha_r,\alpha_c}) = n$. Then the above defines a map

$$\tau(\alpha_r,\alpha_c)\colon W(\alpha_r,\alpha_c)(k) \to L^{cr,co}_{m,n,p}(k) . \tag{6.2.5}$$

6.2.6 Lemma. If $k = \mathbb{R}$ or $\mathbb{C}$ the map $\tau(\alpha_r,\alpha_c)$ is analytic, and algebraic-geometrically speaking the $\tau(\alpha_r,\alpha_c)$ define a morphism of schemes from the affine scheme $W(\alpha_r,\alpha_c)$ into the quasi affine scheme $L^{cr,co}_{m,n,p}$.

6.2.7 Lemma. Let $W(k)$ be the space of all sequences of $p \times m$ matrices $\mathcal{A} = (A_1,A_2,\ldots,A_{2n+1})$ such that $\mathrm{rank}(h_{n+1}(\mathcal{A})) = n = \mathrm{rank}\, h_n(\mathcal{A})$. Let $h\colon L^{cr,co}_{m,n,p}(k) \to W(k)$ be the map $h(F,G,H) = (HG,HFG,\ldots,HF^{2n}G)$. Then $h \cdot \tau(\alpha_r,\alpha_c)$ is equal to the natural embedding of $W(\alpha_r,\alpha_c)(k)$ in $W(k)$. (I.e. $h \cdot \tau(\alpha_r,\alpha_c)$ is the identity of $W(\alpha_r,\alpha_c)(k)$.)

Proof. Let $\mathcal{A} \in W(\alpha_r,\alpha_c)(k)$. By partial realization theory (cf. 6.1 above) we know that $\mathcal{A}$ is realizable, say by $\Sigma' = (F',G',H')$. Then because $\mathcal{A} \in W(\alpha_r,\alpha_c)(k)$ we have that $S = R(F',G')_{\alpha_c}$ is invertible. Let

$$\Sigma = (F,G,H) = \Sigma'^{S^{-1}} = (S^{-1}F'S, S^{-1}G', H'S).$$

Then Σ also realizes $\mathcal{A}$ and $R(F,G)_{\alpha_c} = I_n$. Now observe that the realization algorithm described above simply recalculates precisely these F,G,H from $\mathcal{A}$.

6.2.8. Corollary. Let $k = \mathbb{R}$ or $\mathbb{C}$ and let $h: M^{cr,co}_{m,n,p}(k) \to W(k)$ be the map induced by $h: L_{m,n,p}(k) \to W(k)$. Then h is an isomorphism of analytic manifolds.

6.2.9. Corollary. More generally $h: L^{co,cr}_{m,n,p} \to W$ induces an isomorphism of schemes $M_{m,n,p} \to W$. In particular if k is an algebraically closed field then we have an isomorphism of the algebraic varieties $M_{m,n,p}(k)$ and $W(k)$.

6.3. Realization with parameters.

6.3.1. The topological case. Let $T_a(s)$, $a \in V$ be a family of transfer functions depending continuously on a parameter $a \in V$. For each $a \in V$ write $T_a(s) = A_1(a)s^{-1} + A_2(a)s^{-2} + \ldots$ and for each a let $n(a)$ be the rank of the block Hankel matrix of $\mathcal{A}(a) = (A_1(a), A_2(a), \ldots)$. The question we ask is: does there exist a continuous family of systems $\Sigma(a) = (F(a),G(a),H(a))$ such that the transfer function of $\Sigma(a)$ is $T_a(s)$ for all a. The answer to this is definitely "yes" provided $n(a)$ is bounded as a function of a. Simply take a long enough chunk of the $\mathcal{A}(a)$ of all a and do the usual realization construction by means of block companion matrices and observe that this is continuous in the $A_i(a)$. [True if V is paracompact and normal, one needs partitions of unity (in any case, I do) to find continuous $T_i(a)$ such that $B_{n+1} = T_1B_n + \ldots + T_nB_1$ where B_i is the i-th block column of $h(\mathcal{A})$.] The question becomes much more delicate if we ask for a continuous family of realizations which are all cr and co. This obviously requires that $n(a)$ is constant and provided that the space V is such that all $n = n(a)$ dimensional bundles are trivial this condition is also sufficient. Indeed if $n(a)$ is constant then the $\mathcal{A}(a)$ determine a continuous map $V \to W(\mathbb{R})$ and hence by Corollary 5.2.8 a continuous map $V \to M^{cr,co}_{m,n,p}(\mathbb{R})$. Pulling back the universal family over $M^{cr,co}_{m,n,p}(\mathbb{R})$ to a family over V gives us a family $(E;F,G,H)$ over V such that the transfer function of the system over $a \in V$ is $T_a(s)$ for all a. The bundle E is trivial by hypothesis, so there are continuous sections $e_1,\ldots,e_n: V \to E$ such that $\{e_1(a),\ldots,e_n(a)\}$ is a basis for $E(a)$ for all $a \in V$. Now write out the matrices of F,G,H with respect to these bases to find a continuous family $\Sigma(a)$, which realizes $T_a(s)$ and such that $\Sigma(a)$ is cr and co for all a.

6.3.2. <u>The polynomial case.</u> Let k be a field and $\bar{k}$ its algebraic closure, e.g. $k = \mathbb{R}$ and $\bar{k} = \mathbb{C}$. Let $T_x(s)$ be a transfer function with coefficients in $k[x_1,\ldots,x_q]$, where $x_1,\ldots,x_q$ are indeterminates. We ask whether there exists a realization of $T(s)$ over $k[x_1,\ldots,x_q]$, that is a triple of matrices (F,G,H) with coefficients in $k[x_1,\ldots,x_q]$ such that $T_x(s) = H(sI-F)^{-1}G$. Again the answer is obviously "yes" if we do not require any minimality conditions on the realization (provided $n(x_1,\ldots,x_q)$ the degree of the Hankel matrix of $T(s)$ is bounded for all $(x_1,\ldots,x_q) \in \bar{k}^q$

Now assume that $n(x_1,\ldots,x_q)$ is constant for all $(x_1,\ldots,x_q) \to \bar{k}^q$. Then $(x_1,\ldots,x_q) \mapsto \mathscr{A}(x_1,\ldots,x_q)$ defines a morphism of algebraic varieties $\bar{k}^q \to M^{cr,co}_{m,n,p}(k)$. Pulling back the universal family by means of this morphism we find a family $(E;F,G,H)$ over $\bar{k}^q$ which is defined over k because the morphism $\bar{k}^q \to W(k)$ and the isomorphism with $M^{cr,co}_{m,n,p}(k)$ are defined over k. Thus Σ is defined over k and by the Quillen-Suslin theorem E is trivializable over k. Taking the corresponding sections and writing out the matrices of F,G,H with respect to the resulting bases we find an F,G,H with coefficients in $k[x_1,\ldots,x_q]$ which realize $T_x(s)$ for all $x \in \bar{k}^q$, i.e. such that $T_x(s) = H(sI-F)^{-1}G$. Moreover this system (F,G,H) is cr over $k[x_1,\ldots,x_q]$ meaning that $R(F,G)$: $k[x_1,\ldots,x_q]^{(n+1)m} \to k[x_1,\ldots,x_q]^n$ is surjective; it is also co and even stronger its dual system is also cr (i.e. (F,G,H) is split in the terminology of [So 3]).

6.3.3. <u>Realization by means of delay-differentiable systems.</u> Let $\Sigma = (F(\sigma_1,\ldots,\sigma_q),\ G(\sigma_1,\ldots,\sigma_q),\ H(\sigma_1,\ldots,\sigma_q))$ be a delay differential system with q incommensurable delays. Here σ_i stands for the delay operator $\sigma_i F(t) = F(t-a_i)$, cf. 2.3 above for this notation. The transfer function of Σ is then

$$T(s) = G(e^{-a_1 s},\ldots,e^{-a_q s})(sI-F(e^{-a_1 s},\ldots,e^{-a_q s}))^{-1} H(e^{-a_1 s},\ldots,e^{-a_q s}) \qquad (6.3.4)$$

which is a rational function in s whose coefficients are polynomials in

$$e^{-a_1 s}, \ldots, e^{-a_q s} \quad .$$

Now inversely suppose we have a transfer function $T(s)$ like (6.3.4) and we ask whether it can be realized by means of a delay-differential system $\Sigma(\sigma)$. Now if the a_i are incommensurable then the functions

$$s, e^{-a_1(s)}, \ldots, e^{-a_q s}$$

are algebraically independent and there is precisely one transfer function $T(s;\sigma_1,\ldots,\sigma_q)$ whose coefficients are polynomials in $\sigma_1,\ldots,\sigma_q$ such that $T(s) = T'(s, e^{-a_1 s}, \ldots, e^{-a_q s})$. Thus the problem is mathematically identical with the one treated just above 6.3.2. In passing let us remark that complete reachability for delay-systems in the sense of that the associated system over the ring $\mathbb{R}[\sigma_1,\ldots,\sigma_q]$ is required to be cr seems often a reasonable requirement, e.g. in connection with pole placement, cf. [So 1] and [Mo].

7. THE "CANONICAL" COMPLETELY REACHABLE SUBSYSTEM.

7.1. Σ^{cr} <u>for systems over fields</u>. Let $\Sigma = (F,G,H)$ be a system over a field k. Let X^{cr} be the image of $R(F,G): k^r \to k^n$, $r = m(n+1)$. Then obviously $F(X^{cr}) \subset X^{cr}$, $G(k^m) \subset X^{cr}$, so that there is an induced subsystem $\Sigma^{cr} = (X^{cr}; F', G', H')$ which is called the canonical cr subsystem of Σ. In terms of matrices this means that there is an $S \in GL_n(k)$ such that Σ^S has the form

$$\Sigma^S = \left(\begin{pmatrix} G_1 \\ 0 \end{pmatrix}, \begin{pmatrix} F_{11} & F_{12} \\ 0 & F_{22} \end{pmatrix}, (H_1 \quad H_2) \right) \tag{7.1.1}$$

with $(F_{11}, G_1, H_1) = \Sigma^{cr}$, the "canonical" cr subsystem. The words Kalman "decomposition" are also used in this context. There is a dual construction relating to co and combining these two constructions "decomposes" the system into four parts.

In this section we examine whether this construction can be globalized, i.e. we ask whether this construction is continuous, and we ask whether something similar can be done for time varying linear dynamical systems.

7.2. Σ^{cr} for time varying systems. Now let $\Sigma = (F,G,H)$ be a time varying system, i.e. the coefficients of the matrices F,G,H are allowed to vary, say differentiably, with time. For time varying systems the controllability matrix $R(\Sigma) = R(F,G)$ must be redefined as follows

$$R(F,G) = (G(0) \vdots G(1) \vdots \ldots \vdots G(n)) \tag{7.2.1}$$

where

$$G(0) = G;\ G(i) = FG(i-1) - \dot{G}(i-1) \tag{7.2.2}$$

where the $\cdot$ denotes the differentiation with respect to time, as usual. Note that this gives back the old $R(F,G)$ if F,G do not depend on time. The system is said to be cr if this matrix $R(\Sigma)$ has full rank. These seem to be the appropriate notions for time varying systems; cf. e.g. [We, Haz 5] for some supporting results for this claim.

A time variable base change $x' = Sx$ (with $S = S(t)$ invertible for all t) changes Σ to Σ^S with

$$\Sigma^S = (SFS^{-1} + \dot{S}S^{-1}, SG, HS^{-1}) . \tag{7.2.3}$$

Note that $R(\Sigma)$ hence transforms as

$$R(\Sigma^S) = SR(\Sigma) . \tag{7.2.4}$$

7.2.5. Theorem. Let Σ be a time varying system with differentiably varying parameters. Suppose that rank $R(\Sigma)$ is constant as a function of t. Then there exists a differentiable time varying matrix S, invertible for all t, such that Σ^S has the form (7.1.1) with (F_{11},G_1,H_1) cr.

Proof. Consider the subbundle of the trivial $(n+1)m$ dimensional bundle over the real line generated by the rows of $R(\Sigma)$. This is a vectorbundle because of the rank assumption. This bundle is trivial. It follows that there exist r sections of the bundle, where $r = \text{rank } R(\Sigma)$, which are linearly independent everywhere. The continuous sections of the bundle are of the form $\Sigma a_i(t)z_i(t)$, where $z_i(t),\ldots,z_n(t)$ are the rows of $R(\Sigma)$ and the $a_i(t)$ are continuous functions of t. Let $b_1(t),\ldots,$ $b_r(t)$ be the r everywhere linear independent sections and let $b_j(t) = \Sigma a_{ji}(t)z_i(t)$, $j = 1,\ldots,r$; $i = 1,\ldots,n$.

Let E' be the r dimensional subbundle of the trivial bundle E of dimension n over the real line generated by the r row vectors $a_j(t) = (a_{j1}(t),\dots,a_{jn}(t))$. Because the quotient bundle E/E' is trivial we can complete the r vectors $a_1(t),\dots,a_r(t)$ be a set of n vectors $a_1(t),\dots,a_n(t)$ such that the determinant of the matrix formed by these vectors is nonzero for all t. Let $S_1(t)$ be the matrix formed by these vectors, then $S_1R(\Sigma)$ has the property that for all t its first r rows are linearly independent and that it is of rank r for all t. It follows that there are unique continuous functions $c_{ki}(t)$, $k = r+1,\dots,n$; $i = 1,\dots,r$ such that $z'_k(t) = \Sigma c_{ki}(t)z'_i(t)$, where $z'_j(t)$ is the j-th row of $S_1R(\Sigma)$. Now let

$$S_2(t) = \begin{pmatrix} I_r & 0 \\ -C(t) & I_{n-r} \end{pmatrix}$$

Then $S(t) = S_2(t)S_1(t)$ is the desired transformation matrix (as follows from the transformation formula (7.2.4)).

Virtually the same arguments give a smoothly varying $S(t)$ if the coefficients of Σ vary smoothly in time, and give a polynomial $S(t)$ if the coefficients of Σ are polynomials in t (where in the latter case we need the constancy of the rank also for all complex values of t and use that projective modules over a principal ideal ring are free).

7.3. Σ^{cr} for families. For families of systems these techniques give

7.3.1. Theorem. Let Σ be a continuous family parametrized by a contractible topological space (resp. a differentiable family parameterized by a contractible manifold; resp. a polynomial family). Suppose that the rank of $R(\Sigma)$ is constant as a function of the parameters. Then there exists a continuous (resp. differentiable; resp. polynomial) family of invertible matrices S such that Σ^S has the form (7.1.1) with (F_{11},G_1,H_1) a family of cr systems.

The proof is virtually the same as the one given above of theorem 7.2.5; in the polynomial case one, of course, relies on the Quillen-Suslin theorem [Qu; Sus] to conclude that the appropriate bundles are trivial. Note also that, inversely, the existence of an S as in the theorem implies that the rank of $R(\Sigma)$ is constant.

For delay-differential systems this gives a "Kalman decomposition" provided the relevant, obviously necessary, rank condition is met.

Another way of proving theorem 7.3.1 for systems over certain rings rests on the following lemma which is also a basic tool in the study of isomorphisms of families in [HP] and which implies a generalization of the main lemma of [OS] concerning the solvability of sets of linear equations over rings.

7.3.2. Lemma. Let R be a reduced ring (i.e. there are no nilpotents $\neq 0$) and let A be a matrix over R. Suppose that the rank of $A(p)$ over the quotient field of R/p is constant as a function of p for all prime ideals p. Then $\mathrm{Im}(A)$ and $\mathrm{Coker}(A)$ are projective modules.

Now let Σ over R be such that rank $R(\Sigma(p))$ is constant, and let R be projective free (i.e. all finitely generated projective modules over R are free). Then $\mathrm{Im}\, R(\Sigma) \subset R^n$ is projective and hence free. Taking a basis of $\mathrm{Im}\, R(\Sigma)$ and extending it to a basis of all of R^n, which can be done because $R^n/\mathrm{Im}\, R(\Sigma) = \mathrm{Coker}\, R(\Sigma)$ is projective and hence free, now gives the desired matrix S.

There is a complete set of dual theorems concerning co.

7.4. Σ^{cr} for delay differential systems. Now let $\Sigma(\sigma) = (F(\sigma), G(\sigma), H(\sigma))$ be a delay differential system. Then, of course, we can interpret Σ as a polynomial system over $\mathbb{R}[\sigma] = [\sigma_1,\ldots,\sigma_r]$ and apply theorem 7.3.1. The hypothesis that rank $R(\Sigma(\sigma))$ be constant as a function of $\sigma_1,\ldots,\sigma_r$ (including complex and negative values of the delays) is rather strong though.

Now if we assume that all functions involved in

$$\dot{x}(t) = F(\sigma)x(t) + G(\sigma)u(t), \quad y(t) = H(\sigma)x(t) \tag{7.4.1}$$

are zero sufficiently far in the past, an assumption which is not unreasonable and even customary in this context, then it makes perfect sense to talk about base changes of the form

$$x' = S(\sigma)x \tag{7.4.2}$$

where $S(\sigma)$ is matrix whose coefficients are power series in the delays $\sigma_1,\ldots,\sigma_r$ and which is invertible over the ring of power series $\mathbb{R}[[\sigma_1,\ldots,\sigma_r]]$. Indeed if $\sigma_1\alpha(t) = \alpha(t-a_1)$, $a_1 > 0$ and the function $\beta(t)$ is zero for $t < -Na_1$ then

$$\left(\sum_{i=0}^{\infty} b_i\sigma_1^i\right) \beta(t) = \sum_{i=0}^{N+N'} b_i\beta(t-ia_i)$$

where N' is such that $t < N'a_i$.

Allowing such basis changes one has

7.4.3. Theorem. Let $\Sigma(\sigma)$ be a delay-differential system. Suppose that rank $R(\Sigma(\sigma))$ considered as a matrix over the quotient field $k(\sigma_1,\ldots,\sigma_r)$ is equal to rank $R(\Sigma(0))$ (over $\mathbb{R}$) where $\Sigma(0)$ is the system obtained from $\Sigma(\sigma)$ by setting all σ_i equal to zero. Then there exists a power series base change matrix $S \in GL_n(\mathbb{R}[[\sigma]])$ such that Σ^S has the form (7.1.1) with (F_{11},G_1,H_1) a cr system (over $\mathbb{R}[[\sigma]]$).

The proof is again similar where now, of course, one uses that a projective module over a local ring is free.

Note that $\Sigma(0)$ is not the system obtained from $\Sigma(\sigma)$ by setting all delays equal to zero. For example if $\Sigma(\sigma)$ is the one dimensional, one delay system $\dot{x}(t) = x(t) + 2x(t-1) + u(t) + u(t-2)$, $y(t) = 2x(t) - x(t-1)$, then $\Sigma(0)$ is the system $\dot{x}(t) = x(t) + u(t)$, $y(t) = 2x(t)$ obtained by removing all delay terms.

8. CONCLUDING REMARKS ON FAMILIES OF SYSTEMS AS OPPOSED TO SINGLE SYSTEMS

8.1. Non extendability of moduli spaces $M^{cr}_{m,n,p}$ and $M^{co}_{m,n,p}$. One aspect of the study of families of systems rather than single systems is the systematic investigation of which of the many constructions and algorithms of systems and control theory are continuous in the system parameters (or more precisely to determine, so to speak, the domains of continuity of these constructions). This is obviously important if one wants e.g. to execute these algorithms numerically.

Intimately (and obviously) related to this continuity problem is the question of how a given single system can sit in a family of systems (deformation (perturbation) theory). The fine moduli spaces $M^{cr}_{m,n,p}$ and $M^{co}_{m,n,p}$ answer precisely this question (for a system which is cr or co): for a given cr (resp. co) system the local structure of $M^{cr}_{m,n,p}(\mathbb{R})$ (resp. $M^{co}_{m,n,p}(\mathbb{R})$) around the point represented by the given system describes exactly the most complicated family in which the given system can occur (all other families can up to isomorphism be uniquely obtained from this one by a change of parameters). Thus one may well be interested to see whether these moduli spaces can be extended a bit. In particular one could expect that $M^{cr}_{m,n,p}(\mathbb{R})$ and $M^{co}_{m,n,p}(\mathbb{R})$ could be combined in some way to give a moduli space for all systems which are cr or co. The following example shows that this is a bit optimistic.

8.1.1. Example. Let Σ and Σ' be the two families over $\mathbb{C}$ (or $\mathbb{R}$) given by the triples of matrices

$$\Sigma = \left(\begin{pmatrix} 1 & 1 \\ \sigma & 1 \end{pmatrix}, \begin{pmatrix} 1 \\ 0 \end{pmatrix}, (1,0) \right),$$

$$\Sigma' = \left(\begin{pmatrix} 1 & \sigma \\ 1 & 1 \end{pmatrix}, \begin{pmatrix} 1 \\ 0 \end{pmatrix}, (1,0) \right).$$

Σ is co everywhere and cr everywhere but in $\sigma = 0$, and Σ' is cr everywhere and co everywhere but in $\sigma = 0$. The systems $\Sigma(\sigma)$ and $\Sigma'(\sigma)$ are isomorphic for all $\sigma \neq 0$, but $\Sigma(0)$ and $\Sigma'(0)$ are definitely not isomorphic. This kills all chances of having a fine moduli space for families which consist of systems which are co or cr. There cannot even be a coarse moduli space for such families.

Indeed let $\mathcal{F}$ be the functor which assigns to every space the set of all isomorphism classes of families of cr or co systems. Then a coarse moduli space for $\mathcal{F}$ (cf. [Mu] for a precise definition) consists of a space M together with a functor transformation $\mathcal{F}(-) \to \mathrm{Mor}(-,M)$ which is an isomorphism if $- = pt$ and which also enjoys an additional universality property. Now consider the commutative diagram

$$\begin{array}{ccc} \mathcal{F}(\mathbb{C}\backslash\{o\}) & \to & \mathrm{Mor}(\mathbb{C}\backslash\{o\},M) \\ \uparrow & & \uparrow \\ \mathcal{F}(\mathbb{C}) & \xrightarrow{\alpha} & \mathrm{Mor}(\mathbb{C},M) \\ \downarrow & & \downarrow \\ \mathcal{F}(\{o\}) & \xrightarrow{\sim} & \mathrm{Mor}(\{o\},M) \end{array} .$$

Consider the elements of $\mathcal{F}(\mathbb{C})$ represented by Σ and Σ'. Because Σ and Σ' are isomorphic as families restricted to $\mathbb{C}\backslash\{o\}$ we see by continuity (of the elements of $\mathrm{Mor}(\mathbb{C},M)$) that $\alpha(\Sigma) = \alpha(\Sigma')$. Because $\Sigma(0)$ and $\Sigma'(0)$ are not isomorphic this gives a contradiction with the injectivity of $\mathcal{F}(\{o\}) \to \mathrm{Mor}(\{o\},M)$.

Coarse moduli spaces represent one possible weakening of the fine moduli space property. Another, better adapted to the idea of studying families by studying a maximally complicated example, is that of a versal deformation. Roughly a versal holomorphic deformation of a system Σ over $\mathbb{C}$ is a family of systems $\Sigma(\sigma)$ over a small neighborhood U of 0 (in some parameter space) such that $\Sigma(0) = \Sigma$ and such that for every family Σ' over V such that $\Sigma'(0) = \Sigma$ there is some (not necessarily unique) holomorphic map ϕ (i.e., a holomorphic change in parameters) such that $\phi^!\Sigma \simeq \Sigma'$ in a neighborhood of 0.

For square matrices depending holomorphically on parameters (with similarity as isomorphism) Arnol'd, [Ar], has constructed versal deformations and the same ideas work for systems (in any case for pairs of matrices (F,G), cf. [Ta 2]).

8.2. <u>On the geometry of</u> $M^{cr,co}_{m,n,p}$. From the identification of systems point of view not only the local structure of $M^{co,cr}_{m,n,p}(\mathbb{R})$ is important but also its global structure cr. also [BrK] and [Haz 8]. Thus, for example, if $m = 1 = p$, $M^{co,cr}_{m,n,p}(\mathbb{R}) = \mathrm{Rat}(n)$ decomposes into $(n+1)$ components, and some of these components are of rather complicated topological types, [Br], which argues ill for the linearization tricks which are at the back of many identification procedures. One way to view identification is as finding a sequence of points in $M^{co,cr}_{m,n,p}(\mathbb{R})$ as more and more data come in. Ideally this sequence of points will then converge to something. Thus the question comes up of whether $M^{co,cr}_{m,n,p}(\mathbb{R})$ is compact, or compactifiable in such a way that the extra points can be interpreted as some kind of systems. Now $M^{co,cr}_{m,n,p}(\mathbb{R})$ is never compact. As to the compactification question, there does exist a partial compactification $\bar{M}_{m,n,p}$ such that the extra points, i.e. the points of $\bar{M}_{m,n,p} \setminus M^{cr,co}_{m,n,p}$ correspond to systems of the form

$$\dot{x} = Fx + Bu, \quad y = Hx + J(D)u \tag{8.2.1}$$

where D is the differentiation operator and J is a polynomial in D. This seems to give still more motivation for studying systems more general than $\dot{x} = Fx + Gu$, $y = Hx$ [Ros]. This partial compactification is also maximal in the sense that if a family of systems converges in the sense that the associated family of input/

output operators converges (in the weak topology) then the limit input/output operator is the input/output operator of a system of the form (8.2.1). Cf. [Haz 4] for details.

8.3. Pointwise-local-global isomorphism theorems. One perennial question which always turns up when one studies families rather than single objects is: to what extent does the pointwise or local structure of a family determine its global properties. Thus for square matrices one has e.g. the question studied by Wasov [Wa], cf. also [OS]: given two families of matrices $A(z)$, $A'(z)$ depending holomorphically on some parameters z. Suppose that for each separate value of z, $A(z)$ and $A'(z)$ are similar; does it follow that $A(z)$ and $A'(z)$ are similar as holomorphic families?

For families of systems the corresponding question is: let $\Sigma(\sigma)$ and $\Sigma'(\sigma)$ be two families of systems and suppose that $\Sigma(\sigma)$ and $\Sigma'(\sigma)$ are isomorphic for all values of σ. Does it follow that Σ and Σ' are isomorphic as families (globally or locally in a neighborhood of every parameters value σ).

Here there are (exactly as in the holomorphic-matrices-under-similarity-case) positive results provided the dimension of the stabilization subgroups $\{S \in GL_n(\mathbb{R}) | \Sigma(\sigma)^S = \Sigma(\sigma)\}$ is constant as a function of σ, cf. [HP].

REFERENCES

Ans Ansell, H. C.: 1964, *On certain two-variable generalizations of circuits theory to networks of transmission lines and lumped reactances*, IEEE Trans. Circuit Theory, 22, pp. 214-223.

Ar Arnol'd, V. I.: 1971, *On matrices depending on a parameter*, Usp. Mat. Nauk, 27, pp. 101-114.

BC Bellman, R., and Cooke, K. L.: 1963, *Differential Difference Equations*, Academic Press.

BH Byrnes, C. I., and Hurt, N. E.: 1979, *On the moduli of linear dynamical systems, Studies in Analysis*, Adv. Math. Suppl., Vol. 4, pp. 83-122.

BHMR Byrnes, C. I., Hazewinkel, M., Martin, C., and Rouchaleau, Y.: 1980, *Basic material from algebraic geometry and differential topology for (linear) systems and control theory*, this volume.

BOU Bourbaki, N.: 1961, *Algèbre commutative*, Ch. I, II, Paris, Hermann (ed.).

BR Brockett, R. W.: 1976, *Some geometric questions in the theory of linear systems*, IEEE Trans. AC 21, pp. 449-455.

BrK Brockett, R. W., and Krishnaprasad, P. S.: *A scaling theory for linear systems*, to appear, Trans. IEEE AC.

By Byrnes, C. I.: 1979, *On the control of certain deterministic infinite dimensional systems by algebro-geometric techniques*, Amer. J. Math., 100, pp. 1333-1380.

DDH Deistler, M., Dunsmuir, W., and Hannan, E. J.: 1978, *Vector linear time series models: corrections and extensions*, Adv. Appl. Probl., 10, pp. 360-372.

De Deistler, M.: 1978, *The structural identifiability of linear models with autocorrelated errors in the case of cross-equation restrictions*, J. of Econometrics, 8, pp. 23-31.

DH Dunsmuir, W., and Hannan, E. J.: 1976, *Vector linear time series models*, Adv. Appl. Prob. 8, pp. 339-364.

DS Deistler, M., and Seifert, H. G.: 1978, *Identifiability and consistent estimability in econometric models*, Econometrica, 46, pp. 969-980.

Eil Eilenberg, S.: 1978, *Automata, languages and machines*, Vol. A, Acad. Press.

Eis Eising, R.: 1978, *Realization and stabilization of 2-D systems*, IEEE Trans. AC 23, pp. 793-799.

GWi Golub, S. H., and Wilkinson, J. H.: 1976, *Ill conditioned eigensystems and the computation of the Jordan canonical form*, SIAM Rev. 18, pp. 578-619.

Han Hannan, E. J.: 1971, *The identification problem for multiple equation systems with moving average errors*, Econometrica, 39, pp. 751-765.

Haz 1 Hazewinkel, M.: 1977, *Moduli and canonical forms for linear dynamical systems II: the topological case*, Math. System Theory, 10, pp. 363-385.

Haz 2 Hazewinkel, M.: 1977, *Moduli and canonical forms for linear dynamical systems III: the algebraic geometric case*, In: C. Martin, R. Hermann (eds.), Proc. 1976 Ames Research Centre (NASA) Conf. on Geometric Control Theory, Math Sci Press, pp. 291-336.

Haz 3 Hazewinkel, M.: 1979, *On the (internal) symmetry groups of linear dynamical systems*, In: P. Kramer, M. Dal-Cin (eds.), Groups, Systems and Many-Body Physics, Vieweg.

Haz 4 Hazewinkel, M.: 1979, *Families of linear dynamical systems; degeneration identification and singular perturbation*, NATO-AMS Conf. on Algebraic and Geometric Methods in Linear Systems Theory, Harvard University, June 1979. (Preliminary version: Proc. IEEE CDC New Orleans, 1977, pp. 258-264).

Haz 5 Hazewinkel, M.: 1978, *Moduli and invariants for time varying systems*, Ricerche di Automatica, 9, pp. 1-14.

Haz 6 Hazewinkel, M.: *On the representations of the wild quiver* $\bullet \longrightarrow \circlearrowleft$, in preparation.

Haz 7 Hazewinkel, M.: *A partial survey of the uses of algebraic geometry in systems and control theory*, to appear, Symp. Math. INDAM (Severi centennial conference), Acad. Press.

Haz 8 Hazewinkel, M.: 1979, *On identification and the geometry of the space of linear systems*, Proc. Bonn Jan. 1979 Workshop on Stochastic Control and Stochastic Differential Questions, Lect. Notes Control and Inf. Sciences, 16, Springer, pp. 401-415.

HJK Hadlock, C. K., Jamshidi, M., and Kokotovic, P.: 1970, *Near optimum design of three time scale problems*, Proc. 4th Annual Princeton Conf. Inf. & System Sci., pp. 118-122.

HKa Hazewinkel, M., and Kalman, R. E.: 1976, *Invariants, canonical forms and moduli for linear constant, finite dimensional dynamical systems*. In: G. Marchesini, S. K. Mitter (eds.), Proc. of a Conf. on Algebraic System Theory, Udine 1975, Springer Lect. Notes Economics and Math. Systems, 131, pp. 48-60.

HP Hazewinkel, M., and Perdon, A.-M.: 1979, *On the theory of families of linear dynamical systems*, Proc. MTNS '79, (4th Int. Symp. Math. Theory of Networks and Systems, Delft, July 1979), pp. 155-161.

Kam Kamen, E. W.: 1978, *An operator theory of linear functional differential equations*, J. Diff. Equations, 27, pp. 274-297.

Kap Kappel, F.: 1977, *Degenerate difference-differential equations: algebraic theory*, J. Diff. Equations, 24, pp. 99-126.

KFA Kalman, R. E., Falb, P. L., and Arbib, M. A.: 1969, *Topics in System Theory*, McGraw-Hill.

KKU Kar Keung, D. Y., Kokotovic, P. V., and Utkin, V. I.: 1977, *Singular perturbation analysis of high-gain feedback systems*, IEEE Trans. AC, 22, pp. 931-937.

MK Martin, C., and Krishnaprasad, P. S.: 1978, *On the geometry of minimal triples*, preprint, Case Western Reserve Univ.

Mo Morse, A. S.: 1974, *Ring models for delay-differential systems*, Proc. IFAC, pp. 561-567 (reprinted (in revised form) in Automatica, 12, (1976), pp. 529-531).

Mu Mumford, D.: 1965, *Geometric Invariant Theory*, Springer.

OMa O'Malley, R. E., Jr.: 1974, *Introduction to singular perturbations*, Acad. Press.

OS Ohm, J., and Schneider, H.: 1964, *Matrices similar on a Zariski open set*, Math. Z., 85, pp. 373-381.

Qu Quillen, D.: *Projective modules over polynomial rings*, Inv. Math., 36, pp. 167-171.

RMY Rhodes, J. D., Marston, P. D., and Youla, D. C.: 1973, *Explicit solution for the synthesis of two-variable transmission-line networks*, IEEE Trans. CT, 20, pp. 504-511.

Ros Rosenbrock, H. H.: *Systems and polynomial matrices*, this volume.

Rou Rouchaleau, Y.: 1980, *Commutative algebra in systems theory*, this volume.

So 1 Sontag, E. D.: *Linear systems over commutative rings: a survey*, Ricerche di Automatica, 7, pp. 1-34.

So 2 Sontag, E. D.: *On first order equations for multi-dimensional filters*, preprint, Univ. of Florida.

So 3 Sontag, E. D.: 1978, *On split realizations of response maps over rings*, Inf. and Control, 37, pp. 23-33.

Sus Suslin, A.: 1976, *Projective modules over a polynomial ring*, Dokl. Akad. Nauk SSSR, 26.

Ta 1 Tannenbaum, A.: 1978, *The blending problem and parameter uncertainty in control*, Preprint.

Ta 2 Tannenbaum, A.: *Geometric invariants in linear systems*, in preparation.

Wa Wasow, W.: 1962, *On holomorphically similar matrices*, J. Math. Analysis and Appl., 4, pp. 202-206.

We Weiss, L.: 1969, *Observability and controllability*, In: Evangilisti (ed.), Lectures on observability and controllability (CIME), Editione Cremonese.

Yo Youla, D. C.: 1968, *The syntehsis of networks containing lumped and distributed elements*, In: G. Biorci (ed.), Networks and Switching Theory, Acad. Press, Chapter II, pp. 73-133.

ZW Zakian, V., and Williams, N. S.: 1977, *A ring of delay operators with applications to delay-differential systems*, SIAM J. Control and Opt., 15, pp. 247-255.

GRASSMANNIAN MANIFOLDS, RICCATI EQUATIONS AND FEEDBACK INVARIANTS OF LINEAR SYSTEMS

Clyde Martin*

Department of Mathematics
Case Western Reserve University
Cleveland, Ohio

ABSTRACT

The purpose of these lectures is to present a brief introduction of the role of Grassmannian manifolds in linear control theory. The Riccati equations of linear quadratic optimal control occur naturally as vector fields on the Lagrangian Grassmannian manifolds and exhibit some interesting topological behavior that is discussed in this paper. The feedback structure of linear systems can be deduced through a vector bundle structure on $\mathbb{P}^1(\mathbb{C})$ induced from the "natural bundle" structure on the Grassmannian manifold.

1. INTRODUCTION

My intent in these lectures is to examine the role that the Grassmannian manifolds play in the theory of linear systems. The value in systems theory is partially due to the mathematical intricacies of the manifolds and is probably partially accidental. I will look at this from two different prespectives. First we will examine the matrix Riccati equations that arise in the linear quadratic optimal control problem. We will show that these Riccati equations occur naturally as vector fields on a certain algebraic submanifold and that many properties of the Riccati equations can be easily deduced from the elementary properties of this submanifold. In particular, we will examine in some detail the case of the two-dimensional Riccati equation.

---

*Supported in part by NASA Grant #2384.

C. I. Byrnes and C. F. Martin, Geometrical Methods for the Theory of Linear Systems, 195-211.

The fact that the Grassmannian manifold enters linear system theory through the Riccati equation is fortunate, but seems to me not to be fundamental to the theory of linear systems.

The second occurrence which we will discuss is in the structure of linear controllable systems. In [HM] it was first pointed out that with every linear system and with every transfer function can be associated an algebraic curve in a complex Grassmannian. It was proven in [HM] that the McMillan degree of the transfer function was the intersection number of the curve with a hyperplane. It was shown in the same paper that the Kronecker indices could be obtained as the invariants of a certain vector bundle constructed from the system and that the vector bundle was a complete invariant for the system under the action of the feedback group. In [By] Byrnes showed by a very neat argument that the vector bundle was a complete invariant and further more that the correspondence between the category of vector bundles of $\mathbb{P}^1(\mathbb{C})$ and controllable systems was functorial. This fact leads to a very beautiful calculation that yields the dimensions of the feedback orbits as a special case of the Riemann-Roch theorem.

In these lectures we will mainly concentrate on the single input case and leave to the reader the details of the general case through the two original references [HM] and [By].

In the introductory paper of this volume the basic facts about Grassmannian manifolds are collected. For the purposes of this paper we need the representation as a set of all k dimensional subspaces of an n dimensional vector space and we need the fact that there is a natural action of $G\ell(n)$ on the manifold.

2. RICCATI EQUATIONS

We have seen that there is a natural action of $G\ell(V)$ on $\text{Grass}(p,V)$ and that in local coordinates (with respect to a canonical chart) the action takes the form of a linear fractional transformation. In this section we will examine the flows generated by one parameter subgroups of $G\ell(V)$ on $\text{Grass}(p,V)$. We will also examine the submanifold of Lagrangian subspaces of V. The symplectic group $Sp(n)$ will act naturally on this submanifold and we will see that this is the natural setting in which to study the Riccati equations that arise in the linear-quadratic optimization problems.

Let $A(t)$ be a one parameter subgroup of $G\ell(V)$ with infinitesimal generator B. Let $U(X)$ be a fixed element of $\text{Grass}(p,V)$ written with respect to some fixed canonical chart and let $A(t)$ and B be compatibly partitioned. Then one has that

$$A(t) \cdot U(X) = U((A_{21}(t) + A_{22}(t)X)(A_{11}(t) + A_{12}(t)X)^{-1}) \tag{2.1}$$

or just in terms of local coordinates

$$X \to (A_{21}(t) + A_{22}(t)X)(A_{11}(t) + A_{12}(t)X)^{-1} \tag{2.2}$$

and we have a flow (except for singularities) in the space of $p \times (n-p)$ matrices. Let $P(t) = A(t) \cdot X_0$ and calculate its derivative. Using the fact that B is the infinitesimal generator of $A(t)$ one finds that $P(t)$ is the solution of the matrix Riccati equation

$$\dot{P}(t) = B_{21} + B_{22}P(t) - P(t)B_{11} - P(t)B_{12}P(t)$$
$$P(0) = X_0. \tag{2.3}$$

Every matrix Riccati equation arises in this way as the action of a one parameter subgroup of $G\ell(V)$ on Grass(p,V). Thus one can reduce the study of Riccati equations to the study of flows on Grass(p,V).

Recall that a change of basis in V results in a change of charts in Grass(p,V). At the parameter level this is represented by a linear fractional transformation. Thus we have a similarity transformation of B is equivalent to a linear fractional transformation of $P(t)$. From this we can conclude the following-- the class of Riccati equations is closed under the action of the group of generalized linear fractional transformations and every Riccati equation is equivalent to a linear differential equation of the form

$$\dot{P} = A_1P + PA_2 + E \tag{2.4}$$

where E is zero except possibly for the entry in the lower left hand position. This simply says that every Riccati equation leaves invariant an open dense subset of its defining Grassmannian manifold. It also gives a characterization of the Riccati equations as the smallest set of equations that contains the linear equations and is invariant under linear fractional transformations. There is no really satisfactory characterization of the Riccati equations among the class of quadratic matrix equations. One is tempted to conjecture that the Riccati equations are those equations that remain quadratic under linear fractional transformations but it does not appear to be simple to prove.

To this point we have been quite general and nothing has particularly applied to linear systems theory. I want to now consider a setting for the problem of linear quadratic optimal control.

Let V be 2n-dimensional and let U and W be n-dimensional complementary subspaces with $V = U \oplus W$. Define a bilinear form of V by

$$\omega(x_1 + y_1, x_2 + y_2) = x_1'y_2 - y_1'x_2 \quad . \tag{2.5}$$

An n-dimensional subspace S of V is said to be a Lagrangian subspace if for every pair of elements say x and y in S

$$\omega(x,y) = 0 \ . \tag{2.6}$$

Such subspaces exist in great profusion for consider the space $U(A)$. Choosing two arbitrary elements from $U(A)$ say $u + Au$ and $v + Av$ we have

$$\begin{aligned} \omega(u + Au, v + Av) &= u'Av - u'A'v \\ &= u'(A-A')v \ . \end{aligned} \tag{2.7}$$

So we have that $U(A)$ is Lagrangian iff A is symmetric.

Let H be an element of $G\ell(V)$ that leaves ω invariant that is

$$\omega(Hx,Hy) = (x,y) \quad . \tag{2.8}$$

The set of such H is obviously a closed algebraic subgroup of $G\ell(V)$ and the group is denoted by $Sp(n)$. It is also clear that $Sp(n)$ leaves invariant the set of Lagrangian subspaces of V. In fact it acts transitively on the set of Lagrangian subspaces. A proof can be found in Hermann [HE]. The Lagrangian subspaces form an algebraic subset of $Grass(p,V)$ and is an invariant submanifold for $Sp(n)$.

Consider the linear system

$$\begin{aligned} \dot{x} &= Ax + Bu \qquad\qquad x(0) = x_0 \\ y &= C'x \end{aligned} \tag{2.9}$$

and the quadratic performance index

$$J(u) = \int_0^T x'CC'x + u'Ru \tag{2.10}$$

where R is positive definite and we assume that the system is controllable and observable. As is probably the best known fact in control theory the optimal solution is given by

$$u(t) = -R^{-1}P(t)x(t) \tag{2.11}$$

where $P(t)$ is the solution of the matrix Riccati equation

$$\dot{P}(t) = -P(t)A - A'P(t) + P(t)BR^{-1}B'P(t) - C'C$$
$$P(T) = 0 \tag{2.12}$$

The original proof is, of course, due to R. Kalman. If we attempt to directly minimize $J(u)$ we can construct the system

$$\begin{pmatrix} \dot{x} \\ \lambda \end{pmatrix} = \begin{pmatrix} A & -BR^{-1}B' \\ -C'C & -A' \end{pmatrix} \begin{pmatrix} x \\ \lambda \end{pmatrix} \tag{2.13}$$

with initial and final values, $x(0) = x_0$ and $\lambda(T) = 0$.

The matrix

$$H = \begin{pmatrix} A & -BR^{-1}B' \\ -C'C & -A' \end{pmatrix} \tag{2.14}$$

is exactly the infinitesimal generator of the one parameter group that generates $P(t)$. Unfortunately, the solution, for T finite, is time varying, but the limit as T goes to infinity is constant. Kalman's contribution was to show that the minimizing solution for $T = \infty$ was a constant feedback law and that the gain was determined by the positive definite equilibrium point of the Riccati equation.

The matrix H is symplectic and hence e^{Ht} leaves invariant the form defined by the matrix

$$W = \begin{pmatrix} 0 & I \\ -I & 0 \end{pmatrix} . \tag{2.15}$$

Thus we see that the natural place for the Riccati equation of optimal control to live is on the Lagrangian Grassmannian.

The equilibrium solutions of any Riccati equation can be identified as points in Grassmannian that are left invariant by the flow, i.e. $e^{Bt}X = X$. The following easy proposition was proven in [MA], but it was probably known earlier.

Proposition 2.1: The subspace X defines an equilibrium solution of the Riccati equation 2.3 iff $BX \subseteq X$.

Thus we have explicitly reduced the problem of determining the solutions of the algebraic Riccati equations to a problem in invariant subspaces. Potter essentially observed this in his now classic SIAM paper [PO].

In the case that H has 2n-distinct eigenvalues the problem of invariant subspaces is easily solved. One simply takes all combinations of n eigenvectors and the spaces spanned are the equilibrium points. If one seeks the real equilibrium points, the problem is a little harder, but not significantly, since one simply has to make sure that x is chosen then so is $\bar{x}$. This makes the counting problem tedious.

The more interesting problem is to determine which solutions lie in the Lagrangian submanifold. The following proposition is useful in this calculation.

Proposition 2.2: Let H be a symplectic matrix with distinct eigenvalues with nonzero real parts. If $Hx = \lambda x$ and $Hy = \gamma y$ then $w(x,y) = 0$ if and only if $\lambda + \gamma \neq 0$.

Proof. We can assume that

$$H = \begin{pmatrix} A & 0 \\ 0 & -A' \end{pmatrix}$$

where A is stable, since we have assumed H has distinct eigenvalues with nonzero real parts. Suppose $\lambda + \gamma \neq 0$. Then

$$\begin{aligned} (\lambda+\gamma)w(x,y) &= w(\lambda x,y) + w(x,y) \\ &= w(Hx,y) + w(x,Hy) \\ &= 0 \end{aligned}$$

since H is symplectic. Thus $w(x,y) = 0$. On the other hand, suppose $\gamma + \lambda = 0$. Then we calculate $Ax_1 = \lambda x_1$ $A'x_2 = -\lambda x_2$, $Ay_1 = -\lambda y_1$ $A'y_2 = \lambda y_2$. Either $x_1 = 0$ or $y_1 = 0$ since A is stable so assume $x_1 = 0$. This implies that $y_2 = 0$ and since we have assumed distinct eigenvalues we have that $x_2 = ay_1$ where $a \neq 0$. Thus we calculate

$$\begin{aligned} w(x,y) &= x_1'y_2 - y_1'x_2 \\ &= 0 - ax_2'x_2 \\ &\neq 0 . \end{aligned}$$

Thus, we have proved the proposition.

Proposition 2.2 enables us to count the Lagrangian solutions of the algebraic Riccati equation in the case of distinct eigenvalues. Potter in [PO] observed that the equilibrium point associated with the eigenvectors with eigenvalues with positive real parts was symmetric by essentially this argument.

Kalman was able to prove that equation 2.12 has a positive definite equilibrium in the canonical chart. At the present I don't know how to prove that fact with the methods used here. In fact, it seems unlikely that the global methods of Grassmannian manifolds can reproduce this essentially local result.

Also recall that Kalman's proof was a proof concerning the asymtotic behavior of the Riccati equation. He proved that if equation 2.12 is integrated backwards in time that the solution always converges to the positive definite Lagrangian solution. This result simply doesn't follow from the global methods.

One can show, at least in the generic case, that there is an invariant submanifold of the Lagrangian Grassman manifold of codimension two that contains neither the positive definite nor the negative definite solution of the algebraic Riccati equation.

Let H have distinct eigenvalues and let $X(+)$ be the space spanned by eigenvectors corresponding to eigenvalues with positive real part and $X(-)$ be the space spanned by the eigenvectors associated with eigenvalues with negative real part. These are both real Lagrangian subspaces and, furthermore, they are disjoint and span. Therefore, they determine a chart on the manifold. Now H written with respect to this chart is block diagonal and has the form

$$H = \begin{pmatrix} H_1 & 0 \\ 0 & H_2 \end{pmatrix} .$$

Since H is symplectic we have that $H_2 = -H_1'$. At this point we only know that the eigenvalues of H_1 are the negatives of the eigenvalues of H_2. Actually there are two charts defined by $X(+)$ and $X(-)$ depending on the order in which they are taken. Both charts are invariant under the action of the one parameter group e^{Ht} and both charts contain exactly one equilibrium point. One can show by a direct calculation that one equilibrium point is stable and the other totally unstable. Since both charts are invariant under the action of the one parameter group generated by H their union is invariant and hence the complement of the union is invariant and of codimension two.

Let's leave the general case and consider a very special case--namely the case when $n = 2$. Then H is 4×4 and $Grass(2,\mathbb{R}^4)$ is 4 dimensional. Suppose first that H has complex eigenvalues $\lambda, \bar{\lambda}, -\lambda, -\bar{\lambda}$ and eigenvectors $x,\bar{x},y,\bar{y}$, and assume $Re\lambda > 0$. Now we will construct explicitly the invariant two dimensional submanifold.

Let $V(\theta,\gamma) = \langle \cos\theta \, Re(x) + \sin\theta \, Im(x), \cos\gamma \, Im(y) + \sin\gamma \, Imy \rangle$. Since $x,\bar{x},y,\bar{y}$ are the eigenvectors for H and easy calculation shows that the set of $V(\theta,\gamma)$'s is an invariant submanifold for e^{Ht}. I claim there is no equilibrium point in $V(\theta,\gamma)$, for if there were it would imply that $HV(\theta,\gamma) = V(\theta,\gamma)$ for some θ and γ. One sees that this would lead to either x and $\bar{x}$ being in $V(\theta,\gamma)$ or y and $\bar{y}$. This is an obvious contradiction. Thus $V(\theta,\gamma)$ does not intersect canonical charts associated with $\langle x,\bar{x}\rangle$ and $\langle y,\bar{y}\rangle$ and hence we have a two dimensional submanifold contained in the complement of the union of the stable and unstable chart. The set of $V(\theta,\gamma)$'s is closed and we can conclude that they include the entire complement.

The next thing to consider is whether or not any of the $V(\theta,\gamma)$'s are Lagrangian. The Lagrangian Grassmannian is a three dimensional invariant submanifold of $Grass(2,\mathbb{R}^4)$ and the set of $V(\theta,\gamma)$'s is two dimensional so one would expect for dimension reasons that they intersect. (I would imagine that this could be proved by studying the cohomology.) In fact, if one calculates the conditions for $V(\theta,\gamma)$ to be Lagrangian using the two from w one gets the following equation

$$\begin{aligned}\cos\theta\cos\gamma\, w(Rex,Rey) &+ \cos\theta\sin\gamma\, w(Rex,Imy)\\ &+ \sin\theta\cos\gamma\, w(Imx, Rey)\\ &+ \sin\theta\sin\gamma\, w(Imx,Imy) = 0\end{aligned}$$

that must be satisfied by θ and γ if $V(\theta,\gamma)$ is Lagrangian. It's either satisfied identically or we can calculate θ as a function of γ. Let M denote the set of Lagrangian subspaces of the form $V(\theta,\gamma)$ and let V be some fixed element of M. There exists a $t_o > 0$ such that e^{Ht_o} has real eigenvalues.

It is easy to see that Rex, Imx, Rey, Imy are eigenvectors for e^{Ht_0} and hence that $e^{Ht_0}V = V$. M is invariant and $HV \neq V$ thus $e^{Ht}V \neq V$ for all t. We conclude that M is a periodic orbit of e^{Ht} contained in the Lagrangian Grassmannian. [Jan Willems has given me a very elegant proof that this orbit exists in the space of symmetric matrices using classical Riccati techniques.] Also note that this is the only periodic orbit in the Lagrangian Grassmannian.

This special case turns out to be not so special for the existence of periodic orbits can, in general, be reduced to the 2×2 case. A forthcoming paper by Hermann and Martin will discuss this in some generality.

When H has real distinct eigenvalues the situation is different. It is easy to show that H has four Lagrangian equilibrium points and no periodic orbits.

The situation when H has multiple eigenvalues is complicated. In general there are algebraic sets of equilibria and the geometry is not well understood.

The following theorem can be proven by direct calculation.

Theorem 2.3. Let H be an element $Sp(4,\mathbb{R})$ with distinct eigenvalues with nonzero real part. Then the vector field on the Lagrangian points of $Grass(2, \mathbb{R}^4)$ defined by H satisfies the Morse-Smale conditions.

For the real eigenvalue case this result was proven in Schneider [Sc]. The complex case is proven by calculating the Poincaré map of the periodic orbit and directly verifying the other conditions. A general result will appear in Hermann and Martin.

3. LINEAR SYSTEMS

In this section we continue examining the role of the Grassmannian manifolds in system theory. The basic material comes from Hermann-Martin [HM] and from Byrnes [By]. However, we will take a more pedestrian pace and try to clarify some of the issues raised in the two mentioned papers. I'm indebted to Roger Brockett for pointing out the construction of the sections of the line bundles. We will begin by considering the easiest case--the case of a single input system.

A. Single Input Linear Systems

Consider the system

$$\dot{x} = Ax + Bu \tag{3.1}$$

where x is n-vector, A is the shift matrix and b is the n'th standard unit vector. The system 3.1 is controllable by any one of the many definitions of controllability. However, it will be convenient to use the fact that controllability is equivalent to the pencil

$$[A-sI,b] \tag{3.2}$$

having rank n for all s. We introduce homogeneous coordinates into the pencil in order to discuss the behavior at ∞ and obtain

$$\alpha(s,\lambda) = [\lambda A-sI, b] \tag{3.3}$$

for all $x,\lambda \in \mathbb{C}^2\backslash\{(0,0)\}$. Again controllability we have that the rank of $\alpha(s,\lambda)$ is equal to n. Since the rank of α is n the kernel has demension one for all pairs (s,λ) in the punctured plane. In this way we obtain a map

$$\tau(s,\lambda) = \text{Ker}[\lambda A-sI,\lambda b] \tag{3.4}$$

from $\mathbb{C}^2\backslash\{(0,0)\}$ to the Grassmannian of lines in $\mathbb{C}^{n+1}$. Of course, this is just projective n space, $\mathbb{P}^n$. The function $\alpha(s,\lambda)$ is not constant on lines, but it is easy to see that $\tau(s,\lambda)$ is and so, in fact, τ is a map from the Grassmannian of lines in $\mathbb{C}^2$ to $\mathbb{P}^n$.

We can calculate directly that $\tau(s,\lambda)$ is generated by a homogeneous polynomial of degree n. That is, of course, just the Kronecker polynomial

$$p(s,\lambda) = \lambda^n e_1 + \lambda^{n-1} s e_2 + \ldots + \lambda s^{n-1} e_n + s^n e_{n+1} \tag{3.5}$$

where the e_i are the standard unit vectors in $\mathbb{C}^{n+1}$. Again $p(s,\lambda)$ is not constant on lines, but the one dimensional vector space generated by $p(s,\lambda)$ is constant on lines in $\mathbb{C}^2$. We then have alternatively that

$$\tau(s,\lambda) = \{\alpha p(s,\lambda):\alpha \in \mathbb{C}\} . \tag{3.6}$$

We can decompose $\mathbb{C}^{n+1}$ as $X \oplus U$ where X is the n-dimensional state space and U is one dimensional input space.

This gives, as in the introduction, a canonical chart for $\mathbb{P}^n$--namely the set S_X of lines in $\mathbb{C}^{n+1}$ that have zero intersection with X. Note that $\tau(s,\lambda)$ is in S_X except when $s^n = 0$. A better way to phrase this is to say that $\tau(s,\lambda)$ intersects the complement of the canonical chart iff $s^n = 0$.

Let $X_{\bar{a}}$ be the hyper plane in $\mathbb{C}^{n+1}$ defined by

$$X_{\bar{a}} = \{(x_1,\dots,x_n,u):a_1x_1 + \dots + a_nx_n + u = 0\}$$

and let $S_{\bar{a}}$ be the canonical chart defined by $X_{\bar{a}} \oplus U$. Note that $S_0 = S_X$. The map $\tau(s,\lambda)$ intersects the complement of $S_{\bar{a}}$ iff $p(s,\lambda)$ intersects $X_{\bar{a}}$ which is iff $a_1\lambda^n + a_2s\lambda^{n-1} + \dots + a_ns\lambda^{n-1} + s^n = 0$. If $\lambda = 0$ then $s = 0$ which contradicts the definition τ. So we can assume $\lambda = 1$ and we have that $\tau(s,\lambda)$ intersects the complement of $S_{\bar{a}}$ iff

$$s^n + a_ns^{n-1} + \dots + a_2s + a_1 = 0 \quad . \tag{3.7}$$

Counting multiplicity we have that $\tau(s,\lambda)$ has exactly n points of intersection with every canonical hypersurface in $\mathbb{P}^n$.

The feedback group associated with the system 3.1 is the group generated by the transformations

$$\begin{array}{lll} \text{i)} & x \to Rx & p \in G\ell(n) \\ \text{ii)} & u \to ru & r \in G\ell(1) \\ \text{iii)} & u \to k'x + u & k' \in L(\mathbb{R}^n,\mathbb{R}) \quad . \end{array} \tag{3.8}$$

As is well known, this group has a representation of degree $n + 1$, namely by the matrices of the form

$$\begin{pmatrix} P & 0 \\ k' & r \end{pmatrix} \in G\ell(n+1) \tag{3.9}$$

Since the $G\ell(n+1)$ acts naturally on $\mathbb{P}^n$ by transforming lines in $\mathbb{R}^{n+1}$ we obtain an action of the feedback group on $\mathbb{P}^n$. Let X_0 be defined as before and consider the action of

$$\alpha = \begin{pmatrix} I & 0 \\ -a' & 1 \end{pmatrix}$$

on X_0. X_0 is the hyperplane anniliated by the vector $(\bar{0},1)$. We can calculate exactly the form of αX_0. Let $(x,u) \in X_0$ then $\alpha(x,u) = (x,-a'x + u)$. Suppose (b,s) is dual to αX_0 then $x'b + (-a'x + u)s = 0$ for all x and u and hence $b = a$ and $s = 1$. Thus $\alpha X_0 = X_a$.

Now, on the other hand, α acts on the system defined by 3.1 to yield the system

$$\dot{x} = (A - ba')x + bu \ . \tag{3.10}$$

We define as for 3.1 $\alpha_a(s,\lambda)$ and $\tau_a(s,\lambda)$. We again do a simple calculation to determine the necessary and sufficient conditions for

$$\tau_a(s,\lambda) = \text{Ker } [\lambda(A\text{-}ba')\text{-}sI,\lambda b]$$

to intersect X_0. Now $\tau_a(s,\lambda)$ intersects X_0 iff there is no element in $\tau_a(s,\lambda)$ of the form $(x,1)$. Hence iff $\det\lambda(A\text{-}ba')\text{-}sI) = 0$ but

$$A\text{-}ba' = \begin{bmatrix} 0 & 1 & 0 & \dots & 0 \\ 0 & 0 & 1 & \dots & 0 \\ a_1, & & \dots & & -a_n \end{bmatrix}$$

So we have that $\tau_a(s,\lambda)$ intersects X_0 iff equation 3.7 is satisfied. We obtain in this way the beautiful little theorem of Byrnes.

Let Σ denote the image of $\mathbb{P}^1(\mathbb{C})$ determined by 3.1 and $\alpha \cdot \Sigma$ the image determined by 3.10.

<u>Theorem 3.1</u> $\alpha \cdot \Sigma \cap S_0 = \Sigma \cap \alpha \cdot X_0$. Intuitively, we obtain the same intersections if we fix the system and move the hyperplane or if we fix the hyperplane and move the system. From our point of view the critical matter is that the action of the feedback group in the space of controllable systems is the same as the action of $G\ell(n)$ in the Grassmannian.

There are at least two vector bundles on $\mathbb{P}^n(\mathbb{C})$ that can be considered to arise naturally. The first bundle is the bundle whose fibre at the point $x \in \mathbb{P}^1(\mathbb{C})$ is subspace that defines x. I'm told that this bundle is considered to be the "natural" bundle by differential geometers. The other bundle is a bundle whose fibre at x is the dual of the vector space x. This space (I'm told) is considered to be the "natural" bundle by algebraic geometers. In Hermann and Martin the first bundle was used and in Byrnes the second is used. Since I have already presented the case in Hermann and Martin, I will consider here the second case.

Choose the basis for $\mathbb{C}^{n+1}$ for which the equations 3.1 are written and denote it by $e_1,\ldots,e_{n+1}$. Denote the vector bundle on $\mathbb{P}^n(\mathbb{C})$ by η. Define a section s_i of η by the following: for $a \in x$ let

$$s_i(x)(a) = \langle e_i, a\rangle$$

where $\langle\,,\rangle$ is the standard inner product on $\mathbb{C}^{n+1}$. The sections s_i are holomorphic and independent. Now τ defines an embedding of $\mathbb{P}^1(\mathbb{C})$ into $\mathbb{P}^n(\mathbb{C})$ and we can ask to classify the bundle on $\mathbb{P}^1(\mathbb{C})$ that is defined by restricting η to the image of τ.

Let Σ_1 be a system obtained from Σ by feedback. As we have seen there is an element of $G\ell(n+1)$ that transforms the image of τ of Σ_1 into the image of τ of Σ_2. Since the map is defined by a linear map on the points of $\mathbb{P}^1(\mathbb{C})$ it induces an automorphism of η and an isomorphism of the restricted bundles. The calculation is straight forward and can be found in Hermann and Martin.

Let's consider the restriction of the sections s_i to the bundle η. I claim they are linearly independent for if not then there are constants α_i such that $\Sigma\alpha_i s_i \equiv 0$ on the image of τ. This would imply that

$$\Sigma\alpha_i s_i'(\tau(s,\lambda))(p(s,\lambda)) \equiv 0$$

as a function of s,λ but

$$\begin{aligned}\Sigma\alpha_i s_i(\tau(s,\lambda))(p(s,\lambda)) &= \langle\Sigma\alpha_i e_i, p(s,\lambda)\rangle \\ &= \Sigma\alpha_i \lambda^{n-i+1} s^{i-1} \\ &\equiv 0\,.\end{aligned}$$

Thus all of the α_i's are zero. So the space of sections

$H^o(\mathbb{P}^1(\mathbb{C}),\eta)$ of γ has dimension at least $n + 1$. Byrnes has shown in [ ] that $H^1(\mathbb{P}^1(\mathbb{C}),\eta)$ is zero. If we apply the Riemann-Roch Theorem we obtain

$$\begin{aligned} H^o - H^1 &= c(\eta) + 1 \\ &= \eta + 1 \end{aligned}$$

and hence the dimension of H^o is exactly $n + 1$. (Using the Riemann-Roch Theorem here is definitely overkill.)

Suppose we have two systems Σ_1 and Σ_2 and that we have constructed bundles over $\mathbb{P}^1(\mathbb{C})$ for each, V_1 and V_2. Let V_1 be bundle isomorphic to V_2 by a bundle map T.

$$\begin{array}{ccc} T: V_1 & \simeq & V_2 \\ & \searrow \quad \swarrow & \\ & \mathbb{R}(\mathbb{C}) & \end{array}$$

The space $\mathbb{P}_1(\mathbb{C})$ has a distinguished point which we will denote by ∞. The fibre over ∞ we will denote by U.

The map T induces a map T_* from $H^o(\mathbb{P}^1(\mathbb{C}),V_1)$ to $H^o(\mathbb{P}^1(\mathbb{C}),V_2)$ and T_* is linear. We define a map ρ_∞ from $H^o(\mathbb{P}^1(\mathbb{C}),V_j)$ to U by $\rho_\infty(s_i) = s_i(\infty)$, or ρ_∞ is just the evaluation map at ∞. So we have a canonical map from H^o to U. We can write then

$$H^o(\mathbb{P}^1(\mathbb{C}),V_j) \approx U \oplus X_j$$

where X_j is n dimensional. Now T_* preserves U and hence T_* has the form

$$\begin{pmatrix} * & 0 \\ * & * \end{pmatrix}$$

and so is an element of the feedback group. I leave to the reader the easy construction that shows that T_* takes Σ_1 to Σ_2. In this way we obtain the theorem that the bundle is a complete invariant.

B. Multi-input Systems

The multi-input case is almost identical to the single input case. Obtaining the dimension of the system as a topological invariant is exactly the same except that we now must work in the Grassmannian of m planes in $\mathbb{C}^{n+m}$, $\mathrm{Grass}(m,\mathbb{C}^{n+m})$, instead of $\mathbb{P}^n(\mathbb{C})$.

We can construct the bundle associated with the system as before. We use the natural bundle on the Grassmannian whose fibre at a point x is the dual of the space s and we denote the bundle by η. As before, we construct $n+m$ "canonical" sections of choosing a basis $\{e_i\}$ in $\mathbb{C}^{n+m}$ and we define for $a \in x$

$$s_i(x)(a) = \langle e_i, a \rangle$$

Exactly as before, we can show that the bundle defined by restricting η to the image of $\mathbb{P}^1(\mathbb{C})$ has at least $n+m$ independent sections obtained by restricting the above sections.

In Hermann and Martin it was shown that bundle over $\mathbb{P}^1(\mathbb{C})$ is a complete invariant by the following argument. First, it was shown that if two systems are feedback equivalent then the feedback transformation generates a bundle isomorphism. The proof is the same regardless of the dimension of the input space. Then is was shown that if the system is in Brunovsky canonical form the bundle on $\mathbb{P}^1(\mathbb{C})$ decomposes as the direct sum of line bundles and the degree of the line bundle corresponds to a Kronecker invariant. By the theorem of Grothendiek vector bundles on $\mathbb{P}^1(\mathbb{C})$ decompose uniquely into the direct sum of line bundles and the set of degrees is a complete set of isomorphism invariants. Thus if two bundles, which are determined from systems, are isomorphic then they have the same Grothendiek invariants, the same feedback invariants, and so determine systems which are feedback equivalent. The proof is completely satisfactory in that it is proved that the bundle is a complete invariant. However, the proof is nonfunctorial in that the bundle isomorphism does not produce an element of the feedback group.

In Byrnes the same theorem is proved, but his proof is functorial. The basic difference in the two proofs comes from the fact that in Hermann-Martin the bundle is the bundle that comes from the "natural bundle" whose fibre at a point on the Grassmannian is space x. This bundle has no sections. Hence, there is no natural way to construct the state and input space from the bundle.

From the proof of Byrnes comes a very clever construction. In [Br] Brockett considered the problem of calculating the dimensions of the orbits of the feedback group. His proof was very complicated and involved detailed counting arguments based on the Brunovsky canonical forms. The point of the proof was to construct the stabilizer of the Brunovsky canonical form.

From the proof of Byrnes we know that the stabilizer of the bundle is the same as the stabilizer of the system and, hence, we need only calculate the group of automorphisms of the vector bundle V. So let T be an automorphism of V. Then T is a section of the group bundle over $\mathbb{P}^1(\mathbb{C})$, $\mathrm{Aut}(V)$, which is a subbundle of $\mathrm{End}(V)$. The dimension of the space of sections of $\mathrm{Aut}V$ is the same as the space of sections of $\mathrm{End}V$ and, hence to calculate the dimension of the stabilizer of a system Σ it is only necessary to calculate the dimension of the space of sections of $\mathrm{End}(V)$. But $\mathrm{End}V \approx V \otimes V^*$, where V^* is the dual bundle of V. We know that V is the direct sum of line bundles, $V \approx \Sigma \mathcal{O}(n_i)$ where $\mathcal{O}(n_i)$ has $n_i + 1$ sections $\Sigma n_i = n$ and i runs from 1 to m. Likewise $V^* \approx \Sigma \mathcal{O}(-n_i)$ and $\mathcal{O}(-n_i)$ has no sections at all. Calculating the tensor product we have that

$$\mathrm{End}V \approx \sum_{\substack{i_{ij}=1}}^{m} \mathcal{O}(n_i - n_j) \ .$$

Calculating the dimension of the space of sections we note that there is a contribution from $\mathcal{O}(n_i - n_j)$ iff $n_i - n_j \geq 0$ and hence we obtain

$$\dim \Gamma(\mathrm{End}V) = \sum_{n_i \geq n_j} n_i - n_j + 1$$

Hence the dimension of the orbit of Σ is

$$n^2 + nm + n^2 - \sum_{n_i \geq n_j} n_i - n_j + 1 \ .$$

Having done this calculation using canonical form, I'm tempted to believe that the above formula is justification enough for the introduction of vector bundles into linear system theory.

REFERENCES

[1] Brockett, R.: 1977, *The geometry of the set of controllable linear systems*, Research Report of Automatic Control Laboratory, Nagoya University, v. 24, pp. 1-7.

[2] Byrnes, C. I.: 1978, *On the control of certain deterministic, infinite-dimensional systems by algebro-geometric techniques*, Amer. Jour. Math., vol. 100, pp. 1333-1381.

[3] Hermann, Robert: 1979, *Cartanian Geometry, Nonlinear Waves and Control Theory, Part A*, Math Sci Press, Brookline, Massachusetts.

[4] Hermann, R., and Martin, C.: 1978, *Applications of algebraic geometry to systems theory: The McMillan degree and Kronecker indices of transfer functions as topological and holomorphic system invariants*, SIAM J. Control and Optimization, vol. 16, no. 5.

[5] Potter, J. E.: 1966, *Matrix quadratic solutions*, SIAM J. Appl. Math., vol. 14, pp. 496-501.

[6] Schneider, C. R.: 1973, *Global aspects of the matrix Riccati equation*, Math. Syst. Theory, vol. 7, pp. 281-286.

COMMUTATIVE ALGEBRA IN SYSTEM THEORY

Y. Rouchaleau

E.N.S. Mines de Paris
Centre de Mathematiques Appliqueés
Valbonne, France

PART I

Suppose one wants to describe a mechanical linkage made of rigid bars

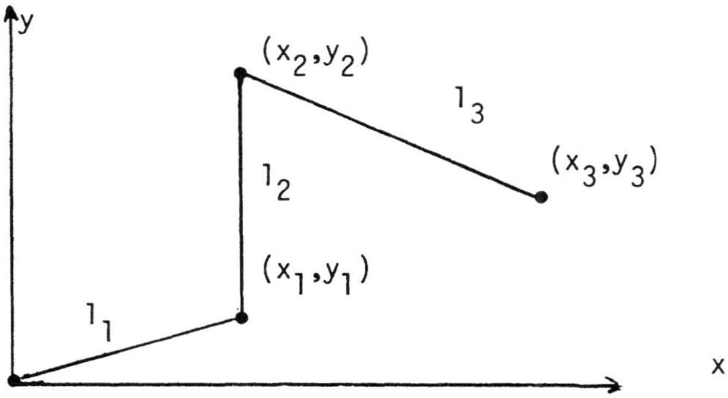

The coordinates of the end points of each bar must satisfy a set of equations

$$(x_1^2 + y_1^2) = l_1^2$$

$$(x_2 - x_1)^2 + (y_2 - y_1)^2 = l_2^2$$

$$(x_3 - x_2)^2 + (y_3 - y_1)^2 = l_3^2$$

C. I. Byrnes and C. F. Martin, Geometrical Methods for the Theory of Linear Systems, 213-231.

So the coordinates of the end points in $\mathbb{R}^6$ cannot take any arbitrary values; they must lie on a surface described by the above algebraic equations.

It is the necessity of finding a suitable framework in which to study systems evolving under similar constraints, but including also other classes previously studied--linear systems, internally or externally bi-linear systems--which has led to investigating discrete-time polynomial systems. The most convenient language for expositing this concept being that of algebraic geometry (or commutative algebra), we shall now introduce the relevant concepts.

1. Algebraic Sets

Let k be an infinite field, and $k[T_1,T_2,\ldots,T_n] = k[T]$ the ring of polynomials in n unknowns over k. When $n = 0$, we shall define $k[T]$ to be the base field k itself. Having assumed the field to be infinite, we shall not make any distinction between formal polynomials and polynomial functions; if $\mathcal{F} \in k[T]$ and $x \in k^n$, then $\mathcal{F}(x)$ will be the specialization of $\mathcal{F}$ at x.

Definition 1.1. *Let Γ be an arbitrary set of polynomials in $k[T_1,\ldots,T_n]$; then we define*

$$V(\Gamma): = \{x \in k^n / \mathcal{F}(x) = 0, \ \forall \ \mathcal{F} \in \Gamma\}$$

Similarly, if S is a set of points in k^n, let

$$I(S): = \{\mathcal{F} \in k[T_1,\ldots,T_n] / \mathcal{F}(x) = 0, \ \forall \ x \in S\}$$

We shall call $V(\Gamma)$ an algebraic set.

Lemma 1.2. *Let $<\Gamma>$ be the ideal generated by Γ. Then*

$$V(\Gamma) = V(<\Gamma>)$$

Proof. Let $\mathcal{F}$, $\mathcal{G} \in \Gamma$; if $\mathcal{F}(x) = \mathcal{G}(x) = 0$, then $a\mathcal{F}(x) - b\mathcal{G}(x) = 0$ for all $a,b \in k$; so $V(\Gamma) \subset V(<\Gamma>)$. But $\Gamma \subset <\Gamma>$ implies $V(\Gamma) \supset V(<\Gamma>)$. Hence the result. ⌑

Lemma 1.3. *For any set $\Gamma \subset k^n$ there exists a set Γ_0 containing a finite number of polynomials such that $V(\Gamma) = V(\Gamma_0)$.*

Proof. By lemma 1.2, $V(\Gamma) = V(<\Gamma>)$. But we saw during the preliminary lectures that $k[T]$ is noetherian, hence that the ideal $<\Gamma>$ is finitely generated by, say, $\{\mathcal{F}_1,\ldots,\mathcal{F}_n\} = \Gamma_0$. So

$$V(\Gamma) = V(<\Gamma>) = V(<\Gamma_0>) = V(\Gamma_0)$$ ⌑

This result, a direct consequence of Hilbert's theorem, is of great practical importance, since it will later on turn out to be the key to establishing that certain system theoretic concepts can be checked in a finite way.

It is easy to check that

$$V(I(S)) \supset S \quad \text{and} \quad I(V(\Gamma)) \supset \Gamma \ ;$$

these inclusions show that we have a Galois connection, and the operator

$$S \to V(I(S)) = \bar{S}$$

is a closure operator: $\bar{\bar{S}} = \bar{S}$.

Furthermore, $V(<0>) = k^n$ and $V(k[T]) = \emptyset$: k^n itself and the empty set are algebraic sets. Also, if

$$S_1 = V(<\Gamma_1>) \quad \text{and} \quad S_2 = V(<\Gamma_2>),$$

then

$$S_1 \cup S_2 = V(<\Gamma_1 \cdot \Gamma_2>) \quad \text{and} \quad S_1 \cap S_2 = V(<\Gamma_1 \cup \Gamma_2>)$$

so a finite union and an intersection of algebraic sets are algebraic sets. This justifies the following:

Definition 1.4. *The closure operator* $S \to V(I(S))$ *induces a topology on* k^n *in which the closed sets are the algebraic sets. This topology is called the Zariski topology. This will also induce a topology on the algebraic sets of* k^n.

Note that this topology is very weak (T_0 only !); indeed, any two open sets have a non-empty intersection. A basis of open sets is given by sets of the form

$$V_{\mathscr{F}} = \{x \in k^n / \mathscr{F}(x) \neq 0\}$$

Weak Nullstellensata 1.5. *Let* k *be an algebraically closed field. Then the maximal ideals in the ring* $k[T]$ *are the ideals*

$$<T_1 - s_1, T_2 - x_2, \ldots, T_n - x_n> \quad , \quad x_1, \ldots, x_n \in k$$

Proof. See Mumford [Introduction to Algebraic Geometry, p. 9]. ⌑

Nullstellensatz (Hilbert) 1.6. $I(V(<\Gamma>))$ *is the radical of* $<\Gamma>$, *i.e.*

$$I(V(<\Gamma>)) = \{\mathscr{F} \in k[T]/\exists n,\ \mathscr{F}^n \in <\Gamma>\}$$

Proof. See Mumford [Ibid., Chapter 1, §2, Th. 1]. ⌑

Lemma 1.7. *The radical of an ideal* $<\Gamma>$ *is the intersection of all prime ideals containing* $<\Gamma>$.

Proof. It is clear by induction that every prime ideal containing $<\Gamma>$ must contain its radical $\text{rad}(\Gamma)$.

Let now $\mathscr{F}$ be an element of $k[T]$ which is not in $\text{rad}(\Gamma)$, and let S be the set of all ideals I such that $\mathscr{F}^n \notin I$ for any integer n. This set is not empty, since it contains $<\Gamma>$. Let p be a maximal element of S. p is a prime ideal.

Indeed, the ideal

$$q = \{g/\exists n,\ g\mathscr{F}^n \in p\}$$

does not contain $\mathscr{F}^n$ (by definition of S), hence is in S; but it contains p, which is maximal, so they must be equal. Let, therefore, $gh \in p$, $g \notin p$; the ideal $<g> + p$ is strictly larger than p. Thus $\mathscr{F}^n \in <g> + p$, and

$$\mathscr{F}^n = rg + m, \quad r \in k[T], \quad m \in p$$

$$\Rightarrow \mathscr{F}^n h = rgh + mh \in p$$

$$\Rightarrow \quad h \in q = p$$

p is therefore a prime ideal. So if $\mathscr{F} \in \text{rad}(\Gamma)$, then there is a prime ideal which does not contain $\mathscr{F}$, yet contains Γ. ⌑

A radical is therefore an intersection of prime ideals. In fact, if we get rid of all the unnecessary elements in this intersection (i.e., those which include another element), then the remaining ideals turn out to be finite in number and uniquely defined. Furthermore, these prime ideals have a clear geometric interpretation.

Definition 1.8. *An algebraic set* $V \subset k^n$ *is irreducible if there do not exist two algebraic sets* V_1 *and* V_2 *such that* $V_1, V_2 \neq V$ *and* $V = V_1 \cup V_2$. *An irreducible algebraic set is called an affine variety.*

Theorem 1.9. V is irreducible if and only if $I(V)$ is a prime ideal.

Proof. Recall that $V_1, V_2 \subset V \Leftrightarrow I(V_1), I(V_2) \supset I(V)$, and $V = V_1 \cup V_2 \Leftrightarrow I(V) = I(V_1) \cap I(V_2)$.

Thus if $I(V)$ is not irreducible, there exist $\mathcal{F}_1 \in I(V_1)$, $\mathcal{F}_2 \in I(V_2)$, $\mathcal{F}_1, \mathcal{F}_2 \notin I(V)$ yet $\mathcal{F}_1 \mathcal{F}_2 \in I(V_1) \cap I(V_2) = I(V)$: $I(V)$ would not be prime.

Conversely, if $I(V)$ is not prime, there exist $\mathcal{F}, \mathcal{G} \notin I(V)$ such that $\mathcal{F}\mathcal{G} \in I(V)$; so taking $V_1 = V(<\mathcal{F}> \cup I(V))$, $V_2 = V(<\mathcal{G}> \cup I(V))$ shows that V is not irreducible. ¤

2. Morphisms of Algebraic Sets

To the algebraic sets can be associated in a very natural manner suitable morphisms, together with which they will form a category.

Definition 2.1. *A polynomial function from an algebraic set* V *into the base field* k *is the restriction to* V *of a polynomial function from* k^n *into* k.

Definition 2.2. *Let* $V \subset k^n$, $W \subset k^m$ *be algebraic sets. Then* $\mathcal{F}: V \to W$ *will be called a morphism if and only if there exist* m *polynomial functions* $\mathcal{F}_1, \ldots, \mathcal{F}_m$ *such that*

$$\mathcal{F}(x) = (\mathcal{F}_1(x), \ldots, \mathcal{F}_m(x)) \;, \; \forall \; x \in V \;.$$

Of particular interest will be the morphisms from the algebraic set V into the ground field k (i.e. the polynomial functions). Each such function comes from a polynomial in $k[T]$, and two polynomials give rise to the same morphism from V into k if and only if they agree on all of V, i.e., if and only if their difference is 0 on V. Thus

Lemma 2.3. *The set of polynomial functions from* V *into* k *is isomorphic to* $k[T]/I(V)$.

Proof. This results from the considerations above, and the fact that the set of polynomials which are 0 on all of V is exactly $I(V)$. ¤

Definition 2.4. *The* k*-algebra* $A(V)$ *of polynomial functions from* V *into* k *is called the coordinate ring of* V.

It is clear that to a morphism $\mathcal{F}: V \to W$ of algebraic sets can be associated a co-morphism of k-algebras $A(\mathcal{F}) : A(W) \to A(V)$ defined by:

$$\forall \mathcal{G} \in A(W) \;, \; (A(\mathcal{F}))(\mathcal{G}) := \mathcal{G} \circ \mathcal{F}$$

We obtain in this manner a category of coordinate rings dual to that of algebraic sets.

This duality is in fact very strong: the knowledge of V is enough to define A(V). But, conversely, we can reconstruct V from A(V)! So both categories contain essentially the same information.

Lemma 2.5. *The coordinate ring of an algebraic set is a finitely generated* k-*algebra with no nilpotent elements, provided the field is algebraically closed.*

Proof. It is the quotient of a polynomial ring in a finite number of unknowns, and $\mathcal{F}^n \in I(V) \Rightarrow \mathcal{F} \in I(V)$, so

$$\mathcal{F}^n \equiv 0 \quad (I(V)) \Rightarrow \mathcal{F} \equiv 0 \quad (I(V)) . \qquad \text{⌑}$$

Theorem 2.6. *If* A *is a finitely generated* k-*algebra with no nilpotent elements, there exists an algebraic set of which it is the coordinate ring.*

Proof. Let $u_1,\ldots,u_n$ be a set of generators of A. For any k-homomorphism Ψ from A into k, let $x_1 = \Psi(u_1),\ldots,x_n = \Psi(u_n)$. Then the set V of points $x = (x_1,\ldots,x_n)$ of k^n obtained for all homomorphisms is such that A = A(V). ⌑

The image of an algebraic set by a morphism is not necessarily an algebraic set. Indeed, look at the projection of the surface of equation $yz-1 = 0$ of k^3 parallel to the z-axis. The morphism is $(\mathcal{F}_1 = x,\ \mathcal{F}_2 = y,\ \mathcal{F}_3 = 0)$, and the image the set $y \neq 0$.

The latter is an open set in the Zariski topology, hence not an algebraic set. However,

Lemma 2.7. *Morphisms are continuous for the Zariski topology.*

Proof. Direct application of the definitions. ⌑

Lemma 2.8.

(i) $\overline{\mathcal{F}(V)} = W \Leftrightarrow A(\mathcal{F})$ is *injective*

(ii) $\mathcal{F}$ *injective* $\Leftarrow A(\mathcal{F})$ *is surjective.*

Proof. $A(\mathcal{F})$ injective means that $\mathcal{G}\circ\mathcal{F} = 0 \Leftrightarrow \mathcal{G} = 0$, $\forall\ \mathcal{G} \in A(W)$, hence the result. ⌑

A counter example to the missing implication is the following: The morphism $\mathcal{F} : y \to y$ from $\{x,y/x = 0\}$ onto

$\{x,y/y^2 - x^3 = 0\}$ is injective, however $k[T^2,T^3]$ is not equal to $k[T]$.

In order to get $A(\mathcal{F})$ surjective, we would have to assume that $\mathcal{F}$ was a closed immersion.

It turns out, in fact, that the language of coordinate rings is easier for introducing certain notions. For example,

Definition 2.9. *The product of two algebraic sets* V *and* W *is an algebraic set the points of which belong to* $V \times W$ *and such that*

$$A(V \times W) = A(V) \otimes A(W)$$

It is important to note that the Zariski topology on the product is not the product topology. There are many more closed sets!

We have seen how to define the Krull dimension of a finitely generated algebra; we shall use this to get:

Definition 2.10. $\dim (V) := \dim A(V)$

If V is an affine variety, $I(V)$ is prime, so $A(V)$ is an integral domain. We then know that $\dim (V)$ is equal to the transcendence degree of $A(V)$ over k. In particular,

$$\dim k^n = \dim k[T_1,\dots,T_n] = n \quad .$$

$\dim A(V)$ is an upper bound for the length of a chain of subvarieties contained in V.

3. Formal Definition of a Polynomial System

Let Y be an algebraic set and U an affine variety over the field k. Then a *discrete time, polynomial dynamical system* with input space U and output space Y will be a quadruple

$$\Sigma = (X,x_o,p,h)$$

where X (the state space) is an algebraic set over k, p (the transition function) and h (the output function) are morphisms from $X \times U$ into U and X into Y respectively, and x_o (the initial state) is an equilibrium state of X (i.e., $\exists\, u_o \in U$, $p(x_o,u_o) = x_o$).

Thus, if x_{t-1}, u_{t-1} are respectively the state and input at time t-1,

$$x_t = p(x_{t-1}, u_{t-1}) \qquad y_t = h(x_t)$$

are respectively the state and output at time t.

The transition function p can be extended recursively to a function $p : X \times U^* \to X$ where U^* is the free monoid on U (i.e., the set of sequences of inputs). Let

$$g : U^* \to X \quad g(w) = p(x_o, w) \ .$$

Definition 3.1. *The input/output function of system* Σ *is the function:* $U^* \to Y : \mathscr{F}_\Sigma = h \circ g$

The restrictions g_t and $\mathscr{F}_t$ of g and $\mathscr{F}$ to U^t are morphisms from U^t into X and Y respectively. In fact, one can even define the notion of morphism from one system to another:

Definition 3.2. A morphism from a system Σ to a system $\tilde{\Sigma}$ is a morphism T of algebraic sets such that the following diagram commutes:

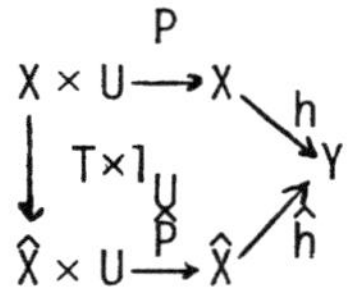

and $T(x_o) = \hat{x}_o$

Note that the existence of such a morphism implies that Σ and $\hat{\Sigma}$ have the same input/output function!

$X_t := g_t(U^t)$ is the set of reachable states at time t. Since we assumed that the system was in an equilibrium state originally, it is clear that the sequence $\{X_t\}$ increases monotically.

We shall now look in turn at the main system theoretic concepts: reachability, observability, canonical representations, and ask about them the most fundamental question from a practician's point of view: when can an answer be obtained after a finite number of operations?

4. Reachability

Let us first start with an example showing that reachability stricto-sensu is not a finite concept, i.e., no matter how large the time t, there can always be some reachable states which have not yet been reached:

Example 4.1. Let $U = X = Y = \mathbb{R}$, $p(x,u) = x+u^2-2u$, $h(x) = x$, $x_o = 0$. Then $X_t = \{x \in \mathbb{R} / x \geq -t\}$.

However, if we look at the closure of the set of reachable states, and call the system *quasi-reachable* if this closure is the whole state space, then we have the following very satisfying results:

Lemma 4.2. If $\bar{X}_t = \bar{X}_{t+1}$, *then* $\bar{X}_{t+1} = \bar{X}_{t+2}$.

Proof. $X_{t+1} = p(X_t \times U)$; since morphisms are continuous, $p(\overline{X_t \times U}) \subseteq \overline{p(X_t \times U)}$. So $\bar{X}_{t+1} = \overline{p(X_t \times U)} \supseteq p(\overline{X_t \times U}) = p(\bar{X}_t \times U) = p(\bar{X}_{t+1} \times U)$. But

$$p(\bar{X}_{t+1} \times U) \supseteq p(X_{t+1} \times U) = X_{t+2} \supseteq X_{t+1}$$

So

$$\bar{X}_{t+1} \supseteq X_{t+2} \supseteq X_{t+1} \Rightarrow \bar{X}_{t+1} = \bar{X}_{t+2}$$ ⌑

Theorem 4.3. *If* $\dim\ \Sigma = n$, *then the set of quasi-reachable states is* $\bar{X}_n$.

Proof. Since U was assumed to be irreducible, U^t is irreducible, so $\bar{X}_t = \bar{g}_t(U^t)$ is irreducible. Consequently, the length of a chain

$$\bar{X}_o \subset \bar{X}_1 \subset \dots$$

cannot exceed $n+1$.

We shall therefore know by time n what the set of quasi-reachable states will be.

5. Identification

Theorem 5.1. *Let* Σ *and* $\hat{\Sigma}$ *be two systems of dimension* $\leq n$. *Then* $\mathcal{F}_\Sigma = \mathcal{F}_{\hat{\Sigma}} \Leftrightarrow \mathcal{F}_{2n} = \hat{\mathcal{F}}_{2n}$

Proof. Let us consider the difference between the output functions of both systems. $\tilde{\mathcal{F}} : w \to \mathcal{F}(w) - \hat{\mathcal{F}}(w)$ is well defined, and can be realized as the input/output function of a system $\tilde{\Sigma}$ such that

$$\tilde{X} := X \times \hat{X}\ ,\quad \tilde{x}_o := (x_o, \hat{x}_o)$$

$$\tilde{p}((x,\hat{x}),u) := (p(x,u),\hat{p}(\hat{x},u))\ ,\quad \tilde{h}(x,\hat{x}) := h(x) - h(\hat{x})$$

The dimension of $\tilde{\Sigma}$ is $\leq 2n$, and we must prove that

$$\tilde{\mathscr{F}}_{2n} \equiv y \in Y \Rightarrow \tilde{\mathscr{F}}_t \equiv y, \quad \forall t > 0. \quad \text{But:}$$

$$\tilde{\mathscr{F}}_t \equiv y \Leftrightarrow \tilde{h}/\tilde{X}_t \equiv y$$

By assumption, $h/X_{2n} \equiv y$; since h is continuous, $\tilde{h}/\bar{\tilde{X}}_{2n} \equiv y$; theorem 4.3 then implies that

$$\bar{\tilde{X}}_t \subseteq \bar{\tilde{X}}_{2n}, \forall t \Rightarrow \tilde{h}/\tilde{X}_t \equiv y, \forall t$$ ⌑

It is impossible in general to get a better bound than $2n$ for characterizing an input/output function: even the linear case requires that length of time!

6. Observability and Isomorphism

Definition 6.1. $x_1, x_2 \in X$ are distinguishable if $\exists\, w \in U^*$ such that $h^w(x_1) \neq h^w(x_2)$, where $h^w(x) := h \circ p(x,w)$.

Theorem 6.2. *For any system* Σ, *there exists a finite set of input sequences* $w_1, \ldots, w_q$ *such that*

$$h^{w_i}(x_1) = h^{w_i}(x_2),$$

$\forall i \Leftrightarrow x_i$ *and* x_2 *are not distinguishable.*

Proof. An input sequence w lets us define a morphism

$$l^w : X \times X \to Y : (x_1, x_2) \to h^w(x_1) - h^w(x_2)$$

If we compose this map with the p coordinate projections $Y \subset k^p \to k$, we get p morphisms $l_i^w : X \times X \to k$. The states which are not distinguishable are exactly that algebraic subset of X given as $V(l_i^w / w \in U^*, \ 1 \leq i \leq p)$. It follows from Hilbert's theorem that finitely many of the l_i^w's suffice to characterize this algebraic set, hence that trying finitely many input sequences w will tell us which states are not distinguishable. ⌑

In fact, the elements of these sequences can be chosen almost at random (i.e., outside a closed set).

Unfortunately, reachability and distinguishability of all the states are not enough to select a system having a given input/output function uniquely up to isomorphism.

Example 6.3. Let $V = k$, $W = \{(x,y)/x^3 - y^2 = 0\}$ $q: z \to (z^2, z^3)$

Define $U = V$, $Y = W$, $\Sigma := (V, x_0, pr_2, q)$,

$$\hat{\Sigma} := (W, q(x_0), q \circ \hat{pr}_2, 1_W)$$

where $x_0 \in V$ is arbitrary and $pr_2 : V \times V \to V$, $\hat{pr}_2 : W \times V \to V$ are the projections on the second factor.

q induces a morphism from Σ to $\hat{\Sigma}$, so $\mathcal{F}_\Sigma = \mathcal{F}_{\hat{\Sigma}}$. q being bijective, both systems are reachable and have distinguishable states. Yet their state spaces are not isomorphic.

If we wish to salvage the isomorphism theorem for canonical systems, then we need to take a stronger notion for observability.

Definition 6.4. *A system* Σ *is algebraically observable if there exist an integer* s *and* $\P = (w_1, \ldots, w_s) \in (U^*)^s$ *such that for any costate* $q \in A(X)$, *there exists* $\hat{q} \in A(Y^s)$ *such that*

$$\hat{q} \circ H^{\P} = q, \quad \textit{where} \quad H^{\P} : X \to Y^s :$$
$$x \to (h^{w_1}(x), \ldots, h^{w_s}(x))$$

This is the same as requiring that $A(H^{\P}) : A(Y^s) \to A(X)$ be surjective. We can now state:

Theorem 6.5. *Let* $\Sigma, \hat{\Sigma}$ *be such that* $\mathcal{F}_\Sigma = \mathcal{F}_{\hat{\Sigma}}$, Σ *and* $\hat{\Sigma}$ *being both quasi reachable and algebraically observable; there exists a unique isomorphism* $T : \Sigma \to \hat{\Sigma}$.

Proof. This follows directly from Zeiger's lemma applied to the coordinate rings, and duality. ⌑

The conditions of this theorem hold for example in the case of so-called "variable structure" systems, in which h is linear and $p(x,u) = p_1(u) + p_2(u)x$; h^w is then affine, and algebraic observability is the same thing as state distinguishability. They hold also in the case of externally bilinear systems.

We have seen how the tools of commutative algebra let us establish for polynomial systems some fundamental results, very close to those of linear system theory. Some results could be obtained for the realization (or minimization) of these systems; but we would need more sophisticated results, beyond the scope of this seminar.

PART II

The motivation for this talk can come from several directions. On the one hand, if we have a delay-differential system

$$S \begin{cases} \dot{x}(t) = \sum_{i=1}^{r} [F_i x(t-\tau_i) + G_i \quad u(t-\tau_i)] \\ y(t) = \sum_{i=1}^{r} H_i \; x(t-\tau_i) \end{cases}$$

where $x(t) \in \mathbb{R}^n$, $u(t) \in \mathbb{R}^m$, $y(t) \in \mathbb{R}^p$, and F_i, G_i, H_i are matrices of suitable dimensions, we can associate with it a transfer function

$$W(s, e^{-s\tau_1}, \ldots, e^{-s\tau_r}).$$

If we write s for the derivative operator and δ_i for the Dirac distribution $\delta_{\tau i}$, we can write S as

$$\Sigma \begin{cases} s * x = (F * x) + (G * u) \\ \qquad y = H * x \end{cases}$$

where F, G, H are polynomial matrices in the unknowns δ_i and $*$ is the convolution product [see Kamen, J. Math. Syst. Th., 1975, pp. 57-74].

Σ is therefore a linear system defined over the polynomial ring $\mathbb{R} = \mathbb{R}[\delta_1, \ldots, \delta_r]$, to which we associate the transfer function

$$W(s) = H(sI - F)^{-1} \; G$$

On the other hand, if we follow the current trend for nostalgia and elect to study the D.D.A. once again, we are faced with the fact that at some point we are dealing with a linear system in which "all quantities are discrete both in amplitude and time" [Sizer, The Digital Differential Analyzer, Chap. 4: Error Analysis]. It is clear therefore that the object to be studied is a linear, discrete-time system Σ the coefficients, inputs and outputs of which are integers.

The same situation would arise if one built a filter around a C.C.D. device, and had the gains generated by computer.

Both the ring of polynomials and the ring of integers are integral domains, hence they have a quotient field (the field of rational functions and that of rationals respectively). One could therefore look at systems Σ as evolving over these fields and apply the standard results of linear system theory. This approach however might introduce elements outside the original rings, and lead to systems not physically realizable; a proper rational function, for example, would involve inverses of the delay operators, i.e. ideal predictors.

We must therefore take into account the specificity of linear systems over rings, and, true to our program so far, see how their properties relate to those they would have if studied over a field. We shall concentrate in this talk on the issues of realizability and stabilizability by state feedback. This approach has a double advantage: since the answer to these questions over a field is well known not only from a theoretical but also from a computational point of view, it pays to know what can be lifted from the quotient field back to the ring; conversely, and to use for once the language of geometry, concepts which are equivalent over the field (e.g., minimal and canonical, observability and reachability of the dual, etc. . .) can be separated over the ring; we get a resolution of the concept, so to speak.

1. Realization of Linear Systems over Noetherian Integral Domains

A remarkable feature of the study of linear systems over rings is that a lot of the familiar arguments for systems over fields go through.

Let $A_1, A_2, \ldots, p\times m$ matrices over R, be an input-output sequence, $\Sigma = (F,G,H)$ a realization--if one exists--of that sequence with state module X.

All of abstract realization theory goes through without any change: construction of a realization from an input-output function via Nerode equivalence, reduction of a realization to a canonical one, uniqueness up to isomorphism of canonical realizations (via Zieger's lemma still), equivalence between the existence of a realization and that of a monic recursion among the matrices of the input-output sequence:

$$A_{n+k} = \alpha_1 A_{n+k-1} + \ldots + \alpha_n A_k , \quad \forall k > 0$$

(via the use of the companion matrix realization), or also with the existence of a transfer function with monic denominator.

In the standard case, we know that this is equivalent to the behaviour matrix

$$B = \begin{pmatrix} A_1 & A_2 & A_3 & \cdots \\ A_2 & A_3 & A_4 & \cdots \\ \cdot & \cdot & \cdot & \cdots \\ \cdot & \cdot & \cdot & \cdots \end{pmatrix}$$

having finite rank; the proof of this does not generalize directly, since either the Kalman-Ho algorithm or Silvermann's formulas involves dividing by determinants, hence may take us out of the ring. However, even this result remains true:

Theorem 1.1. *Let* R *be a noetherian ring,* K *its quotient field,* $A_1, A_2, \ldots$ *an input/output sequence over* R. *This sequence has a realization over* R *if it has one over* K.

Proof. It is easy to check that there is a realization over R, if the module X generated by the columns of B is finitely generated: in this case, taking this module as state module, F as a matrix representation of the shift operator, G and H as specified by Silvermann's formulas, one gets a realization which is in fact canonical.

The theorem therefore amounts to proving that this module X is finitely generated whenever it is of finite rank (i.e., the vector space $X \otimes_R K$ is finite dimensional).

Assume that $v_1, \ldots, v_n$ are a set of basis columns for B over K. Any column of B can then be written as

$$v = \sum_{i=1}^{n} \alpha_i(v) v_i \ , \quad \alpha_i(v) \in K.$$

The coefficients $\alpha_i(v)$ can be obtained using Cramer's formulas

$$\alpha_i(v) = \frac{\Delta_i(v)}{\Delta} \ , \quad \Delta_i(v) \ , \qquad \Delta \in R$$

(since these determinants are obtained by adding and multiplying elements of B, they are indeed in R). Note that the denominator does not depend on v. So, writing $u_i = v_i/\Delta$, we have:

$$v = \sum_{i=1}^{n} \Delta_i(v) \, u_i \ , \quad \Delta_i(v) \in R, \quad \forall v \in B \ .$$

The R-module generated by the columns of B is therefore contained in the finitely generated R-module generated by

$u_1,\dots,u_n$. Since R is noetherian, it is finitely generated. ☐

The next step in classical linear system theory is to assert that the notions of canonical realizations and minimal realizations are equivalent, and that the dimension of such a realization is equal to rank B. This is where we must part company.

The notion of minimality is directly related to that of size of the realization, i.e. to the number of differential equations (or integrators) in the case of the delay-differential example, or, equivalently, to the number of generators of the state module to which it is equal. To get a canonical system from any other, one must take quotients--which leave the number of generators unchanged--and subobjects. If we could guarantee that a submodule of a minimal state module has fewer generators, then we would be able to say that canonical implies minimal; this is true when R is a P.I.D. (in the case of the integers for example):

Theorem 1.2. *Let* R *be a P.I.D. Then a canonical system is always minimal.*

Proof. Since a finitely generated module is always the quotient of a free module on the same generators, any minimal realization can be lifted to another minimal realization having a free state module. But we know that over a P.I.D. a submodule of a free module is free, hence cannot have more generators than the larger module. Thus the canonical realization is free, and must be minimal too. ☐

The converse is not true, even in the case of P.I.D.'s.

Example 1.3. Let $X = \mathbb{Z}$, $F = 1$, $G = 2$, $H = 1$, and $\hat{X} = \mathbb{Z}$, $\hat{F} = 1$, $\hat{G} = 1$, $\hat{H} = 2$; they have the same input-output function, are both minimal, yet the first one is not reachable.

Since the canonical realization of a linear system over a P.I.D. is free, its dimension is equal to the rank of the behaviour matrix B. The existence of minimal realizations of dimension equal to the rank of B can be guaranteed in general in one other case only:

Lemma 1.4. *A minimal realization of an input/output sequence* S *over a ring* R *will always have dimension equal to the rank of the behaviour matrix associated with* S *iff every finitely generated reflexive module over* R *is free.*

Proof. *Sufficiency.* Let (F,G,H) be a canonical realization of S with state module X. The latter is finitely generated, and its dual and bidual X^* and X^{**} are reflexive. So

X and X^{**} have the same dual X^*, and the map $H : X \to R^p$ can be viewed as a map $H^{**} : X^{**} \to R^p$. The map $F : X \to X$ induces canonically a map $F^* : X^* \to X^*$ hence a map $F^{**} : X^{**} \to X^{**}$. So the system (F,G,H) induces canonically a system (F^{**},G^{**},H^{**}) with state module X^{**} and the same input/output function S. It follows from the assumption that X^{**} is free; since we can write--abusively, perhaps--

$$X \subset X^{**} \subset X \otimes_R K$$

and the latter is the state space of a canonical realization of S over K, the dimension of X^{**} is equal to the rank of B.

Necessity. Let X be the state module of a canonical realization of a system, M that of the minimal realization of dimension equal to rank B. We have seen that M could be assumed free. So $M \otimes_R K$ is the state space of a minimal--hence canonical--realization of the system over K. It follows that the realization of state-module M must be observable (a map $\mathscr{F} : M \to R^p \times R^p \times \ldots$ is injective if $\mathscr{F} \otimes K : M \otimes K \to K^p \times K^p \times \ldots$ is.

So, by Zeiger's lemma, there is an injection $X \to M$. Hence $X \subset M \subset X \otimes_R K$, and $M^* \subset X^*$ [see Bourbaki: Commutative Algebra, Chap. 7, Sec. 4, no. 1 and 2].

Now, given any X, finitely generated, torsion free module over T, let us construct a canonical system with state module X. Since the ring R is noetherian, X^* is finitely generated, by, say, $\{u_1,\ldots,u_p\}$. X being finitely generated by, say, m generators, there is a projection

$$R^m \xrightarrow{G} X \to 0$$

Take R^m and G as the input side of the system, the identity matrix for F and the matrix having for rows the linear forms $[u_1,\ldots u_p]$ for H.

If the lemma is true, there is a system $(\hat{F},\hat{G},\hat{H})$ with state module M, and the same input/output function such that $\dim M = \operatorname{rank} B$.

Zeiger's lemma shows that the map $H : X \to R^p$ extends to the map $\hat{H} : M \to R^p$. Since in our construction the rows of H $[u_1,\ldots,u_p]$ generate X^*, these generators extend to linear maps $M \to R$, and so $X^* \subset M^*$. It follows that $X^* = M^*$. If X is reflexive, then $X = M$ and so X is free. ∎

Theorem 1.5. *Among polynomial rings, those in one or two variables are the only ones such that the conditions of* (1.3) *always hold.*

Proof. Since reflexive modules must be free, they must at least be projective. But every finitely generated reflexive R-module is projective if the global dimension of R is inferior or equal to 2 [Faith: Rings, Modules and Categories], and we know that the dimension of a polynomial ring in n unknowns is n. So we cannot go beyond two unknowns. However, Seshadri proved that over a polynomial ring in two unknowns every finitely generated projective module was free. ⌑

If the system had only one input or one output, this could be generalized to many other rings (noetherian integrally closed).

In the case of delay-differential systems, if one wishes to avoid this limitation to one or two noncommensurable delays, one must consider a slightly more general class of systems, of the form:

$$\begin{cases} \dot{x}_o(t) = F_{oo}x_o(t) + \dots + F_{or}x_r(t) + G_o\, u(t) \\ x_1(t+\alpha_1) = F_{1o}x_o(t) + \dots + F_{1r}x_r(t) + G_1\, u(t) \\ \vdots \\ x_r(t+\alpha_r) = F_{ro}x_o(t) + \dots + F_{rr}x_r(t) + G_r\, u(t) \\ y(t) = H_o x_o(t) + \dots + H_r x_r(t) \end{cases}$$

where $x_i(t) \in \mathbb{R}^{1_i}$, $i = 0,\dots,r$ and $\alpha_i > 0$, $\forall i$ [Sontag and Rouchaleau, Journées de l'Optimisation. Montreal, 1978].

2. Stabilization by State Feedback

The classic result in this domain is that the poles of a linear system over a field can be modified arbitrarily if, and only if, the system is completely reachable.

It is not overly surprising that this condition remains necessary over rings:

Theorem 2.1. *If the poles of a system* (F,G,H) *can be modified arbitrarily by state-feedback, then the pair* (F,G) *is completely reachable.*

Proof. To say that the system is reachable is to say that its reachability map is onto. We know that this is a local property, i.e., that it is true if and only if it is true at each stalk of the structure sheaf of the ring.

Let therefore R be a local ring, M its maximal ideal. A map $\mathscr{F}$ between two modules A and B over R is surjective, if

$$\mathscr{F} \otimes R/M : A \otimes R/M \to B \otimes R/M$$

is surjective (this is a direct consequence of Nakayama's lemma, see for example [M. Artin's M.I.T course notes, Commutative Rings, Chap. 5, §F]).

Since if one can modify arbitrarily the poles of a system, one can certainly do the same for the systems induced over the fields, it follows that each local system is reachable, hence that the global system is too. ⌑

Morse [Ring models for delay-differential system, 1974, IFAC symp., Manchester] has proved the converse in the case of a P.I.D. The assumption of complete reachability of the ring model is very stringent, however, and Sontag [Linear Systems over Commutative Rings, Recerce di Automatica, Vol. 7, no. 1] has suggested a generalization of the class of feedback under consideration to get a stable system under more general conditions.

Consider the set R_s of rational functions $p(\sigma)/q(\sigma)$ with real coefficients such that:

$$q(\sigma) \text{ has no zeroes in } \{z \in \mathbb{C},\ |z| \leq 1\}$$

$R_s[\sigma]$ is a P.I.D. which contains $R[\sigma]$. It follows from Morse's result that if the pair (F,G) is reachable over $R_s[\sigma]$ there is a matrix K (over $R_s[\sigma]$) such that $\rho(e^{-s},s)$ has the real part of its zeroes bounded from above by a negative number, where $\rho(\sigma,z) = \det(zI - F + GK)$.

We can interpret $p(\sigma)/q(\sigma)$ as the transfer function of a system in which the integrators would be replaced by delays (compare with the system at the end of paragraph 1) and which would be stable (because of the restriction on the zeroes of $q(\sigma)$). If

$$\begin{cases} x(t+\sigma) = \hat{F}\, x(t) + \hat{G}\, u(t) \\ y(t) = \hat{H}\, x(t) + \hat{J}\, u(t) \end{cases}$$

is a minimal realization of K, then the system

$$\dot{x}(t) = (F(\sigma) - G(\sigma)\hat{J})\, x(t) - G(\sigma)\, \hat{H}\nu(t)$$

$$\nu(t+\sigma) = \hat{F}\nu(t) + \hat{G}\, x(t)$$

is uniformly asymptotically stable.

REFERENCES

Artin, M.: 1966, *Commutative Algebra,* Notes for course 18.732, M.I.T.

Bourbaki: 1964, *Algebre*, Chap. 7, (Hermann).

Bourbaki: 1965, *Algebre Commutative,* Chaps. 1,2,8, (Hermann).

Dieudonne: 1974, *Cours de Geometrie Algebrique,* Vols. 1 and 2, P.U.F.

Kalman: 1968, *Lecture on Controllability and Observability,* CIME Lecture Notes.

Mumford: 1968, *Introduction to Algebraic Geometry,* Harvard Lecture Notes.

Zariski and Samuel: 1965, *Commutative Algebra,* Vols. 1 and 2, Van Nostrand.

SYSTEMS AND POLYNOMIAL MATRICES

H. H. Rosenbrock

University of Manchester Institute of Science and Technology
Manchester, England

1. INTRODUCTION

The usual starting-point for a theory of time-invariant, finite-dimensional linear systems is

$$\left.\begin{aligned} \dot{x} &= Ax + Bu \\ y &= Cx \end{aligned}\right\} \qquad (1)$$

but for certain purposes this is too restrictive.

(i) Several subsystems may be described as in (1), and they may be connected together to form a composite system. This does not have its equations in the form (1); they may be brought to this form, but only by losing the identity of the subsystems in the final description of the system.

(ii) In many other circumstances the system equations do not arise in the form (1), and have to be brought subsequently to this form; a common example is when nonlinear equations are linearised by a perturbation technique.

(iii) Equations (1) give rise to a strictly proper transfer function matrix G. If we wish to consider non-strictly-proper G, we can replace the second of equations (1) by

$$y = Cx + D(p)u \qquad (2)$$

where p is the differential operator. But then the strictly proper part $C(sI-A)^{-1}B$ of G, and the remaining

C. I. Byrnes and C. F. Martin, Geometrical Methods for the Theory of Linear Systems, 233-255.

part $D(s)$ are treated in quite different ways. In particular, $D(s)$ lacks information about structure of the system at infinite frequencies, whereas for finite frequencies this information is contained in A,B,C. So a satisfactory theory of structure at infinite frequency becomes impossible.

For these reasons we need a more general description, and we use

$$\left.\begin{aligned} T(p)\xi &= U(p)u \\ y &= V(p)\xi + W(p)u \end{aligned}\right\} \qquad (3)$$

Here ξ is an r-vector and r may be larger or smaller than the order n of the system (which is yet to be defined). The matrices T, U, V, W are polynomial matrices in p, respectively $r \times r$, $r \times \ell$, $m \times r$, $m \times \ell$. Equations (3) can be written more compactly

$$\begin{pmatrix} T(p) & U(p) \\ -V(p) & W(p) \end{pmatrix} \begin{pmatrix} \xi \\ -u \end{pmatrix} = \begin{pmatrix} 0 \\ -y \end{pmatrix} \qquad (4)$$

and if we Laplace-transform with zero initial conditions the form of the equation is unchanged. We simply replace p by s, and ξ, u, y by their Laplace transforms; these will be denoted by the same letters, the ambiguity being resolved by the context.

The appropriate definition of the order of the system (4) is not immediately apparent. One approach [Rosenbrock, 1970a, pp. 48, 49] is to investigate the number of arbitrary initial conditions in the complementary function, which is the solution of

$$T(p)\xi = 0 \qquad (5)$$

using a technique described by Gantmacher [1959]. The result is that n is equal to the degree of the determinant of $T(p)$ or $T(s)$, $n = \delta(|T(s)|)$. Alternatively, we may assume this result, investigate its consequences, and show that they are consistent with the definition of order for (1).

If $r < n$ in (4), we augment the equations to make $r = n$, giving after Laplace transformation

$$\begin{pmatrix} I_{n-r} & 0 & 0 \\ 0 & T(s) & U(s) \\ 0 & -V(s) & W(s) \end{pmatrix} \begin{pmatrix} \xi_0 \\ \xi \\ -u \end{pmatrix} = \begin{pmatrix} T_1(s) & U_1(s) \\ -V_1(s) & W_1(s) \end{pmatrix} \begin{pmatrix} \xi_1 \\ -u \end{pmatrix} = \begin{pmatrix} 0 \\ -y \end{pmatrix} \qquad (6)$$

The new variables introduced in this way satisfy the equation $\xi_0 = 0$ which is simply adjoined to the original equations. This is an operation (a transformation of the system equations) which we should ordinarily regard as leaving the system substantially unaltered. It could be included in the definition of strict system equivalence below, but the procedure adopted here is somewhat more convenient.

The matrix

$$\begin{pmatrix} T_1(s) & U_1(s) \\ -V_1(s) & W_1(s) \end{pmatrix} \tag{7}$$

which has been augmented as in (6) if necessary, now has $r \geq n$. We call it the (polynomial) system matrix because it summarises the complete set of system equations.

2. STRICT SYSTEM EQUIVALENCE

Given the system matrix, $P(s)$, or the corresponding equations

$$\begin{pmatrix} T(s) & U(s) \\ -V(s) & W(s) \end{pmatrix} \begin{pmatrix} \xi \\ -u \end{pmatrix} = \begin{pmatrix} 0 \\ -y \end{pmatrix} \tag{8}$$

in which now $r \geq \delta(|T(s)|)$, one of the things which we shall wish to do is to bring them to state-space form. For this purpose, a number of operations are ordinarily permitted. For example, we may make a change of variables, we may differentiate one of the equations in $T(p)\xi = U(p)u$ and add it to any other equation, and so on. All these operations which are normally allowed can be summarised by three kinds of elementary operations on $P(s)$, carried out either on rows or columns.

(i) Multiply any one of the first r rows (columns) of $P(s)$ by a nonzero constant.

(ii) Add a multiple, by a polynomial in s, of any of the first r rows (columns) in $P(s)$ to any other row (column).

(iii) Interchange any two among the first r rows (columns) of $P(s)$.

In turn, these operations can be summarised by the operation of strict system equivalence (sse) defined by

$$\begin{pmatrix} M(s) & 0 \\ X(s) & I_m \end{pmatrix} \begin{pmatrix} T(s) & U(s) \\ -V(s) & W(s) \end{pmatrix} \begin{pmatrix} N(s) & Y(s) \\ 0 & I_\ell \end{pmatrix} = \begin{pmatrix} T_1(s) & U_1(s) \\ -V_1(s) & W_1(s) \end{pmatrix} \tag{9}$$

in which M, N, X, Y are polynomial matrices, while M,N are unimodular.

As an example, consider

$$\left.\begin{aligned} \dot{\xi}_1 + \dddot{\xi}_2 &= -\xi_1 \\ \dot{\xi}_2 &= -\xi_2 + u \\ y &= \xi_1 \end{aligned}\right\} \tag{10}$$

Differentiate and change sign in the second equation twice and then add to give

$$\begin{aligned} \dot{\xi}_2 &= -\xi_2 + u \\ -\ddot{\xi}_2 &= \dot{\xi}_2 - \dot{u} \\ \dddot{\xi}_2 &= -\ddot{\xi}_2 + \ddot{u} \end{aligned}$$

$$\dddot{\xi}_2 = -\xi_2 + u - \dot{u} + \ddot{u} \tag{11}$$

Substitute this into the original equations, giving

$$\left.\begin{aligned} \dot{\xi}_1 &= -\xi_1 + \xi_2 - u + \dot{u} - \ddot{u} \\ \dot{\xi}_2 &= -\xi_2 + u \end{aligned}\right\} \tag{12}$$

and substitute

$$\left.\begin{aligned} x_1 &= \xi_1 + \dot{u} - 2u \\ x_2 &= \xi_2 \end{aligned}\right\} \tag{13}$$

to give the state-space equations

$$\left.\begin{aligned} \dot{x}_1 &= -x_1 + x_2 - 3u \\ \dot{x}_2 &= \quad\; - x_2 + u \\ y &= x_1 - \dot{u} + 2u \end{aligned}\right\} \tag{14}$$

All these transformations can be summarised by

$$\left(\begin{array}{cc|c} 1 & -(s^2-s+1) & 0 \\ 0 & 1 & 0 \\ \hline 0 & 0 & 1 \end{array}\right)\left(\begin{array}{cc|c} s+1 & s^3 & 0 \\ 0 & s+1 & 1 \\ \hline -1 & 0 & 0 \end{array}\right)\left(\begin{array}{cc|c} 1 & 0 & s-2 \\ 0 & 1 & 0 \\ \hline 0 & 1 & 1 \end{array}\right)$$

$$= \left(\begin{array}{cc|c} s+1 & -1 & -3 \\ 0 & s+1 & 1 \\ \hline -1 & 0 & -s+2 \end{array}\right) \tag{15}$$

It is now easy to prove [Rosenbrock, 1970a, pp. 53,54].

<u>Theorem 1</u> Any polynomial system matrix may be brought by strict system equivalence to the form

$$\left(\begin{array}{cc|c} I_{r-n} & 0 & 0 \\ 0 & sI-A & B \\ \hline 0 & -C & D(s) \end{array}\right) \tag{16}$$

where $D(s)$ is an $m \times \ell$ polynomial matrix.

It can also be shown that two system matrices P, P_1 with $r = r_1$ are sse iff any matrix of the form (16) which is sse to P, and any matrix of the form (16) (but with A replaced by A_1, etc.) which is sse to P_1, satisfies

$$\begin{pmatrix} H^{-1} & 0 \\ 0 & I \end{pmatrix}\begin{pmatrix} sI-A & B \\ -C & D(s) \end{pmatrix}\begin{pmatrix} H & 0 \\ 0 & I \end{pmatrix} = \begin{pmatrix} sI-A_1 & B_1 \\ -C_1 & D(s) \end{pmatrix} \tag{17}$$

for some real nonsingular matrix H [Rosenbrock, 1970a, 1977a]. In other words, the state-space representations are related by a change of basis in the state space, which we may also call system similarity (ss). This shows that sse is a natural generalisation of ss.

For the truth of the last result, it is essential to have $r \geq n$, as is shown by the following example. The system matrices

$$\left(\begin{array}{cc|c} 1 & 0 & 0 \\ 0 & (s+2)^2 & -1 \\ \hline 0 & s+1 & 0 \end{array}\right), \quad \left(\begin{array}{cc|c} 1 & 0 & 0 \\ 0 & (s+2)^2 & s+1 \\ \hline 0 & -1 & 0 \end{array}\right)$$

can be shown [Rosenbrock, 1970a, p. 107] to be strictly system equivalent. But if we try to find m, n, x(s), y(s) such that

$$\begin{pmatrix} m & 0 \\ x(s) & 1 \end{pmatrix} \begin{pmatrix} (s+2)^2 & -1 \\ s+1 & 0 \end{pmatrix} \begin{pmatrix} n & y(s) \\ 0 & 1 \end{pmatrix} = \begin{pmatrix} (s+2)^2 & s+1 \\ -1 & 0 \end{pmatrix} \tag{18}$$

and m, n are unimodular (i.e. nonzero constants) we have

$$x(s)\ (s+2)^2 b + (s+1)n = 0 \tag{19}$$

whence $x(s) = 0$, and so $n = 0$, and (18) cannot be satisfied.

This fact, which was the reason for satisfying the condition $r \geq n$ before defining sse, was initially somewhat puzzling. Light has been thrown on the difficulty by Fuhrmann [1977] who has shown that with an appropriate abstract definition of sse, the most general explicit relation which embodies it is not (9) but

$$\begin{pmatrix} M(s) & 0 \\ X(s) & I_m \end{pmatrix} \begin{pmatrix} T(s) & U(s) \\ -V(s) & W(s) \end{pmatrix} = \begin{pmatrix} T_1(s) & U_1(s) \\ -V_1(s) & W_1(s) \end{pmatrix} \begin{pmatrix} N(s) & Y(s) \\ 0 & I \end{pmatrix} \tag{20}$$

in which M,N need no longer be unimodular, nor even square, and we no longer need $r = r_1$, nor $r \geq n$, nor $r_1 \geq n$. We do need M, T_1 left comprime and T,N right coprime. That is, whenever we express

$$T(s) = T_o(s)Q(s), \quad N(s) = N_o(s)Q(s) \tag{21}$$

we have $|Q(s)|$ a nonzero constant, and similarly for M,T_1. For example, though (18) is frustrated, we have

$$\begin{pmatrix} -(s+1) & 0 \\ 0 & 1 \end{pmatrix} \begin{pmatrix} (s+2)^2 & -1 \\ s+1 & 0 \end{pmatrix} = \begin{pmatrix} (s+2)^2 & s+1 \\ -1 & 0 \end{pmatrix} \begin{pmatrix} -(s+1) & 0 \\ 0 & 1 \end{pmatrix} \tag{22}$$

where $-(s+1)$, $(s+2)^2$ are left coprime (which for scalars means simply coprime) and $(s+2)^2$, $-(s+1)$ are right coprime. Furhmann's proof of the above result is not couched in terms of polynomial matrices, but Pugh and Shelton [1978] have subsequently given a polynomial matrix proof.

We now seem to have enlarged the transformation of sse in (20) beyond that in (9). Since (9) incorporates all the operations we generally allow, this seems embarrassing. The reconciliation [Rosenbrock, 1977a] is obtained by showing that anything which may be achieved by (20) may also be achieved by (9) if we first make $r = r_1 \geq n$ as in (6). That is, the transformations generate the same equivalence classes (subject to the proviso) even though they have different explicit forms. This is fortunate, because (20) is not in the form of operations on the original system matrix leading to the second. If we do convert it to this form we obtain for example from (22)

$$\begin{pmatrix} -(s+1) & 0 \\ 0 & 1 \end{pmatrix} \begin{pmatrix} (s+2)^2 & -1 \\ s+1 & 0 \end{pmatrix} \begin{pmatrix} -(s+1)^{-1} & 0 \\ 0 & 1 \end{pmatrix} = \begin{pmatrix} (s+2)^2 & s+1 \\ -1 & 0 \end{pmatrix} \tag{23}$$

which represents the insertion of a new pole into the system at $x = -1$, and its subsequent removal. It is interesting that this is permitted under Fuhrmann's sse, but it is something which most control engineers would no doubt wish to avoid.

Under sse, it is easy to check that the order of the system is unchanged, because (9) gives

$$|T_1| = |M|\,|T|\,|N| \tag{24}$$

and as $|M|$, $|N|$ are nonzero constants, $|T_1|$, $|T|$ have the same degree. Also if we use Theorem 1 we find that

$$M(s)\; T(s)\; N(s) = \begin{pmatrix} I_{r-n} & 0 \\ 0 & sI-A \end{pmatrix} \tag{25}$$

whence it follows that apart from some unit entries, the invariant polynomials of $T(s)$ are those of $sI-A$. The zeros of these polynomials are therefore the poles of the system.

3. CONTROLLABILITY, OBSERVABILITY

Having made the generalisation from (1) to (3), it is highly desirable to associate with (3) some property corresponding to controllability of (1). The appropriate property is coprimeness; iff in a polynomial system matrix $P(s)$ we have $T(s)$, $U(s)$ left coprime, then any system matrix (16) which is sse to $P(s)$ will have A, B a controllable pair. From this (since any system matrix can be transformed into itself by the identity transformation) it follows that $sI-A$, B are left coprime polynomial matrices iff

$$(B,\; AB,\; \ldots,\; A^{n-1}B)$$

has full rank. This is an interesting algebraic result which is easily proved. A convenient general test is that $T(s)$, $U(s)$ are left coprime iff the matrix $(T(s)\ U(s))$ has full rank r at each root of $|T(s)| = 0$.

If T, U are not left coprime, there will exist a polynomial matrix $Q(s)$ such that $T(s) = Q(s)T_0(s)$, $U(s) = Q(s)U_0(s)$, and $\delta(|Q(s)|) \geq 1$. If $Q_1(s)$ satisfies this condition, and is such that any other $Q(s)$ satisfying the condition has $|Q(s)|$ a divisor of $|Q_1(s)|$, then $Q_1(s)$ is called a greatest common left divisor of $T(s)$, $U(s)$. We thus have T, U left coprime iff their greatest common left divisor is unimodular.

The zeros of $|Q_1(s)|$, where $Q_1(s)$ is as just defined, are zeros of $|T(s)|$, and so are poles of the system. However, when we form

$$\begin{aligned} G(s) &= V(s)T^{-1}(s)U(s) + W(s) \\ &= V(s)[Q_1(s)T_o(s)]^{-1}Q_1(s)U_o(s) + W(s) \\ &= V(s)T_o^{-1}(s)U_o(s) + W(s) \end{aligned} \tag{26}$$

we see that $Q_1(s)$ cancels. It will be no surprise to find that these zeros of $|Q_1(s)|$ are the poles corresponding to the uncontrollable part of the system. It is straightforward to check that sse leaves these poles unchanged.

All of the above, with appropriate changes, applies also to observability. We can also, with a little care, separate out those poles of the system corresponding to the uncontrollable and observable part in Kalman's canonical decomposition [1962]. What we do is to see which of the unobservable poles (to use a loose but convenient terminology) disappear when we eliminate the uncontrollable poles [Rosenbrock, 1970a, p. 65].

This way of approaching controllability and observability throws some new light on these properties. For example, if $s = s_o$ makes the rank of

$$\begin{pmatrix} T(s_o) \\ -V(s_o) \end{pmatrix} \tag{27}$$

less than r, then clearly $|T(s_o)| = 0$, so that s_o is a pole of the system. But it is also a zero of the system in the sense that

$$\begin{pmatrix} T(s_o) & U(s_o) \\ -V(s_o) & W(s_o) \end{pmatrix} \begin{pmatrix} \xi(s_o) \\ -u(s_o) \end{pmatrix} = \begin{pmatrix} 0 \\ -y(s_o) \end{pmatrix} \tag{28}$$

permits nonzero vectors $\xi(s_0)$ with $u(s_0) = 0$, $y(s_0) = 0$. We may regard the zero as located at the output and coincident with the pole, decoupling it from the output of the system. For this reason such zeros were named "output decoupling zeros." A corresponding definition can be given for "input decoupling zeros" and for "input-output decoupling zeros." The latter are zeros which (without being duplicated) decouple from both input and output. With some care [Rosenbrock, 1973, 1974a] it is possible to give a direct characterisation of the zeros of a system which agrees with these definitions. But there is more than one way of defining the zeros of a system [MacFarlane and Karcanias, 1976; Rosenbrock, 1977] and more may await discovery on this question.

4. LEAST-ORDER SYSTEMS

If T, U are not left coprime, then (26) shows that the same transfer function can arise from a lower-order system, and similarly if T, V are not right coprime. Consequently if a system has least order (in the sense that no system of lower order can give the same G(s)) it follows that T, U are left coprime and T, V are right coprime. The converse is contained in the following basic result [Rosenbrock, 1970a, p. 106].

Theorem 2 Let P, P_1 be two system matrices having no decoupling zero. Then P, P_1 are sse iff they give the same G.

By the result stated in connection with (17), if P, P_1 have state-space form, sse in the theorem can be replaced by ss to recover a well-known result.

Suppose now that we start with a given G(s) and try to find a least-order representation for it. Let d(s) be the least common denominator of the elements of G(s), which can therefore be written

$$G(s) = N(s)/d(s) \tag{29}$$

A system matrix* giving rise to G(s) is then

$$P(s) = \begin{pmatrix} d(s)I & N(s) \\ -I & 0 \end{pmatrix} \tag{30}$$

This P(s) has no output decoupling zero, because

*Strictly we should add an identity matrix and bordering zeros, but these play no part in the subsequent calculation.

$$\begin{pmatrix} d(s)I \\ -I \end{pmatrix} \tag{31}$$

has full rank for all s. There may, however, be input decoupling zeros, and we can check this by inspecting the rank of $(d(s)I \;\; N(s))$ at each root of $d(s) = 0$. If there is a real input decoupling zero s_0, then $(d(s_0)I \;\; N(s_0))$ will have less than full rank, and there will be some real nonsingular matrix K such that

$$K(d(s_o)I \quad N(s_o)) \tag{32}$$

has one row zero. Thus

$$K(d(s)I \quad N(s)) \tag{33}$$

will have the corresponding row divisible by $s-s_o$, and we may write

$$(d(s)I \quad N(s)) = Q(s)(T_1(s) \quad N_1(s)) \tag{34}$$

where $|Q(s)| = s-s_o$, $T_1(s)$ and $N_1(s)$ are polynomial, and the system matrix

$$\begin{pmatrix} T_1(s) & N_1(s) \\ -I & 0 \end{pmatrix} \tag{35}$$

has $\delta(|T_1(s)|) = \delta(|d(s)I|) - 1$, while still giving the same $G(s)$. A similar and only slightly more complicated reduction will eliminate a complex conjugate pair if s_0 is complex [Rosenbrock, 1970a, pp. 60-62]. Proceeding in this way we shall eventually arrive at a least-order system matrix

$$\begin{pmatrix} T_2(s) & N_2(s) \\ -I & 0 \end{pmatrix} \tag{36}$$

in which T_2, N_2 are left coprime. Then we have

$$G = T_2^{-1} N_2 \tag{37}$$

which is usually called a "matrix fraction description" of G. It is interesting because it generalises the idea of a rational function as the ratio of two polynomials; it was apparently first found by Belevitch [1963, 1968]. An entirely analogous development allows G to be written

$$G = N_3 T_3^{-1} \tag{38}$$

5. MINIMAL INDICES

With the exception of the transformation of sse, most of what has been said above is a restatement in other terms of known results. We now come to some new results, which link up in a satisfactory way with work by Forney, Kalman, and others.

Let $sI-A, B$ be left coprime. In the terminology of Kronecker [Gantmacher, 1969]

$$(sI-A \quad B) \tag{39}$$

is a singular pencil of matrices, and it therefore has a set of minimal indices λ_i for the rows. One way of defining these is as follows. Let $v_1(s)$ be a polynomial vector of lowest degree satisfying

$$(sI-A \quad B)\, v_1(s) = 0 \tag{40}$$

Among all polynomial vectors linearly independent (over the polynomials) of $v_1(s)$, let $v_2(s)$ be a vector of lowest degree satisfying (40). Continue in this way until no more linearly independent vectors can be found. Then the degrees of the polynomial vectors generated in this way are the minimal indices. They will be ℓ in number (the number of columns in B) and their sum will be n. Kronecker shows that although the vectors $v_i(s)$ are not unique, the minimal indices are uniquely defined. It is easy to show [Rosenbrock, 1970a, pp. 96-99] that the λ_i are the sizes of the blocks in Luenberger's standard forms. They are also a complete set of invariants found by Brunovsky [1970] for the transformation

$$\left.\begin{aligned} &\text{(i)} & x_1 &= Hx \\ &\text{(ii)} & u_1 &= Ku \\ &\text{(iii)} & u_2 &= u + Fx \end{aligned}\right\} \tag{41}$$

and are called the controllability indices; the identity of these with the Kronecker indices was pointed out by Kalman [1972].

It is instructive to consider, instead of (1), the discrete-time system

$$\left.\begin{aligned} x_{k+1} &= Ax_k + Bu_k \\ y_k &= Cx_k + Du \end{aligned}\right\} \tag{42}$$

which we assume to have least order and which on z-transformation becomes

$$\left.\begin{aligned}(zI-A)x &= Bu \\ y &= Cx + Du\end{aligned}\right\} \qquad (43)$$

Note that the term Du in (43) is the most general which can occur; a term $D_1 zu$ for example would make the output y depend upon future inputs.

Now the first equation in (43) is just

$$(zI-A \quad B)\begin{pmatrix} x(z) \\ -u(z)\end{pmatrix} = 0 \qquad (44)$$

and clearly the degree of $u(z)$ is greater than that of $x(z)$. Hence a solution of (40) gives an input sequence $u(z)$ of minimum length such that the corresponding state vector $x(z)$ is polynomial. The last condition implies that the state sequence (which is shorter than the input sequence) is zero after time zero. In other words, the $u(z)$ generated from (40) takes the state from its initial value zero at time $-\lambda_1$, moves it away from zero, and returns it to zero at time $t = 1$. Since u and x are both zero for $t > 0$, y will also be zero for $t > 0$. The minimal indices are therefore the lengths of a minimum set of linearly independent inputs [Rosenbrock, 1970a, p. 148] having this property.

Now if $D = 0$ in (42), Kalman [1969] shows how a module structure can be associated with the system. The input-output behaviour is represented as a mapping $f : \Omega \to \Gamma$. Here Ω is a finite free module over the ring of polynomials in z, and its elements may be identified with the z-transforms of finite input sequences terminating at time zero. Also, Γ is a module over the ring of polynomials. Each element of Γ may be regarded as the z-transform of that part of the output which occurs after $t = 0$. Then [Rosenbrock, 1970b] the interpretation of what was said in connection with (44) is that the kernel of f is the submodule of Ω which is generated by the vectors $u_1(z), u_2(z), \ldots, u_\ell(z)$.

Using the idea of the Nerode equivalence class, Kalman identifies the states of the system with equivalence classes of inputs; two inputs are in the same equivalence class iff their difference lies in the kernel of f. In view of what was said above, this means that their difference is an input sequence which generates the zero state at $t = 1$.

It is possible in fact [Rosenbrock, 1972] to give an explicit representation of the state generated by an input sequence $\omega(z)$. Let $T(z)$ be the matrix having the $u_i(z)$ as its columns

$$T(z) = (u_1(z), u_2(z), \ldots, u_\ell(z)) \tag{45}$$

which is nonsingular because the u_i are linearly independent [Rosenbrock, 1970a, p. 148]. Then the polynomial vector $w(z)$

$$w(z) = T(z) \text{ [strictly proper part of } T^{-1}(z)\omega(z)] \tag{46}$$

represents the state generated by $\omega(z)$. This is verified by showing that two inputs $\omega_1(z)$, $\omega_2(z)$ generate the same $w(z)$ iff they belong to the same Nerode equivalence class. An alternative representation of the state has been given by Salehi [1978].

These are connections between the theory of minimal indices and Kalman's module theory. On the other hand, there are connections with the theory of linear vector spaces over the rational functions. This was developed and applied by Forney [1975] while some results in this area can also be found in Wedderburn [1934]. Consider the $(m+\ell) \times \ell$ matrix over the rational functions

$$\begin{pmatrix} G(s) \\ I \end{pmatrix} \tag{47}$$

The columns span a linear vector space V over the rational functions, and by multiplying each column by an appropriate polynomial we may generate a polynomial basis for V. Among polynomial bases, some will be more complicated than others, and Forney calls a basis minimal if it has the following two properties

(i) The greatest common divisor of all $\ell \times \ell$ minors is 1.

(ii) The sum of the column degrees has the smallest possible value, namely the highest degree occurring among the $\ell \times \ell$ minors.

If a given polynomial basis is not minimal, then Forney shows how a minimal basis may be obtained from it. Moreover, Forney shows that the column degrees of all minimal bases for a given V are invariant within reordering: they are properties of V and Forney calls them the invariant dynamical indices.

Now multiply (47) on the right by $T(z)$ defined by (45). For each column $u_i(z)$ of T, Gu_i is $Cx_i + Du_i$, where x_i is the corresponding polynomial vector in (44). It follows that GT is polynomial, and has each column of no higher degree than the corresponding column of T. Now in

$$\begin{pmatrix} G \\ I \end{pmatrix} T = \begin{pmatrix} V \\ T \end{pmatrix} \tag{48}$$

we have the column degrees equal to the λ_i, and $\Sigma\lambda_i = n$. The degree of $|T(z)|$ cannot exceed $\Sigma\lambda_i$, by (45). It follows that $\delta(|T|) = n$, and V and T are right coprime, for if either of these conditions were not true, G could arise from a system having order less than that of (42), whereas (42) has least order. From this it follows that Forney's two conditions are satisfied, and (48) defines a minimal basis. Hence Forney's invariant dynamical indices are the minimal indices.

Forney further shows that the sum of the dynamical indices of the dual basis $V^{\perp}$ orthogonal to V is the same as the sum of the dynamical indices of V. The vectors v_i in (40) therefore span the basis $V^{\perp}$ orthogonal to the rows of $(sI-A \quad B)$ and it follows at once that $\Sigma\lambda_i = n$, as was stated before.

6. NONPROPER SYSTEMS

In Section 5 we assumed that G was proper, or strictly proper. The second assumption is essential in Kalman's module theory, which cannot distinguish two systems having $G = 0$, $G = 1$. Forney's work does not require either restriction, but clearly if $G(s)$ is not proper, the dynamical indices can no longer be the same as the minimal indices (i.e. controllability indices). For the sum of the latter is always n, the order of the system. On the other hand,

$$G = VT^{-1} = V\,\mathrm{diag}(s^{-\lambda_i})[T\,\mathrm{diag}(s^{-\lambda_i})]^{-1} \tag{49}$$

and because $\delta(|T|) = n$, it follows that G is non-proper only if $V\,\mathrm{diag}(s^{-\lambda_i}) \to \infty$ as $s \to \infty$. So the sum of the column degrees in (48), that is, the sum of Forney's dynamical indices, will now be greater than n. We should expect Forney's indices to have some system-theoretic importance, and it is therefore interesting to investigate what this may be [Rosenbrock and Hayton, 1974].

A first result of the investigation is that the sum of the dynamical indices is the McMillan degree $\delta(G)$ of G. This is a satisfactory generalisation; when G is proper, the McMillan degree is just the least order associated with G. When G is non-proper, the McMillan degree takes account not only of finite poles, but also of poles at infinity.

Secondly, [Rosenbrock, 1974b] we may generalise the state-space description of a system to

$$\begin{pmatrix} sE-A & B \\ -C & 0 \end{pmatrix} \begin{pmatrix} x \\ -u \end{pmatrix} = \begin{pmatrix} 0 \\ -y \end{pmatrix} \tag{50}$$

in which E is singular. Using a result of Kronecker, we can find real nonsingular matrices M,N such that

$$\begin{pmatrix} M & 0 \\ 0 & I \end{pmatrix} \begin{pmatrix} sE-A & B \\ -C & 0 \end{pmatrix} \begin{pmatrix} N & 0 \\ 0 & I \end{pmatrix} = \left|\begin{array}{cc|c} sI-A & 0 & B_1 \\ 0 & I+sJ & B_2 \\ \hline -C_1 & -C_2 & 0 \end{array}\right| \tag{51}$$

where in J all elements are zero except perhaps for unit entries in the superdiagonal. The transformation (51) is a restricted form of sse, which leaves G unchanged, so

$$G(s) = C_1(sI-A_1)^{-1}B_1 + C_2(I+sJ)^{-1}B_2 \tag{52}$$

The first term on the right is strictly proper, while the second is polynomial. Hence (50) can give rise to non-proper G. The non-proper part of G is related to an underlying structure which can be investigated in a parallel way to the investigation of the strictly proper part. Contrast this with (2), in which the polynomial part of G is represented only by its transfer function $D(s)$, without reference to the structure of the system from which it has arisen.

Subject to appropriate conditions we now find that the dynamical indices of G are the minimal indices of

$$(sE-A \quad B) \tag{53}$$

The conditions in Rosenbrock [1974b] are related to the decoupling zeros defined above, and to a new set of infinite decoupling zeros which play a parallel role for that part of the system giving rise to the polynomial part of G. Just as in Kalman's canonical decomposition we can divide this part of the system into four, only one of which contributes to G. We can also show that two system matrices of the form in (51), each having no finite or infinite decoupling zero, give rise to the same G iff they are related by the restricted form of sse defined by (52).

All of this work concerning structure at infinity has some defects, and Verghese [1978] has recently proposed changes which improve it. Work still remains to be done before the structure of such systems can be regarded as fully understood.

A third result [Hayton, 1975] is that the minimal indices of (53) are invariant under the transformation on (5) represented by

$$\left.\begin{array}{ll}\text{(i)} & x_1 = Hx \\ \text{(ii)} & u_1 = Ku \\ \text{(iii)} & u_2 = u + Fx\end{array}\right\} \qquad (54)$$

which may be compared with (41).

7. COMPOSITE SYSTEMS

Systems in practice are often constructed by interconnecting subsystems. If the subsystems are of some specified type, and the interconnections are also carried out in some specified way, the properties of the overall system will be determined largely by the general properties of the subsystems and the way they are interconnected. This is a relatively unexplored area in general, though the special area of electrical network theory is highly developed.

We may represent the subsystems which are to be interconnected by

$$\left.\begin{aligned} T_i(s)\xi_i &= U_i(s)u_i \\ y_i &= V_i(s)\xi_i + W_i(s)u_i \end{aligned}\right\} \qquad (55)$$

An interesting type of interconnection is

$$\left.\begin{aligned} u_i &= -\sum_i F_i^{(t)} y_t + K_i u \\ y &= \sum_i L_i y_i \end{aligned}\right\} \qquad (56)$$

where u is the overall system input and y is its output. The scheme is general enough to include all interconnections which can be represented by block diagrams, or which can be set up on an analogue computer.

For such a composite system we can investigate a wide range of properties [Rosenbrock and Pugh, 1974; Pugh, 1974]. There are not many surprises; for example the composite system cannot have least order unless every subsystem has least order. In general, the decoupling zeros of the composite system consist of those of all the subsystems, plus a set of decoupling zeros which is determined by the interconnections and by the transfer function matrices of the subsystems. However, contrary to what one might conjecture, no result of this kind is true for input decoupling zeros separately (or for the other two types).

Similarly, if every subsystem has a strictly proper transfer function matrix, the order of the composite system is the sum of the orders of the subsystems. But if any subsystem has a non-strictly-proper transfer function matrix, this additive property of orders is lost.

More surprisingly, the interconnection scheme is not general enough for electrical network theory. Two capacitors can be represented as two subsystems (55) with

$$\begin{pmatrix} T(s) & U(s) \\ -V(s) & W(s) \end{pmatrix} \begin{pmatrix} \xi \\ -u \end{pmatrix} = \begin{pmatrix} sC_k & 1 \\ -1 & 0 \end{pmatrix} \begin{pmatrix} v_k \\ -i_k \end{pmatrix} = \begin{pmatrix} 0 \\ -v_k \end{pmatrix} \tag{57}$$

where the inputs are the currents i_k, the outputs are the voltages v_k, and (because the equations are in state-space form) the voltages v_k are also the states. If these capacitors are connected in parallel, this imposes a constraint $v_1 = v_2$ on the two states. When the equations for the composite system are put into state-space form, they have first order. This type of constraint on states is impossible with (56), or with the normal block-diagram conventions.

If a resistor is inserted in the capacitor-loop, the system immediately becomes second-order. As the resistance decreases, one of the system poles goes off to infinity. It is therefore tempting to believe that the type of redundancy represented by a capacitor loop corresponds to a "decoupling zero at infinity." This was the motivation of the work mentioned in the previous section, which as stated there is not in a completely satisfactory form.

An appropriate interconnection scheme for an LCR multiport is represented [Rosenbrock, 1974c] by the composite system matrix

$$\left(\begin{array}{cc:c} \begin{pmatrix} W_{RL}(s) & 0 \\ 0 & W_C(s) \end{pmatrix} & I & 0 \\ \begin{pmatrix} -A_1 & 0 \\ 0 & -B_1 \end{pmatrix} & \begin{pmatrix} 0 & A_2 \\ B_2 & 0 \end{pmatrix} & K \\ \hdashline -L_1 & -L_2 & 0 \end{array}\right) \tag{58}$$

Here $W_{RL}(s)$ contains information about those subsystems which are resistors or inductors, and $W_C(s)$ about those which are capacitors. The matrices, A_1, A_2 are partitions of the incidence matrix, and B_1, B_2 are partitions of the circuit matrix. The matrices K and L_1, L_2 define the way in which inputs and outputs of the composite system are connected to the sub-systems.

One of the interesting results in the theory of LCR multiports is that certain transfer functions (sc. admittances or impedances) can be synthesised using LCR (without ideal transformers) only if more reactive elements are used than the least order of the impedance would seem to require. For example, the one-port impedance

$$G(s) = R \frac{s^2+a_1 s+a_o}{s^2+b_1 s+b_o} \tag{59}$$

with a_o, a_1, b_o, b_1 real and positive and

$$a_1 b_1 = (\sqrt{a_o} - \sqrt{b_o})^2 \tag{60}$$

can be realised by an LCR network with five reactive elements (instead of two). Except in special cases, no way of realising (59) with less than five reactive elements is known. Then of what kind is the redundancy?

It can be shown that (58) can be brought to a symmetric form by sse. As sse leaves the decoupling zeros unchanged, this shows that for every i.d. zero of (58) there is also an equal o.d. zero. Hence an LCR multiport has only two possible kinds of redundancy due to uncontrollability or unobservability,

(i) An uncontrollable and unobservable subspace.

(ii) A subspace of dimension k which is uncontrollable but observable, accompanied by a subspace of dimension k which is controllable but unobservable, both subspaces having the same eigenvalues associated with them.

For the Bott and Duffin synthesis of (59) when $a_o > b_o$ it has been shown that there are three redundant poles, of which one is uncontrollable and unobservable, one is uncontrollable but observable, and one is controllable but unobservable, the last two being coincident.

Uncontrollability and unobservability are not the only possible kinds of redundancy in LCR multiports. We may also have capacitor loops or inductor nodes, and these can play the same role as unobservability or uncontrollability in making possible an LCR realisation. This kind of redundancy, however, is not invariant under sse, and we do not know what transformation should be used instead. A full investigation is therefore not available at present.

8. POLE ASSIGNMENT

The methods of polynomial matrix theory can be applied to a wide range of problems in linear systems theory. As would be expected, they are superior to other methods in some problems and not in others. One problem for which they are particularly apt is pole assignment.

Let the controllable pair A, B be given. Under what conditions can we find C such that the invariant polynomials of sI-A+BC (i.e. the closed-loop invariant polynomials) are given polynomials $\phi_i(s)$, with $\phi_i(s)$ dividing $\phi_{i-1}(s)$? This is a generalisation of the usual pole assignment problem.

We need first a lemma which is established by a constructive proof in Rosenbrock and Hayton [1978].

Lemma 1 Let the integers α_i, β_i be given, with $\alpha_1 \geq \alpha_2 \geq \ldots \geq \alpha_\ell \geq 0$, $\beta_1 \geq \beta_2 \geq \ldots \geq \beta_\ell \geq 0$. Also let $\phi_1, \phi_2, \ldots, \phi_\ell$ be given monic polynomials with ϕ_i dividing ϕ_{i-1}, $i = 2,3,\ldots,\ell$. Then a necessary and sufficient condition for the existence of an $\ell\times\ell$ polynomial matrix $\Phi(s)$ having Smith form $\mathrm{diag}(\phi_\ell(s), \phi_{\ell-1}(s),\ldots,\phi_1(s))$ and satisfying

$$\lim_{s\to\infty}(\mathrm{diag}(s^{-\alpha_i})\ \Phi(s)\ \mathrm{daig}(s^{-\beta_i})) = I \tag{61}$$

is

$$\sum_{i=1}^{k} \delta(\phi_i) \geq \sum_{i=1}^{k} (\alpha_i+\beta_i),\ k = 1,2,\ldots,\ell \tag{62}$$

with equality holding when $k = \ell$.

We can now show that a sufficient condition for the existence of a C such that the closed-loop system has the required invariant polynomials is

$$\sum_{i=1}^{k} \delta(\phi_i) \geq \sum_{i=1}^{k} \lambda_i,\ k = 1,2,\ldots,\ell \tag{63}$$

with equality holding when $k = \ell$.

First, sI-A,B are left coprime, which implies [Rosenbrock, 1970a, p. 71] that there exist polynomial matrices X(s), Y(s) such that the square matrix

$$\begin{bmatrix} sI-A & B \\ X(s) & Y(s) \end{bmatrix} \tag{64}$$

is unimodular. With (T)s as in (45), and V(s) formed similarly from the $x_i(s)$,

$$\begin{pmatrix} sI-A & B \\ X(s) & Y(s) \end{pmatrix} \begin{pmatrix} V(s) \\ T(s) \end{pmatrix} = \begin{pmatrix} 0 \\ M(s) \end{pmatrix} \tag{65}$$

where M(s) is some polynomial matrix. But from (65)

$$(sI-A)^{-1}B = VT^{-1}$$

and the least order corresponding to the left-hand transfer function is n. Since T has its columns of degree $\lambda_1, \lambda_2, \ldots, \lambda_\ell$, $\delta(|T|) \leq \Sigma\lambda_i = n$. But this implies $\delta(|T|) = n$, and V,T right coprime. Then in (64) the first matrix on the left is unimodular, and so nonsingular for all s, while the second has full rank ℓ for all s. Hence M(s) has full rank for all s and so is unimodular.

In Lemma 1, put $\alpha_i = 0$, $\beta_i = \lambda_i$ and construct Φ which is possible by (63). Then from (65)

$$\begin{pmatrix} sI-A & B \\ \Phi M^{-1}X & \Phi M^{-1}Y \end{pmatrix} \begin{pmatrix} V \\ T \end{pmatrix} = \begin{pmatrix} 0 \\ \Phi \end{pmatrix} \tag{67}$$

and by subtracting suitable multiples of the first n rows from the last ℓ we can reduce the degree of $\Phi M^{-1}X$ to zero, giving

$$\begin{pmatrix} sI-A & B \\ -C_1 & Q(s) \end{pmatrix} \begin{pmatrix} V(s) \\ T(s) \end{pmatrix} = \begin{pmatrix} 0 \\ \Phi(s) \end{pmatrix} \tag{68}$$

But the high-order coefficient matrix of T(s) (i.e. the real matrix having as column i the coefficient of s^{λ_i} in the ith column of t) is nonsingular because $\delta(|T|) = \Sigma\lambda_i$. From this, and (61), and the fact that each column of V has lower degree than the corresponding column of T, it follows that Q(s) is real and nonsingular, say D. Now the matrix

$$\begin{pmatrix} sI-A & B \\ -C & I \end{pmatrix} \tag{69}$$

where $C = D^{-1}C_1$ is obtained from

$$\begin{pmatrix} I_n & 0 \\ 0 & \Phi \end{pmatrix} \tag{70}$$

by operating on it with unimodular matrices, and so (69) has the same Smith form as (70), and so do

$$\begin{pmatrix} sI-A+BC & 0 \\ C & I \end{pmatrix} \quad \text{and} \quad \begin{pmatrix} sI-A+BC & 0 \\ 0 & I \end{pmatrix} \tag{71}$$

Hence the closed-loop system has the required invariant polynomials. A separate proof that (63) is necessary will be found in Rosenbrock and Hayton [1978].

This result is obtained in a remarkably simple and efficient manner. It can be generalised to deal with a dynamic compensator, in which case the best result available (loc. cit.) replaces (63) by the sufficient condition

$$\sum_{i=1}^{k} \delta(\phi_i) \geq \sum_{i=1}^{k} (\lambda_i+\mu_1-1), \quad k = 1,2,\ldots,\ell \tag{72}$$

with equality for $k = \ell$. Here μ_1 is the largest of the observability indices of G. It is conjectured that μ_1 in (72) might be replaced by μ_i, giving a sharper and more symmetrical condition. Though some improvement of (72) can be made, the conjecture is at present still open.

REFERENCES

Belevitch, V.: 1963, *Developments in network theory*, edited by S. R. Deards, P. 19 (Pergamon).

Belevitch, V.: 1968, *Classical network theory*, (Holden-Day).

Brunovsky, P.: 1970, *A classification of linear controllable systems*, Kybernetika (Praha), 3, pp. 173-187.

Forney, G. D.: 1975, *Minimal bases of rational vector spaces, with applications to multivariable linear systems*, J. Siam Control, 13, pp. 493-520.

Fuhrmann, P. A.: 1977, *Strict system equivalence and similarity*, Int. J. Control, 25, pp. 5-10.

Gantmacher, F. R.: 1959, *The theory of matrices*, Vols. 1, 2 (Chelsea Publishing Co.).

Hayton, G. E.: 1975, *Properties of dynamical indices*, Int. J. Control, 22, pp. 289-293.

Kalman, R. E.: 1962, *Canonical structure of linear dynamical systems*, Proc. Nat. Acad. Sci., 48, pp. 596-600.

Kalman, R. E.: 1969, *Topics in mathematical system theory*, with P. L. Falb and M. A. Arbib (McGraw-Hill).

Kalman, R. E.: 1972, *Kronecker invariants and feedback*, in Ordinary Differential Equations, edited by Weiss, pp. 495-471 (Academic Press).

MacFarlane, A. G. J. and Karcanias, N.: 1976, *Poles and zeros of linear multivariable systems: a survey of the algebraic, geometric and complex-variable theory*, Int. J. Control, 24, pp. 33-74.

Pugh, A. C.: 1974, *The relationship between order, degree and complexity in the hierarchical theory of systems*, Int. J. Control, 20, pp. 713-719.

Pugh, A. C. and Shelton A. K.: 1978, *On a new definition of strict system equivalence*, Int. J. Control, 27, pp. 657-672.

Rosenbrock, H. H.: 1970a, *State-space and multivariable theory*, (Nelson-Wiley).

Rosenbrock, H. H.: 1970b, *Further properties of minimal indices*, Electronics Letters, 6, p. 450.

Rosenbrock, H. H.: 1972, *Modules and the definition of state*, Int. J. Control, 16, pp. 433-435.

Rosenbrock, H. H.: 1973, *The zeros of a system*, Int. J. Control, 18, pp. 297-299.

Rosenbrock, H. H.: 1974a, *Correction to "The zeros of a system,"* Int. J. Control, 20, pp. 525-527.

Rosenbrock, H. H.: 1974b, *Structural properties of linear dynamical systems*, Int. J. Control, 20, pp. 191-202.

Rosenbrock, H. H.: 1974c, *Non-minimal LCR multiports*, Int. J. Control, 20, pp. 1-16.

Rosenbrock, H. H.: 1977a, *The transformation of strict system equivalence*, Int. J. Control, 25, pp. 11-19.

Rosenbrock, H. H.: 1977b, *Comments on "Poles and zeros of linear multivariable systems, etc.,"* Int. J. Control, 26, pp. 157-161.

Rosenbrock, H. H. and Hayton, G. E.: 1974, *Dynamical indices of a transfer function matrix*, Int. J. Control, 20, pp. 177-189.

Rosenbrock, H. H. and Hayton, G. E.: 1978, *The general problem of pole assignment*, Int. J. Control, 27, pp. 837-852.

Rosenbrock, H. H. and Pugh, A. C.: 1974, *Contributions to a hierarchical theory of systems*, Int. J. Control, 19, pp. 845-867.

Salehi, S. V.: 1978, *On the state-space realizations of matrix fraction descriptions for multivariable systems*, IEEE Trans. AC, AC-23, pp. 1054-1057.

Verghese, G.: 1978, *Infinite frequency behaviour in generalised dynamical systems*, Ph.D. dissertation, Stanford University.

Wedderburn, J. H. M.: 1934, *Lectures on matrices*, (New York, American Mathematical Society).

FUNCTIONAL MODELS, FACTORIZATIONS AND LINEAR SYSTEMS

Paul A. Fuhrmann

Ben Gurion University of the Negev
Beer Sheva, Israel

1. INTRODUCTION

In this paper we attempt to describe a circle of ideas which makes possible a unified exposition of a large part of linear algebra, operator theory and both finite and infinite dimensional multivariable linear systems. As the title suggests the unifying concepts will be those of functional models, module theory and various factorizations of polynomial and analytic matrix valued functions. The full exposition of the ideas presented here will be the theme of a forthcoming monograph [20].

While in general one may expect that finite dimensional results will always precede an infinite dimensional generalization, this is not the case. A variety of problems in linear algebra can sometimes be solved by brute force methods depending very much on finite dimensionality, the availability of matrix representations and the use of elementary transformations. In operator theory however, one is generally forced to state his results and give his proofs in a basis independent language. Thus while a functional model may seem natural to the operator theorist people who use matrices will generally balk at replacing them by a polynomial model.

In the study of operator theory, and we will restrict ourselves to the more developed theory of operators in Hilbert spaces, it became clear very early in the development of the theory that infinite matrix representations, besides being cumbersome, will

Partially supported by National Science Foundation under Grant NSF ENG77-28444.

C. I. Byrnes and C. F. Martin, Geometrical Methods for the Theory of Linear Systems, 257-282.

not do in general for the study of operators. In fact one of the early successes of operator theory was the proof of the spectral theorem for self-adjoint operators and the accompanying theory of spectral representations and spectral multiplicity. This yielded also a complete set of unitary invariants for self-adjoint operators analogous to the invariant factors of a finite dimensional linear transformation. Thus a study of a self-adjoint operator could be replaced by the study of any of its, unitarily equivalent, spectral representations.

It was to be expected that a representation theory for some classes of nonself-adjoint operators will follow. This indeed was the case and it can probably be traced to Livsic's introduction of characteristic functions and triangular models of operators [29]. The use of shift operators as models dates back to a simple but important paper of Rota [33], work of deBranges and Rovnyak [3], Helson's beautiful book on invariant subspaces [24] arising from his work on prediction theory, the Lax-Phillips approach to scattering theory [28] and the extensive study of contractions by Sz.-Nagy and Foias [35].

With the development of so many results on contractions, contractive semigroups and functional models the time seemed ripe for an application of these ideas to the study of infinite dimensional systems. This occurred to a number of people more or less simultaneously and within a relatively short time a realization theory for a fairly wide class of transfer functions emerged, spectral properties of the realization analyzed, various isomorphism results as well as a partial generalization of McMillan degree theory obtained. We cite [1,6,11,12,13,25,26] as a sample of papers dealing with these matters.

The elegance of the operator theoretic methods stated usually, by necessity, in a basis free language was such that it seemed proper to try and recast the finite dimensional theory in these terms. This was done by the author in a series of papers [14,15,16,17] which showed the power of polynomial methods in integrating and compactifying a variety of results. One side benefit of this was the clarification of the connections between state space theory, module theoretic techniques and the theory of polynomial system matrices.

However, it would be highly misleading to imply that finite dimensional theory has no intuition to offer for the study of infinite dimensional systems. In fact in spite of the progress in infinite dimensional theory there is one area, that of feedback design, which remains largely unknown. On the other hand feedback has been extensively studied in the finite dimensional theory with major contributions by Rosenbrock [32] and Wonham [37]. A straightforward generalization of reachability and observability indices is not possible for the same reason that McMillan degree theory is not generalizable in a simplistic way. Thus what seems called for is a recasting of finite dimensional results on feedback in terms which are generalizable to the infinite dimensional case. Specifically one would want to study feedback, moving away from canonical forms, in polynomial terms. The theory of polynomial models is particularly well suited for this. What seems reassuring is that Toeplitz operators play a significant role. Connections between Kronecker indices and Wiener-Hopf factorization indices are established and contact is made with geometric control theory. One would hope that with a generalization of classical Fredholm theory the road would be open to the study of feedback in infinite dimensional systems. It seems that this will become possible in the not too distant future.

2. OPERATORS AND SYSTEMS IN HILBERT SPACE

We shall review briefly the relevant operator theoretic results that have a bearing on the study of linear systems. Probably the single most useful idea is that of a model for an operator. Naturally, in the study of operators in Hilbert, there are very few nontrivial, useful results of complete generality. One instance is of course the closed graph theorem, but in general one must pose some restrictions. Thus let us assume we have a contraction operation T in a Hilbert space H for which $\lim_{n\to\infty} T^n x = 0$ for each vector x in H. Of course, if T is any bounded operator than αT would satisfy this condition for $|\alpha| < \|T\|^{-1}$, but this is not a very useful observation.

For each $x \in H$ we define a map W by $Wx = (x, Tx, T^2x, \ldots)$. One would like to renorm Wx so that the map W is isometric. We define a new norm on vectors in H by the equality

$$\|x\|^2 = \sum_{n=0}^{\infty} \|T^n x\|_1^2$$

which readily implies

$$\|x\|_1^2 = \|x\|^2 - \|Tx\|^2 = \|(I-T^*T)^{1/2}x\|^2 .$$

Define a new Hilbert space $\mathscr{D}$ by $\mathscr{D} = \overline{\text{Range } D_T}$ where $D_T = (I-T^*T)^{1/2}$ and then it is easily checked that W is an isometry of H onto a subspace M of $\ell^2(0,\infty;\mathscr{D})$ which is invariant under the left shift operator S^* given by $S^*(x_0, x_1, \ldots) = (x_1, x_2, \ldots)$. Moreover, we have the equality $WT = (S^*|M)W$ which shows that a left shift restricted to a left invariant subspace can serve as a model for a wide class of contractions. The dimensions of $\mathscr{D}$ is called the multiplicity of the shift.

Naturally in most cases the multiplicity is infinite which carries with it a lot of technical difficulties as well as some pathological behavior [24]. It is a lucky break that most of linear system theory can be done with shifts of finite multiplicity where the use of matrix valued functions and determinants facilitates their study. We note that by passing to a model we have replaced the general operator by a universal one. The properties of the original operator are only preserved in the structure of the left invariant subspace.

The next step is of course the study of the structure of left invariant subspaces. To obtain some feeling for this, let us characterize the 1-dimensional subspaces. These are spanned by an eigenfunction of the left shift, say $(x_0,x_1,\dots)$ corresponding to an eigenvalue $\bar{\alpha}$. Thus

$$S^*(x_0,x_1,\dots,) = (x_1,x_2,\dots) = \bar{\alpha}(x_0,x_1,\dots) .$$

This yields

$$x_{k+1} = \bar{\alpha}^k x_0$$

or the subspace is spanned by $(1,\bar{\alpha},\bar{\alpha}^2,\dots)$ with $|\alpha| < 1$ necessarily. Since sums of left invariant subspaces are also such, we can characterize the finite dimensional ones. Even this is a bit tricky as these may be generalized eigenfunctions corresponding to the kernel of $(S^*-\bar{\alpha}I)^n$. Given now α_i, $i = 1,2,\dots$ we can define

$$M = \operatorname{span}\{(1,\bar{\alpha}_i,\bar{\alpha}_i^2,\dots) \,\big|\, i = 1,2,\dots\} .$$

It is not at all clear, however, whether in this case M will be a proper subspace of $\ell^2(0,\infty)$ or just might coincide with it. This is one of the clearest cases for trying to find a better language for handling these problems. Thus one turns to a functional model. The change, from an operator theoretic point of view is trivial, since all Hilbert spaces, of the same cardinality, are isomorphic, but the implications are far reaching. The passage from $\ell^2(0,\infty)$ to a functional Hilbert space is by way of Fourier series. There is a classical unitary equivalence between $\ell^2(-\infty,\infty)$ and $L^2(T)$ the space of Lebesgue square integrable functions on the unit circle given by

$$f \to \left\{\frac{1}{2\pi}\int f(e^{it})e^{-int}\,dt\right\}_{n=-\infty}^{\infty}$$

and

$$\{a_n\}_{n=-\infty}^{\infty} \to \sum_{n=-\infty}^{\infty} a_n\, e^{int} .$$

The image of $\ell^2(0,\infty)$ will be denoted by H^2 and the important fact is that H^2 can be identified with the set of boundary value functions of a space of analytic functions in the unit disc. Under this map the eigenfunctions $(1,\bar{\alpha},\bar{\alpha}^2,\ldots)$ are mapped into

$$\sum_{n=0}^{\infty} \bar{\alpha}^{-n} e^{int} = (1 - \bar{\alpha}e^{it})^{-1}$$

whose analytic extensions into the disc are given by $(1 - \bar{\alpha}z)^{-1}$. To get an orientation for what is to come let us characterize the orthogonal complement of this one-dimensional space. If $f \in H^2$ then Cauchy's formula implies that $f \perp (1 - \bar{\alpha}z)^{-1}$ if and only if $f(\alpha) = 0$. It is now clear that $\text{span}\{(1 - \bar{\alpha}_i z)\}$ is going to be a proper subspace of H^2 if and only if there exists a nonzero function f in H^2 which satisfies $f(\alpha_i) = 0$. The question of how many zeroes an H^2 function has in the unit disc has been solved at the turn of the century. If α_i are the zeroes then

$$\sum_{i=1}^{\infty} (1-|\alpha_i|)$$

must converge and this is also sufficient. The simplest function, actually it is bounded, i.e., in H^∞, is just the Blaschke product

$$B(z) = \prod_{i=1}^{\infty} \frac{z-\alpha_i}{1-\bar{\alpha}_i z} \frac{-\alpha_i}{|\alpha_i|} .$$

Thus BH^2 is a nontrivial subspace of H^2. This is still not the most general one but it is close, the general one has been characterized by Beurling [2] and is of the form qH^2 where q is an inner function, that is a function H^∞ satisfying $|q(e^{it})| = 1$, a.e. Now a subspace qH^2 is also invariant under multiplication by all H^∞ functions and thus invariant subspaces of H^2 coincide with H^∞-submodules of H^2 where H^2 is considered as a module over H^∞. The left invariant subspaces can be therefore considered as quotient H^∞-module. There are other points one may want to stress. A module structure is always intimately connected with a functional calculus. In this particular case the operator in the shift, in the "multiplication by z" form, and the functions are in H^∞, thus $(\phi(S)f)(z) = \phi(z)\cdot f(z)$. But this by itself is not as significant as when one turns to the quotient structure namely to

$$H(q) = \{qH^2\}^\perp = H^2 \ominus qH^2 .$$

If $P_{H(q)}$ is the orthogonal projection of H^2 onto $H(q)$ then $H(q)$ is still an H^∞-module with the action being $\phi\cdot f = P_{H(q)}(\phi f)$.

Denoting $S(q)$ the operator that corresponds to $\phi(\lambda) = \lambda$ we have here an operational calculus for H^∞-functions which is a special case of a more general calculus constructed by Sz.-Nagy and Foias [35].

Now given a functional calculus the natural question arises whether a spectral mapping theorem exists or, put in other words, what is the spectrum of a function of an operator in terms of the spectrum of the operator and the analytic behavior of the function. In the case of the shift S in H^2 and $\phi \in H^\infty$ $\phi(S)$ is boundedly invertible if and only if ϕ is an invertible element of H^∞ which is equivalent to it being bounded away from zero. If one is interested in the range then $\phi(S)$ has dense range if and only if ϕ is outer [2]. Passing to $H(q)$ naturally those conditions are sufficient but certainly not necessary as a moment's reflection will show. This is of course very closely related to ideal theory in H^∞ and in fact the solution rests on a deep result of Carleson [5]. Thus $\phi(S(q))$ is boundedly invertible if and only if for some $\delta > 0$ we have $\phi|(z)| + |q(z)| \geq \delta$ for all $z \in D$. In particular this implies the weaker necessary condition that ϕ and q have no common zeroes. This last condition is easier to grasp. In fact if $\phi(\alpha) = q(\alpha) = 0$ for some α in the open unit disc D then $g_\alpha(z) = q(z)/z-\alpha$ is an eigenfunction of $S(q)$ corresponding to the eigenvalue α. But then clearly $\phi(S(q))g_\alpha = \phi(\alpha)g_\alpha = 0$. This argument can be generalized to yield the following. $\phi(S(q))$ has a nontrivial kernel if and only if ϕ and q have a nontrivial common inner factor. In that case we say ϕ and q are coprime. The stronger Carleson relation will naturally be called strong coprimeness.

We can specialize the previous analysis to the case $\phi(z) = z-\alpha$. If $|\alpha| < 1$ the Carleson condition is violated if and only if $q(\alpha) = 0$, i.e., we have α as an eigenvalue. If $|\alpha| = 1$ then α is in the spectrum if and only if q is not bounded away from zero. This is equivalent to q not having an analytic continuation at α.

Next we consider for a moment the question of duality. From the global viewpoint this looks bad as the right and left shifts in H^2 have widely differing spectral characteristics. For example S has no eigenvalues whereas the point spectrum of S^* coincides with the open unit disc. The situation changes drastically when we pass to the left invariant subspace $H(q)$. If we define $\tilde{q}$ by $\tilde{q}(z) = \overline{q(\bar{z})}$ then it can be shown that $S(q)^*$ and $S(\tilde{q})$ are unitarily equivalent. The unitary map $\tau_q : H(q) \to H(\tilde{q})$ which intertwines $S(q)^*$ and $S(\tilde{q})$ is given by

$$(\tau_q f)(e^{it}) = e^{-it}\tilde{q}(e^{it})f(e^{-it}).$$

This is an extremely useful map and facilitates many considerations.

We treat next the question of cyclicity, and coprimeness conditions again play an important role. This is not surprising inasmuch the shift in H^2 is cyclic and its cyclic vectors are the H^2 outer functions. Similarly a function f in $H(q)$ is cyclic for $S(q)$ if and only if f and q are coprime. The question of characterizing cyclic vectors for $S(q)^*$, which is cyclic because the unitary equivalent $S(\tilde{q})$ is, follows. Every $f \in H(q)$ has a representation of the form

$$f(e^{it}) = e^{-it}q(e^{it})\overline{h(e^{it})},$$

for some $h \in H^2_0$. By unitary equivalence f is cyclic for $S(q)^*$ if and only if $\tilde{h} = \tau_q f$ is cyclic for $S(\tilde{q})$ which is the case if and only if $\tilde{h}$ and $\tilde{q}$ are coprime. The last condition is equivalent to the coprimeness of h and q. This last result has an interpretation in terms of Hankel operators. The function f, assumed now bounded, is cyclic for $S(q)^*$ if and only if the vectors $S(q)^{*n}f$ span $H(q)$. If

$$f(z) = \sum_{i=0}^{\infty} f_i z^i$$

then in terms of the Hankel matrix $(h_{ij}) = (f_{i+j})$ the subspace $H(q)$ is the image, under the Fourier transform, of the range of the Hankel operator. In fact we may define the functional Hankel operator induced by f and by

$$H_f g = P_{H^2} fJ(g)$$

where

$$(Jg)(e^{it}) = g(e^{-it}).$$

While this does not add much in the scalar case Hankel operators are certainly useful in the multivariable theory.

One extra point to note is related to the commutant of the operators we have studied. The commutant of S, i.e., the set of all operators commuting with S, coincides with all multiplication operations by H^∞ functions, this is a relatively easy observation. Surprisingly the same is true for $S(q)$ where, however, we have to project after multiplying, namely if $XS(q) = S(q)X$ then $X = \phi(S(q))$ for some $\phi \in H^\infty$ which further can be taken so that $\|\phi\|_\infty = \|X\|$. This beautiful result is due to Sarason [34] and is the basis for the much more general lifting theorem proved by Sz.-Nagy and Foias.

One last observation concerns the relation between cyclic vectors for $S(q)^*$ and their singularities. Since f is cyclic for $S(q)^*$ if and only if

$$f(e^{it}) = e^{-it}q(e^{it})\overline{h(e^{it})}$$

with h and q coprime we can easily construct a msomorphic extension of f into the exterior of the unit disc. In fact

$$e^{-it}\overline{h(e^{it})}$$

is the boundary value of a function analytic outside the unit circle. The inner function has an extension by the reflection principle, the one given by

$$\hat{q}(z) = \overline{q(\bar{z}^{-1})}^{-1}$$

which is clearly meromorphic. Now $\hat{q}$ is a bona fide analytic continuation of q if and only if the unit circle is not the natural boundary of q, clearly a pathological case.

From the point of view of system theory it is clear that in essence we have done realization theory and spectral analysis of realizations for functions in H^∞ whose left translates span a proper subspace of H^2 and these functions were characterized by having factorization representations of the form $f = \bar{z}q\bar{h}$. The inner function q in this representation gives a measure of the size of the left invariant subspace that is generated by f and hence is a candidate for the equivalent of the McMillan degree.

When passing to the multivariable case many of the technical difficulties increase because of the noncommutative situation. A subspace of $H^2(\mathbb{C}^n)$ is right invariant if and only if it is an H^∞-submodule and has the representation $QH^2(\mathbb{C}^n)$ where Q is now a rigid function that is analytic, $\|Q\|_\infty \leq 1$ and a.e. $Q(e^{it})$ a partial isometry having a fixed initial space. This is the essence of the Lax-Halmos [27,22] generalization of Beurling's theorem. An inner function Q is one for which $Q(e^{it})$ are unitary a.e. and invariant subspaces that correspond to inner function are called full, as $Q(e^{it})$ has full range a.e. Thus Q in inner is equivalent to $QH^2(\mathbb{C}^n)$ being "large" or dually $H(Q) = \{QH^2(\mathbb{C}^n)\}^\perp$ being "small" in some sense. A full invariant subspace determines the inner function up to a constant unitary factor on the right.

A few things should be said about the geometry of invariant subspaces. Given two full invariant subspaces $Q_1H^2(\mathbb{C}^n)$ and $Q_2H^2(\mathbb{C}^n)$, then their intersection and their span are also invariant subspaces. If

$$PH^2(\mathbb{C}^n) = Q_1H^2(\mathbb{C}^n) \cap Q_2H^2(\mathbb{C}^n)$$

then P is the least common right inner multiple of both Q_1 and

Q_2. Similarly if

$$RH^2(\mathbb{C}^n) = Q_1H^2(\mathbb{C}^n) \vee Q_2H^2(\mathbb{C}^n)$$

then R is the greatest common left inner factor of Q_1 and Q_2. Also invariant subspace inclusion is reflected in factorization of inner functions. In fact

$$Q_2H^2(\mathbb{C}^n) \subset Q_1H^2(\mathbb{C}^n)$$

if and only if $Q_2 = Q_1R$ for some inner function R. An H^∞-submodule of $H(Q)$ is of the form $PH(R)$ for some factorization $Q = PR$ of Q into the product of two inner functions. More precisely we have $H(Q) = H(P) \oplus PH(R)$.

As in the scalar case one is interested in the H^∞-module structure of $H(Q)$, moreover one is interested in the relation between different models. In the scalar case $S(q)$ and $S(q_1)$ are similar if and only if $q = q_1$ modulo a constant factor of absolute value one. This of course, is no longer true in the high multiplicity case.

What is needed in general is a concrete characterization of operator X that intertwine two contractions, i.e., for which $XT = T_1X$. Furthermore we want criteria that guarantee invertibility properties of such X. The best approach to this problem is one which goes through the unitary dilation theory, in fact isometric dilations suffice for our purposes. An isometric operator V in a Hilbert space $K \supset H$ is called a strong isometric dilation of T if

$$T^n = P_HV^n|H$$

for all $n \geq 0$. The Sz-Nagy-Foias lifting theorem states that given a bounded operator X that intertwines T and T_1, then there exists a bounded operator $\bar{X}$ intertwining the isometric dilations (not necessarily minimal) V and V_1 of T and T_1, respectively, such that

$$\bar{X}(K \ominus H) \supset K_1 \ominus H_1\,, \qquad \|\bar{X}\| = \|X\|$$

and $X = P_{H_1}\bar{X}|H$.

The utility of this theorem lies in the fact that for a wide class of operators there is a simple isometric dilation. In fact if Q is an inner function then the shift S in $H^2(\mathbb{C}^n)$ is an isometric dilation. Moreover, operators intertwining two shifts can be described as multiplication operator by bounded matrix valued functions. This yields the following result. Let $S(Q)$ and $S(Q_1)$ be the compressions of the shift to the subspaces

$H(Q)$ and $H(Q_1)$. Thus $S(Q)^* = S^*|H(Q)$ and $S(Q)f = P_{H(Q)}zf$ for each $f \in H(Q)$. If $X: H(Q) \to H(Q_1)$ intertwines $S(Q)$ and $S(Q_1)$ then there exist matrix valued analytic functions Ξ and Ξ_1 satisfying $\Xi Q = Q_1 \Xi_1$ such that $\|X\| = \|\Xi\|_\infty$ and $Xf = P_{H(Q_1)}\Xi f$.

The spectral analysis of such operators obviously depends only on the analytic behavior of Ξ, Ξ_1, Q and Q_1. In fact the injectivity of X is equivalent to Ξ_1 and Q having no common right inner factor. X has dense range and is equivalent to the left coprimeness of Ξ and Q_1. To obtain bounded one sided inverses each of the coprimeness relations has to be replaced by a stronger one that generalized the Carleson condition in the scalar case [9], thus two bounded matrix valued functions A and B are strongly left coprime if there exists a $\delta > 0$ such that

$$\inf_{|z|<1} \{\| A(z)\xi \| + \| B(z)\xi \| \mid \|\xi\| = 1\} \geq \delta .$$

The operators of the form $S(Q)$, with Q an inner function, play the central role in realization theory and so one would want to develop a structure theory of these operators, a decomposition into more elementary components. This in fact turns out to be possible and a multiplicity theory for these operators can be developed, analogous in many ways to the finite dimensional theory and to spectral multiplicity theory for self adjoint operators. Naturally, these are technical difficulties that arise out of the fact that the sum of two nonintersecting closed subspaces in a Hilbert space need not be closed and thus some of the results are of a weaker nature. A sample of what can be achieved is the following. Given an inner function Q then $S(Q)$ is quasisimilar to $S(Q_J)$ with $Q_J = \mathrm{diag}(q_1,\ldots,q_n)$ and q_i inner functions with $q_i | q_{i-1}$. Two operators A and B are quasisimilar if there exist two injective operators X and Y with injective adjoints satisfying $XA = BX$ and $YB = AY$. $S(Q_J)$ splits naturally into the direct sum of cyclic components $S(q_i)$ and naturally one calls $S(Q_J)$ a Jordan operator associated with $S(Q)$. The Jordan operator is a quasisimilarity invariant of $S(Q)$. The inner function Q_J plays the role of the Smith form in finite dimensional theory.

Going to the question of realization with operators of the form $S(Q)$ as generators we need to concentrate on those functions, and we put in a boundedness constraint for mathematical simplicity, for which the range of the induced Hankel operator is a subspace whose orthogonal complement has full range. If $H_T f = P_{H^2} T(Jf)$ then Range $H_T \subset H(Q)$ if and only if T has a factorization of the form

$$T(e^{it}) = e^{-it} Q(e^{it}) H(e^{it})^*$$

with H another H^{∞} matrix valued function. Range H_T is dense in $H(Q)$ if and only if in the previous factorization H and Q are right coprime and Range $H_T = H(Q)$ if and only if Q and H are strongly right coprime. As in the scalar case the class of functions, called strictly noncyclic, having the range of the induced Hankel operator associated with an inner function can also be characterized in terms of their meromorphic extension into the exterior of the unit disc. The meromorphic extension is actually an analytic extension of T if and only if the unit circle is not the natural boundary of T which is equivalent to some point of the unit circle being in the resolvent set of $S(Q)$.

The Hankel operator H_T induced by T can be viewed as the reachability operator of the following realization with $H(Q)$ as state space. We take $A = S(Q)^*$, $(B\xi)(z) = T(z)\xi$ and $Cf = f(0)$ for every $f \in H(Q)$. This is called the shift realization. As in the scalar case the spectrum of $S(Q)^*$, which is unitarily equivalent to $S(\tilde{Q})$, is completely determined by Q. Similarly the singularities of

$$T(e^{it}) = e^{-it}Q(e^{it})H(e^{it})^*$$

are all outside of the open unit disc and there is a one-to-one correspondence between these singularities and the spectrum of $S(Q)^*$.

With the previous factorization of J there is associated also a left coprime factorization

$$T(e^{it}) = e^{-it}H_1(e^{it})^*Q_1(e^{it})$$

and thus we obtain the equality $H_1Q = Q_1H$ which yields a map $X:H(Q) \to H(Q_1)$ defined by $Xf = P_{H(Q_1)}H_1f$ which, due to the assumed coprimeness conditions, is injective and has dense range. A special consequence of this fact is that Q and Q_1 are associated with the same Jordan model Q_J and in particular $\det Q = \det Q_1$.

In fact for left invariant subspaces associated with inner function, the determinant of the inner function provides a generalization of the dimension function in finite dimensions. We have the following implications, with P and R assumed inner. $H(P) \subset H(R)$ implies $\det P|\det R$. if $H(Q) = H(P) \vee H(R)$, or equivalently

$$QH^2(\mathbb{C}^n) = PH^2(\mathbb{C}^n) \cap RH^2(\mathbb{C}^n) ,$$

then $\det Q|\det P\cdot\det R$ with equality if and only if P and R are left coprime which geometrically means that $H(P) \cap H(R) = \{0\}$.

With the availability of the determinant function one can try to develop a generalization of McMillan degree theory to the class of strictly noncyclic functions. Given such a function with the coprime factorizations $T = \bar{z}QH^* = \bar{z}H_1^*Q_1^*$ we define $\Delta(T) = \det(Q) = \det(Q_1)$, equality is always taken modulo a constant factor of absolute value one. We have immediately $\Delta(T_1 + T_2)|\Delta(T_1)\cdot\Delta(T_2)$ as well as $\Delta(T_1T_2)|\Delta(T_1)\cdot\Delta(T_2)$.

These division relations are dependent on an extensive study of Hankel operators [12]. Replacement of the division relation by equality is dependent on no "zero-pole" cancellation which can be precisely given as certain coprimeness condition. The details can be found in [12,13].

All of the development so far has been for the discrete time case. Many of the results can be transformed to the case of continuous time systems with a judicious use of Cayley transform techniques and the Paley-Wiener theorem which is the key, in the continuous time case, to the passage from the time domain to the frequency domain. However, going beyond realization theory lies the whole area of feedback in infinite dimensional systems which so far remains largely unexplored.

3. ALGEBRAIC SYSTEM THEORY

In many cases the algebraic theory is much simpler to develop due to the assumption of finite dimensionality. However, various methods which depended heavily on duality considerations were simpler in the Hilbert space context. We begin by introducing models for linear transformation.

Suppose we want to write down a linear transformation in an n-dimensional vector space over a field F, which is cyclic and has a preassigned characteristic polynomial d. One way of course is simply to write down the companion matrix to this polynomial. The model approach uses essentially the same construction by considering the induced action of the identity polynomial λ in the quotient ring $F[\lambda]/dF[\lambda]$ which can be identified with the set of remainders modulo d.

This approach will now be formalized. Consider the following modules $F^n((\lambda^{-1}))$, the set of all truncated Laurent series with F^n coefficients, $F^n[\lambda]$ the vector polynomials and $\lambda^{-1}F^n[[\lambda^{-1}]]$ the set of formal power series in λ^{-1} with vanishing leading coefficient. Clearly we have $F^n((\lambda^{-1})) = F^n[\lambda] \oplus \lambda^{-1}F^n[[\lambda^{-1}]]$ and we denote by π_+ and π_- the projections of $F^n((\lambda^{-1}))$ on $F^n[\lambda]$ and $\lambda^{-1}F^n[[\lambda^{-1}]]$ respectively.

Let now $D(\lambda)$ be a nonsingular $m \times m$ polynomial matrix. Using projections π_+ and π_- we introduce two new projections acting in $F^m[\lambda]$ and $\lambda^{-1}F^m[[\lambda^{-1}]]$ respectively by $\pi_D f = D\pi_- D^{-1} f$ and $\pi^D h = \pi_- D^{-1}\pi_+ Dh$. It is easily checked that π_D and π^D are indeed projections with $\text{Ker}\,\pi_D = DF^m[\lambda]$ a submodule of $F^m[\lambda]$ and $L_D = \text{Range}\,\pi^D$ a submodule of $\lambda^{-1}F^m[[\lambda^{-1}]]$, both taken with their natural $F[\lambda]$-module structure. The range of π_D which we denote by K_D can be given the induced module structure which makes it isomorphic to $F^m[\lambda]/DF^m[\lambda]$.

If we denote by S_+ and S_- the shifts, i.e., the action of λ, in $F^m[\lambda]$ and $\lambda^{-1}F^m[[\lambda^{-1}]]$, respectively, then we have the induced maps in L_D and K_D, respectively defined by $S^D = S_-|L_D$ and $S_D = \pi_D S_+|K_D$.

Let us see now how a linear transformation A can be represented, that is be similar to, in terms of these models. To this end let $D(\lambda) = \lambda I - A$ which is clearly nonsingular. In this case $K_D = \{\xi | \xi \in F^m\}$ and $L_D = \{(\lambda I-A)^{-1}\xi | \xi \in F^m\}$. Clearly $S_D\xi = A\xi$ for all ξ whereas

$$S^D(\lambda I-A)^{-1}\xi = S^D \sum_{j=0}^{\infty} \frac{A^j\xi}{\lambda^{j+1}} = \sum_{j=1}^{\infty} \frac{A^j\xi}{\lambda^j} = \sum_{j=0}^{\infty} \frac{A^j(A\xi)}{\lambda^{j+1}}$$

$$= (\lambda I-A)^{-1}A\xi \;.$$

This can be interpreted as the algebraic analogue of Rota's theorem. So far no new insight has been gained by the introduction of these models. The power of the method is in the operations one can perform on D which keep S_D in the same similarity class. Specifically one wants to develop a theory of similarity for two models S_D and S_{D_1}, that is to say classify the intertwining operators and give conditions on invertibility. Now a map $X:K_D \to K_{D_1}$ that satisfies $XS_D = S_{D_1}X$ is nothing but a module homomorphism. We note that, adapting the language of the previous section, S_+ is a dilation of S_D whereas S_- is a dilation of S^D. Thus one would hope to be able to prove lifting theorems analogous to the one's described previously. Just as in the Hilbert space case an $F[\lambda]$-homomorphism $X:K_D \to K_{D_1}$ can be lifted to a homomorphism

$$\bar{X}:F^m[\lambda] \to F^{m_1}[\lambda]$$

such that $\bar{X}\,\text{Ker}\,\pi_D \subset \text{Ker}\,\pi_{D_1}$ and $Xf = \pi_D\bar{X}f$. The homomorphism $\bar{X}$ is expressable as a multiplication by a polynomial matrix Ξ and the fact that

$$\bar{X}DF^m[\lambda] \subset D_1F^{m_1}[\lambda]$$

means that $\Xi D = D_1\Xi_1$ for some polynomial matrix Ξ_1. Again invertibility of X is determined through coprimeness conditions, namely X is injective if and only if D and Ξ_1 are right coprime and surjective if and only if Ξ and D_1 are left coprime. The analogous results can be derived in the L_D setting. In fact the map $\rho_D : L_D \to K_D$ defined by $\rho_D y = Dy$ is an invertible map that preserves the module structure, i.e.,

$$\rho_D S^D = S_D \rho_D .$$

This yields immediately the characterization of homomorphisms from L_D to L_{D_1}. $Y : L_D \to L_{D_1}$ is such a homomorphism if and only if for some polynomial matrices Ξ and Ξ_1 satisfying $\Xi D = D_1 \Xi_1$ we have $Yh = \pi_-(\Xi_1 h)$. We note that throughout this discussion D and D_1 were not assumed to be of the same size. Invertibility conditions are the same as in the polynomial model case.

One case where the coprimeness conditions necessary for invertibility are trivially satisfied is when D_1 is equivalent to D namely $UD = D_1 V$ for some unimodular polynomial matrices U and V. Thus equivalence of D and D_1 provides for similarity of S_D and S_{D_1}. As a trivial corollary we have, given two matrices A and A_1, that A and A_1 are similar if and only if $\lambda I - A$ and $\lambda I - A_1$ are equivalent. This also opens up an easy way to the structure theory of linear transformation. Given a linear transformation A we can reduce $\lambda I - A$ to $\mathrm{diag}(\alpha_1, \ldots, \alpha_m)$ by the invariant factor algorithm, i.e., $\mathrm{diag}(\alpha_1, \ldots, \alpha_m)$ is the Smith form of $\lambda I - A$. Then A is similar to $S_{\alpha_1} \oplus \ldots \oplus S_{\alpha_m}$ which is the polynomial way of writing the rational canonical form of A.

The geometry of polynomial models is directly related to some factorizations. Thus given D a subset M of K_D is a submodule, if and only if $M = D_1 K_{D_2}$ for some factorization $D = D_1 D_2$. The same holds for submodules of L_D. $N \subset L_D$ is a submodule if and only if $N = L_{D_2}$ with $D = D_1 D_2$. In general $K_D = K_{D_1} \oplus D_1 K_{D_2}$ will not hold, it is true however when D_2^{-1} is a proper rational function.

If $D_1 K_{E_1}$ and $D_2 K_{E_2}$ are submodules of K_D so are $D_1 K_{E_1} + D_2 K_{E_2}$ and $D_1 K_{E_1} \cap D_2 K_{E_2}$. Let $D_0 E_0$ and $D_m E_m$ be the corresponding factorizations of D, then D_0 is a greatest common left divisor of D_1 and D_2 and D_m is the least common right multiple of D_1 and D_2.

One way to obtain direct sum decompositions of K_D is the following. Let $d(\lambda) = \det D(\lambda)$ and let $d = d_+ d_-$ be a factorization of d into coprime factors d_+ and d_-. This factorization induces factorizations $D = D_+E_- = D_-E_+$ with $\det D_\pm = \det E_\pm = d_\pm$, and a corresponding direct sum decomposition

$$K_D = D_+K_{E_-} \oplus D_-K_{E_+} .$$

We adopt now the Kalman view of an external description of a system as an $F[\lambda]$-homomorphism $\bar{f}:F^m((\lambda^{-1})) \to F^p((\lambda^{-1}))$ where causality is built in by the assumption that $f(F^m[[\lambda^{-1}]]) \subset \lambda^{-1}F^p[[\lambda^{-1}]]$. What we have in essence is a Laurent operator expressible as multiplication by a function in $\lambda^{-1}F^{p\times m}[[\lambda^{-1}]]$. Classically, in a Hilbert space context which is easily carried over to this setup, a general Laurent operator L_A given by $L_Af = Af$ induces two closely related operators, the Hankel operator $H_A:F^m[\lambda] \to \lambda^{-1}F^p[[\lambda^{-1}]]$ and $T_A:F^m[\lambda] \to F^p[\lambda]$ defined by

$$H_A = \pi_-L_A|F^m[\lambda]$$

and

$$T_A = \pi_+L_A|F^m[\lambda].$$

Whereas the role of the Hankel operator in realization theory has long been known the Toeplitz operator has not been too widely used in the system theoretic context. In the sequel we will see its role in analyzing problems of feedback.

The invertibility properties of Toeplitz operators are intimately related to some factorizations. Given an element $A \in F^{p\times m}((\lambda^{-1}))$ we say $A(\lambda) = \Gamma(\lambda)D(\lambda)U(\lambda)$ is a right Wiener-Hopf factorization at infinity of A if U is unimodular in $F^{m\times m}[\lambda]$, Γ is an invertible element of $F^{p\times p}[[\lambda^{-1}]]$, commonly also called a bicausal isomorphism, and

$$D = \mathrm{diag}(\lambda^{\kappa_1},\ldots,\lambda^{\kappa_r},0,\ldots,0).$$

The set of indices $\kappa_1,\ldots,\kappa_r$ assumed in decreasing order are are called the right factorization indices. The Toeplitz operator T_A is injective if and only if all κ_i are nonnegative and surjective if and only if all κ_i are nonpositive. One may wonder what is the role of the left factorization indices in the study of invertibility of Toeplitz operators. This becomes clear if we define a second type of Toeplitz operator

$$T^A:\lambda^{-1}F^m[[\lambda^{-1}]] \to \lambda^{-1}F^p[[\lambda^{-1}]]$$

by $T^Ah = \pi_-Ah$. Then injectivity and surjectivity of T^A are dependent on the left factorization indices being nonpositive and nonnegative, respectively. It might happen that T_A is invertible without T^A being invertible. One simple example is obtained if

$$U(\lambda) = \begin{pmatrix} -\lambda & 1 \\ 1 & 0 \end{pmatrix}$$

and

$$\Gamma(\lambda) = \begin{pmatrix} 1 & \lambda^{-1} \\ 0 & 1 \end{pmatrix} \quad .$$

then

$$A(\lambda) = U(\lambda)\Gamma(\lambda) = \begin{pmatrix} -\lambda & 0 \\ 1 & \lambda^{-1} \end{pmatrix} = \begin{pmatrix} -1 & 0 \\ \lambda^{-1} & 1 \end{pmatrix} \begin{pmatrix} \lambda & 0 \\ 0 & \lambda^{-1} \end{pmatrix} \quad .$$

Thus the right factorization indices are both zero whereas the left factorization indices are 1, -1, and so T^A cannot be invertible.

In the Kalman development of linear system theory realization theory is effectively identified with a factorization of the restricted input/output map $f : F^m[\lambda] \to \lambda^{-1}F^p[[\lambda^{-1}]]$ into a product of injective and surjective maps that can be identified with the reachability and observability maps. Here $f = \pi_+ \bar{f} | F^m[\lambda]$ and as $f = L_A$ f is essentially the Hankel operator H_A. For a variety of purposes one would want a more concrete version of realization theory and to this end the polynomial models seem especially well qualified as the realizations are both concrete and coordinate free till the moment one wants to choose a basis.

Let us assume a rational transfer function G is given which we assume is strictly proper. Such G have representations of the form

$$G = ND^{-1} = D_1^{-1}N_1$$

with D and D_1 nonsingular. If N and D are assumed right coprime then such a factorization is unique modulo a right unimodular factor and similarly for left coprime factorizations. This freedom of applying a unimodular matrix to D, or D_1, can be utilized to reduce the matrix to some special form, say column proper, raw proper, Hermite canonical form, etc. Making the right choice of basis this can yield a variety of realizations with the operators in some special form.

Actually relating realizations to factorization can be done in a somewhat more general way. Let us assume G has the representation

$$G(\lambda) = V(\lambda)D(\lambda)^{-1}U(\lambda) + W(\lambda) \quad .$$

With such a representation one associates the Rosenbrock polynomial system matrix

$$\begin{pmatrix} D & U \\ -V & W \end{pmatrix}$$

as well as a realization given by the triple (A,B,C) acting in the state space K_D where (A,B,C) are given by $A = S_D$, $B\xi = \pi_D U\xi$ and $Cf = (VD^{-1}f)_{-1}$ with h_{-1} denoting the coefficient of λ^{-1} in the expansion of h. It has been shown in [15] that this is indeed a realization of G which is reachable if D and U are right coprime and observable if V and D are left coprime.

Given two such representations of G with realizations (A,B,C) and (A_1,B_1,C_1) one naturally would like to obtain a characterization of the intertwining maps $X:K_D \to K_{D_1}$, i.e., the maps satisfying $B_1 = XB$, $XA = A_1X$ and $C = C_1X$ and furthermore conditions for the invertibility of these maps. The availability of a representation for the module homomorphisms $X:K_D \to K_{D_1}$ makes this problem tractable. Two such realizations are isomorphic if and only if there exist polynomial matrices M, M_1, K and L satisfying

$$\begin{pmatrix} M & 0 \\ K & I \end{pmatrix}\begin{pmatrix} D & U \\ -V & W \end{pmatrix} = \begin{pmatrix} D_1 & U_1 \\ -V_1 & W_1 \end{pmatrix}\begin{pmatrix} M_1 & -L \\ 0 & I \end{pmatrix}$$

with the further assumption that M and D_1 are left coprime and M_1 and D right coprime. This relation between two polynomial system matrices is called strict system equivalence.

A different instance of the relation between factorizations and system theoretic properties concerns the notion of simulation. We say an input/output map $f:U[\lambda] \to \lambda^{-1}Y[[\lambda^{-1}]]$ is simulated by $f_1:U_1[\lambda] \to \lambda^{-1}Y_1[[\lambda^{-1}]]$ if there exist two $F[\lambda]$-module homomorphisms $\phi:U[\lambda] \to U_1[\lambda]$ and $\phi:\lambda^{-1}Y_1[[\lambda^{-1}]] \to \lambda^{-1}Y[[\lambda^{-1}]]$ such that $f = \psi f_1\phi$. This is a transitive relation which is associated with a partial order among transfer functions. We say G divides G_1 if $G = \Psi G_1\Phi + \Pi$ for some polynomial matrices Ψ, Φ and Π. Not surprisingly we have the following result [16] that f_1 simulates f if and only if G divides G_1.

Our next object is to analyze the effects of state feedback in terms of polynomial models. This results in a different derivation of a characterization obtained previously by Hautus and Heymann [23] who took an input/output view of the problem.

Given a reachable pair (A,B) then application of state feedback, allowing for similarity transformations in the input and state space, transforming it into $(R^{-1}AR + R^{-1}BK, R^{-1}BP)$. Two pairs (A,B) and (A_1,B_1) are feedback equivalent if there is a feedback transformation mapping one into the other. It is convenient to let the pairs have different state spaces, which are naturally isomorphic. Suppose $H(\lambda)D(\lambda)^{-1}$ is a right coprime factorization of the input to state transfer function $(\lambda I-A)^{-1}B$ then with the factorization HD^{-1} is associated a realization with the pair (S_D,π_D) being actually isomorphic to (A,B). A feedback equivalent pair (A_1,B_1) has a coprime factorization $RH(\lambda)(D(\lambda) + Q(\lambda))^{-1}P^{-1}$ where Q is a polynomial matrix for which QD^{-1} is strictly proper. Rougly speaking state feedback allows for addition of low order terms in the denominator matrix.

To get some intuitive feeling about this result we first obtain a coordinate free representation of the modules L_D and K_D which generalizes the single input control canonical form. If $D(\lambda) = D_0 + D_1\lambda + \ldots + D_s\lambda^s$ is nonsingular then as $\pi^D\lambda^{-(s+1)}F^m[[\lambda^{-1}]] = 0$ the module L_D is spanned by vectors of the form $\pi^D\lambda^{-j}\xi$, $j = 1,\ldots,s$. Now $\pi_+D\xi\lambda^{-j} = (D_j + D_{j+1}\lambda + \ldots + D_s\lambda^{s-j})\xi$ and so if we define $E_0 = 0$ and $E_j(\lambda) = D_j + \ldots + D_s\lambda^{s-j}$ for $j = 1,\ldots,s$ then

$$L_D = \{\sum_{j=1}^{s} \pi_-D^{-1}E_j\xi_j \mid \xi_j \in F^m\}$$

and

$$K_D = \{\pi_D\sum_{j=1}^{s} E_j\xi_j \mid \xi_j \in F^m\}$$

and we call these the control representations of L_D and K_D, respectively. The reason for that lies in the extremely simple action of S_D on these elements, namely we have

$$S_D\pi_DE_j(\lambda)\xi = \pi_DE_{j-1}(\lambda)\xi - \pi_DD_{j-1}\xi .$$

If D is column proper or just such that D^{-1} is proper then the projection π_D can be dropped altogether. Assuming that $D_1 = D+Q$ and QD^{-1} strictly proper then $D_1D^{-1} = I + QD^{-1}$ is a bicausal isomorphism and hence the induced Toeplitz operator $X = T_{D_1D^{-1}}$ is invertible. Clearly X maps $DF^m[\lambda]$ onto $D_1F^m[\lambda]$ and so one suspects that X maps K_D onto K_{D_1}. Not only is this true but X preserves the control representations. In fact if E_j' are the polynomials derived from D_1 in the same way as the E_j from D then one has the equality $X\pi_D E_j\xi = \pi_{D_1}E'_j\xi$, $j = 1,\ldots,S$. Let $Y: K_{D_1} \to K_D$ be the inverse of X, clearly the Toeplitz operator induced by DD_1^{-1}, then (S_{D_1},π_{D_1}) and (S_D,π_D) are feedback equivalent if and only if for some K we have $S_D - YS_{D_1}Y^{-1} = BK_1$. To prove this it suffices to show that the range of $S_DY - YS_{D_1}$ is included in the range of B, which is equal to $\{\pi_D\xi | \xi \in F^m\}$, for some invertible transformation Y. We show it for $Y = T_{DD_1^{-1}}$ and it is the consequence of the equality $(S_DY - YS_{D_1})\pi_{D_1}E_j\xi = -\pi_D(D_{j-1}-D'_{j-1})\xi$.

If we associate with the pair (S_D,π_D) the polynomial matrix $(D\ \ I)$ which is just the first row of the polynomial system matrix associated with D then $(D\ \ I)$ and $(N(D+Q)M\ \ NP)$ are associated with feedback equivalent pairs. Here N and M are unimodular matrices P constant invertible and Q such that QD^{-1} is strictly proper. The unimodular matrices M and N can be chosen in a way that simplifies the structure and exhibits clearly the invariants of the feedback group.

What seems to be the most effective way to achieve this end is to choose M so that DM is column proper [35]. Let P be the high coefficient column matrix and $N = P^{-1}$. If $\kappa_1 \geq \ldots \geq \kappa_m$ are the column degrees then the new high coefficient column matrix is I and adding low order terms we are reduced to the pair $(\Delta\ \ I)$ with

$$\Delta(\lambda) = \mathrm{diag}(\lambda^{\kappa_1},\ldots,\lambda^{\kappa_m}) .$$

This is the polynomial version of the Brunovsky canonical form. In fact $K_\Delta = K_{\lambda^{\kappa_1}} \oplus \ldots \oplus K_{\lambda^{\kappa_m}}$ and choosing in $K_{\lambda^{\kappa_j}}$ the basis $\{\lambda^\nu)0 \leq \nu \leq \kappa_j - 1\}$ the matrix representation of (S_Δ,π_Δ) is just the usual matrix version of the Brunovsky form.

Here we can easily make contact with Wiener-Hopf factorizations. Let $D(\lambda)$ be a nonsingular polynomial matrix. If U is unimodular and chosen so that DU is column proper with column degrees $\kappa_1 \geq \ldots \geq \kappa_m$ then $DU = \Gamma\Delta$ with Γ a bicausal isomorphism and

$$\Delta(\lambda) = \mathrm{diag}(\lambda^{\kappa_1},\ldots,\lambda^{\kappa_m}).$$

This implies the right Wiener-Hopf factorization at infinity $D = G_-\Delta G_+$ with $G_- = \Gamma^{-1}$ and $G_+ = U$. Thus if G is a strictly proper rational function with a right coprime factorization ND^{-1} and a minimal realization (A,B,C) then the right factorization indices of D are equal to the reachability indices of the pair (A,B).

Of course some natural questions arise immediately as to the existence of Wiener-Hopf factorizations for arbitrary rational functions as well as to their system theoretic interpretation. The existence of factorizations for arbitrary rational functions follows the previous outline of a proof in which reduction to column proper form was the crucial step. It should be noted however that the same ideas were implicit already in the much earlier work of Gohberg and Krein [21]. One should note that given $D = G_-\Delta G_+$ then the factor G_- corresponds to state feedback whereas G_+ to similarity so the right Wiener-Hopf factorization at infinity can be interpreted as a reduction of a reachable pair (A,B) by state feedback to the Brunovsky canonical form. Of course this reduction, and hence also the right Wiener-Hopf factorizations are not unique in general. Thus one expects the non-uniqueness of the Wiener-Hopf factorizations to correspond to the stabilizer subgroup of the feedback group which consists of all elements of the feedback group which leave (A,B) invariant. This group has been analyzed by Brockett [4] as well as by Münzner and Prätzel-Wolters [31]. The elements of this group correspond to the solutions of the equation $\Gamma\Delta = \Delta U$ with unimodular U and bicausal isomorphisms Γ. The solution is given by all unimodular matrices $(u_{ij}(\lambda))$ with $\deg u_{ij} \leq \kappa_j - \kappa_i$ for $\kappa_j \geq \kappa_i$ and $u_{ij} = 0$ if $\kappa_j < \kappa_i$.

To get back to the problem of a system theoretic interpretation of the right factorization indices of a rational transfer function we have to introduce the notion of feedback reducibility. We say that G_1 is feedback reducible to G_2 if G_2 is the transfer function of a system that is state feedback equivalent to a canonical realization of G_1. In particular the McMillan degree of G_2 is majorized by that of G_1. A rational function G is feedback irreducible if whenever G is feedback irreducible to G_1 they have the same McMillan degree. The study of feedback irreducibility is closely related to some questions of geometric control theory. Before proceeding to that we just mention that given a strictly proper rational function G then the right factorization indices of G at infinity are equal to the negatives of the reachability indices of any canonical realization of any feedback irreducible G' to which G can be feedback reduced [18].

To conclude we discuss some basic concepts of geometric control theory as developed by Wonham [37] in terms of polynomial models. The results are based on joint work with J.C. Willems and the full details are given in [19]. Much of what follows has been based on previous results of Emre [7] and Emre and Hautus [8]. Again polynomial models seem a good setting for clear characterization of geometric concepts and one would hope that further development of these models will make possible a complete interpretation of the results of [37] which probably would also result in some simplifications and greater accessibility.

A subspace V of the state space is called (A,B)-invariant if for some map K $(A+BK)V \subset V$. Given a nonsingular D it induces a reachable pair (S_D,π_D) acting in the state space K_D. We have seen before that an S_D-invariant subspace V is nothing but a submodule of K_D and as such it has the representation $V = EK_F$ for some factorization $D = EF$ into nonsingular factors. Now a subspace V of K_D which is (A,B)-invariant is similar to an invariant subspace of a feedback equivalent system. These are, up to similarity, of the form (S_{D_1},π_{D_1}) with DD_1^{-1} a bicausal isomorphism. The similarity transformation that exhibits the feedback property is given by a Toeplitz map induced by DD_1^{-1}. Thus we obtain a representation

$$V = T_{DD_1^{-1}}(E_1K_{F_1})$$

with $D_1 = E_1F_1$ a factorization into nonsingular factors and DD_1^{-1} being a bicausal isomorphism. Loosely stated (A,B)-invariant subspaces are images under some Toeplitz map of submodules in a feedback equivalent module.

Given nonsingular D we would like to characterize those F that can be left multiplied by a polynomial matrix E so that $D_1 = EF$ and DD_1^{-1} is a bicausal isomorphism. The answer can be clearly stated in terms of Wiener-Hopf factorization. This is equivalent to the nonnegativeness of all right factorization indices of DF_1^{-1}. The results can be stated also in terms of the rational model L_D. The reachable pair is now given by (S^D,π_-D^{-1}) and a subpsace V of L_D is (A,B)-invariant if and only if $V = \pi^D L$ for some submodule L of $\lambda^{-1}F^m[[\lambda^{-1}]]$. This result shows the close relation between Toeplitz operators

$$T_{DD_1^{-1}}:K_{D_1} \to K_D$$

and projections

$$\pi^D:L_{D_1} \to L_D\ ,$$

specifically we have the equality

$$\rho_D\pi^D = T_{DD_1^{-1}}\rho_{D_1}\quad .$$

A generalization of this is also valid in the context of H^2-theory.

For a general transfer function G let $G = T^{-1}U$ be a left coprime factorization and let us consider an (A,B)-invariant subspace of the realization associated with it. These subspaces have the form

$$V = T\pi_{-}N\pi^{D}L_{F_1}$$

with $D_1 = E_1F_1$ and DD_1^{-1} a bicausal isomorphism.

To get some grasp at the structure of (A,B)-invariant subspaces we consider the 1-dimensional ones. In case of (S_D,π_D) the one dimensional subspaces are spanned by

$$\pi_D\left(\frac{D(\lambda) - D(\alpha)}{\lambda - \alpha}\ \xi\right)$$

where the projection π_D can be dropped whenever D^{-1} is proper. In that case we have also

$$\frac{D(\lambda) - D(\lambda)}{\lambda - \alpha}\ \xi = \sum_{j=1}^{s}\alpha^{j-1}E_j(\lambda)\xi\ .$$

In the case of the pair $(S_T,\pi_T U)$ the one dimensional (A,B)-invariant subspaces are spanned by vector polynomials of the form

$$\frac{T(\lambda)N(\alpha) - U(\lambda)D(\alpha)}{\lambda - \alpha}\ \xi$$

with $\alpha \in F$ and $\xi \in F^m$.

Next we pass to the characterization of the (A,B)-invariant subspaces that are included in the kernel of C. We start with the two coprime factorizations ND^{-1} and $T^{-1}U$ of a transfer function G. These are related to nonsingular right factors of the numerator polynomials N and U. For intuition let us consider the 1-dimensional subspaces. Such a subspace of K_D is of the form

$$\pi_D\left(\frac{D(\lambda) - D(\alpha)}{\lambda - \alpha}\right)\xi\ .$$

Applying the readout map C to it we obtain $N(\alpha)\xi$. So it is in $\operatorname{Ker} C$ if and only if $\xi \in \operatorname{Ker} N(\alpha)$. But this is equivalent, in a properly chosen basis, to $\operatorname{diag}(\lambda-\alpha,1,\dots,1)$ being a right factor of $N(\lambda)$. The same result is true in general.

To see that we note that a submodule of K_{D_1}, with $G = ND_1^{-1}$

and the realization the one associated with this factorization, a submodule $E_1K_{F_1}$ is in Ker C if and only if F_1 is a nonsingular right factor of N. This yields immediately the following characterization. A subspace $V \subset K_D$ is an (A,B)-invariant subspace included in Ker C if and only if

$$V = T_{DD_1^{-1}}(E_1K_{F_1})$$

with $D_1 = E_1F_1$, DD_1^{-1} a bicausal isomorphism and F_1 a right factor of N. This correspondence between nonsingular right factors of N and (A,B)-invariant subspaces in Ker C has first been established by Emre [7]. One notes immediately that there exists a nontrivial (A,B)-invariant subspace of K_D included in Ker C if and only if $G = ND^{-1}$ is feedback reducible. Thus connection is made with the work of Morse [30].

Inclusion relation between two (A,B)-invariant subspaces of K_D is connected with a division relation of matrix polynomials. Thus given two (A,B)-invariant subspaces V_1, V_2 associated with the nonsingular right factors F_1, F_2 of N then $V_1 \subset V_2$ if and only if F_1 is a right factor of F_2.

We conclude with the study of (A,B)-invariant subspaces included in Ker C where the realization is the one associated with the left coprime factorization $G = T^{-1}U$.

Given a nonsingular polynomial matrix $D \in F^{m\times m}[\lambda]$ then the elements of the module K_D can be characterized in either of the following equivalent ways

$$K_D = \{f \in F^m[\lambda] \mid D^{-1}f \in \lambda^{-1}F^m[[\lambda^{-1}]]\}$$

or $$K_D = \{f \in F^m[\lambda] \mid f = Dh, \quad h \in \lambda^{-1}F^m[[\lambda^{-1}]]\}$$

but whereas the first definition makes sense only for nonsingular D the second one makes sense for any rectangular polynomial matrix [8].

One expects a characterization of the (A,B)-invariant subspaces to be related to nonsingular factors of U in analogy with the previous situation. In fact if $U = E_1U_1$ is a factorization of U with E_1 nonsingular then it can be shown that $U = E_1K_{U_1}$ is an (A,B)-invariant subspace contained in Ker C. In particular, taking $E_1 = I, K_U$ is such a subspace. However, not all such subspaces have such a form. The characterization is again in terms of nonsingular right factors of U. If $U = U_0G_0$ with E_0 nonsingular then $V = U_0K_{E_0}$ is an (A,B)-invariant subspace contained in Ker C and each such subspace admits of such a representation. The full details can be found in [19].

To summarize we have seen a large number of analogies between finite and infinite dimensional system theories. While in finite dimensional theory polynomial methods have proved to be of considerable use in a variety of design problems and may prove even more so once a full connection is made with geometric control theory the serious study of feedback in an infinite dimensional setting has yet to begin. One hopes that with the finite dimensional theory put in terms that make sense to an operator theorist it will give enough insight as well as incentive for making progress in this difficult area.

REFERENCES

[1] Baras, J. S., and Brockett, R. W.: 1975, *H^2 functions and infinite dimensional realization theory*, SIAM J. Control, 13, pp. 221-241.

[2] Beurling, A.: 1949, *On two problems concerning linear transformations in Hilbert space*, Acta Math. 81, pp. 239-255.

[3] deBranges, L., and Rovnyak, J.: 1964, *The existence of invariant subspaces*, Bull. Amer. Math. Soc., 70, pp. 718-721.

[4] Brockett, R. W.: 1977, *The geometry of the set of controllable linear systems*, Reserach Report of Automatic Control Laboratory, Nagoya University, v. 24, pp. 1-7.

[5] Carleson, L.: 1962, *Interpolation by bounded analytic functions and the corona problem*, Ann. of Math., (2), 76, pp. 547-559.

[6] Dewilde, P.: 1976, *Input-output description of roomy systems*, SIAM J. Control, 14, pp. 712-736.

[7] Emre, E.: *Nonsingular factors of polynomial matrices and (A,B)-invariant and reachability subspaces*, Memorandum COSOR 78-19, Eindhoven University of Technology.

[8] Emre, E., and Hautus, M. L. J.: *A polynomial characterization of (A,B)-invariant and reachability subspaces*, Memorandum COSOR 78-19, Eindhoven University of Technology.

[9] Fuhrmann, P. A.: 1968, *On the corona theorem and its applications to spectral problems in Hilbert space*, Trans. Amer. Math. Soc., 132, pp. 55-66.

[10] Fuhrmann, P. A.: 1968, *A functional calculus in Hilbert space based on operator valued analytic functions*, Israel J. Math. 6, pp. 267-278.

[11] Fuhrmann, P. A.: 1975, *Realization theory in Hilbert space for a class of transfer functions*, J. Funct. Anal. 18, pp. 338-349.

[12] Fuhrmann, P. A.: 1975, *On generalized Hankel operators induced by sums and products*, Israel J. Math., 21, pp. 279-295.

[13] Fuhrmann, P. A.: 1976, *On series and parallel coupling of infinite dimensional linear systems*, SIAM J. Control, 14, p. 339-358.

[14] Fuhrmann, P. A.: 1976, *Algebraic system theory; An analyst's point of view*, J. Franklin Inst., 301, pp. 521-540.

[15] Fuhrmann, P. A.: 1977, *On strict system equivalence and similarity*, Int. J. Control, 25, pp. 5-10.

[16] Fuhrmann, P. A.: 1978, *Simulation of linear systems and factorization of matrix polynomials*, Int. J. Control, 28, pp. 689-705.

[17] Fuhrmann, P. A.: *Linear feedback via polynomial models*, Int. J. Control, to appear.

[18] Fuhrmann, P. A., and Willems, J. C.: *The factorization indices for rational matrix functions*, to appear.

[19] Fuhrmann, P. A., and Willems, J. C.: *A study of (A,B)-invariant subspaces by polynomial models*, to appear.

[20] Fuhrmann, P. A.: 1979, *Linear systems and operators in Hilbert space*, McGraw-Hill.

[21] Gohberg, I. C., and Krein, M. G.: 1960, *Systems of integral equations on a half line with kernels depending on the difference of arguments*, English translation, A.M.S. Translations (2), 14, pp. 217-287.

[22] Halmos, P. R.: 1961, *Shifts on Hilbert spaces*, J. Reine Angew. Math. 208, pp. 102-112.

[23] Hautus, M. L. J., and Heymann, M.: 1978, *Feedback--an algebraic approach*, SIAM J. Control and Optimization, 16, pp. 83-105.

[24] Helson, H.: 1964, *Lectures on invariant subspaces*, Academic Press, New York.

[25] Helton, J. W.: 1974, *Discrete time systems, operator models and scattering theory*, J. Funct. Anal., 16, pp. 15-38.

[26] Helton, J. W.: 1976, *Systems with infinite dimensional state space: The Hilbert space approach*, Proc. of the IEEE, 60, pp. 145-160.

[27] Lax, P. D.: 1959, *Translation invariant subspaces*, Acta Math. 101, pp. 163-178.

[28] Lax, P. D., and Phillips, R. S.: 1967, *Scattering theory*, Academic Press, New York.

[29] Livsic, M. S.: 1960, *On a class of linear operators in Hilbert space*, Amer. Math. Soc. Transl. (2), 13, pp. 61-83.

[30] Morse, A. S.: *Systems invariants under feedback and cascade control*, in Mathematical Systems Theory, Lecture Notes in Economics and Mathematical Systems, v. 131, Springer, pp. 61-74.

[31] Münzner, H. F., and Prätzel-Wolters, D.: 1978, *Minimalbasen polynomialer Modulen und Brunovsky-Transformationen*, Mathematik Arbeitspapiere, Nr. 12, Univ. Poremen.

[32] Rosenbrock, H. H.: 1970, *State space and multivariable theory*, J. Wiley, New York.

[33] Rota, G. E.: 1960, *On models for linear operators*, Comm. Pure Appl. Math., 13, pp. 469-472.

[34] Sarason, D.: 1967, *Generalized interpolation in* H^∞, Trans. Amer. Math. Soc., 127, pp. 179-203.

[35] Sz.-Nagy, B., and Foias, C.: 1970, *Harmonic analysis of operators on Hilbert space*, North Holland, Amsterdam.

[36] Wolovich, W. A.: 1974, *Linear multivariable systems*, Springer.

[37] Wonham, W. M.: 1974, *Linear multivariable control*, Springer.

STOCHASTIC SYSTEMS AND THE PROBLEM OF STATE SPACE REALIZATION

J. C. Willems* and J. H. van Schuppen**

*Mathematics Institute, University of Groningen
9700 AV Groningen, The Netherlands
**Mathematical Centre, Tweede Boerhaavestraat 49
1091 AL Amsterdam, The Netherlands

ABSTRACT

The purpose of this paper is to give an exposition of an approach to the problem of stochastic realization theory. We will introduce this problem through the concept of splitting relations and splitting random variables and show in detail how one can construct all minimal splitters for gaussian random vectors. With these ideas in mind, we then introduce the relevant definitions of (autonomous) stochastic dynamical systems and the problem of stochastic realization theory and of white noise representation as they arise naturally in this context. The case of gaussian random processes is worked out in detail.

1. INTRODUCTION

It may be argued that from the theoretical (and certainly from the pedagogical) point of view one of the most outstanding contributions of mathematical system theory has been the axiomatization of the concept of an abstract dynamical system and of the concept of state, in the context of systems with inputs and outputs. This has not only provided a long overdue generalization of the autonomous case, which has been studied in great detail in topological dynamics, but it also gives a very nice and useful axiomatic framework for the study of many problems in control theory, (recursive) signal processing, digital computation and automata theory, etc.

One of the central and completely new problems which has arisen quite naturally in this framework is the so-called *problem*

C. I. Byrnes and C. F. Martin, Geometrical Methods for the Theory of Linear Systems, 283-313.

of state space realization. This concerns the questions of representation of an input/output map as a system in state space form. For deterministic systems it may be said that the theory (far from being closed) is quite advanced and completely worked out, both on the abstract and on the algorithmic level, for linear time-invariant systems. However, this is not the case for stochastic systems where, other than some partial results for finite state stochastic stochastic automata and a fairly complete theory for finite dimensional gaussian processes, very little research has been done on these problems. In fact, a conceptual framework in which to treat these questions is still very much absent.

In the present paper we will attempt to give a systematic exposition of some of the main problems and results in this area. In view of the space limitation, we are unable to include proofs. These are either well-documented in the literature, or will appear elsewhere. We have concentrated our efforts for a great deal on some original aspects which involve introducing these problems via deterministic relations and splitting random variables and giving some general definitions of the abstract notion of a stochastic dynamical system and introducing the realization problem from this point of view. We will also give a rather complete description of the situation with gaussian processes. Unfortunately, we have not been able to include a review of some recent results on finite state processes (see [1] for a good exposition and [2,3] for some more recent results), nor have we been able to cover some recent work on a σ-algebraic approach to these problems [4].

Especially the realization of gaussian processes has been given a great deal of attention in the recent system theory literature. Although DOOB [5] already posed some questions in this direction, it is particularly since the work of KALMAN [6] that one has seen some significant progress in this area. Particularly important in this development has been the work of FAURRE [7,8,9] and of ANDERSON [10] who basically solved what we will call the "weak" or "measure theoretic" version of this problem and also showed its relation to the classical problem of spectral factorization. There has been some recent progress in this area through the work of LINDQUIST and PICCI [11,12,13,14,15] and of RUCKEBUSCH [16,17,18,19] which has culminated in a solution of what we will call the "strong" or the "output-induced" stochastic realization problem. In addition, the neat geometric approach followed by these authors has provided a very important and useful approach to this problem. In closing this introduction, we would like to point out some related work by AKAIKE [20] which fits rather well in the approach which we take in our paper. For a discussion of the relevance of stochastic realization theory in Kalman filtering, see [9 and 21].

ACKNOWLEDGEMENT. *The authors would like to acknowledge the help of Cees van Putten of the Mathematical Centre in Amsterdam.*

2. SPLITTING RELATIONS AND SPLITTING RANDOM VARIABLES

It turns out that many of the problems which one encounters in stochastic realization theory are already apparent in the seemingly trivial case of a time index set $T = \{1,2\}$. The idea of "state" then becomes that of a splitting random variable and the basic problem is to find an efficient way for constructing splitting random variables. Moreover, as far as we are aware of, it is now known that these concepts and problems also have very natural analogues in the context of deterministic systems. We will introduce the problems from this point of view since we believe that it gives one a very clear and sharp introduction to this area.

2.1 Relations

A relation is simply a subset of a product set. In our context it is best to think of the two components of this product set as representing the "past" and the "future" of some dynamical phenomenon. In a splitting relation one should think of the splitting variable as the state which contains the information in the past which is relevant for the future (and vice-versa!) With this intuitive picture in mind, we now proceed with the formal development:

A *relation* R on the product space $Z := Z_1 \times Z_2 \times \ldots \times Z_n$ is simply a subset $R \subset Z$. The subset of Z_i defined by

$$\{z_i \in Z_i \mid \exists z_1,\ldots,z_{i-1},z_{i+1},\ldots,z_n \text{ such that } (z_1,\ldots,z_{i-1},z_i,z_{i+1},\ldots,z_n) \in R\}$$

is called the *projection* of R on Z_i and will be denoted by $P_{Z_i}R$. The relation on $Z_1 \times \ldots \times Z_{i-1} \times Z_{i+1} \times \ldots \times Z_n$ defined by

$$R_{\{z_i=a_i\}} := \{(z_1,\ldots,z_{i-1},z_{i+1},\ldots,z_n) \mid (z_1,\ldots,z_{i+1},a_i,z_{i+1},\ldots,z_n) \in R\}$$

is called the *relation* R *conditioned by* $\{z_i=a_i\}$. Obvious generalization to the case $z_i \in A_i \subset Z_i$, or conditioning to more than one of the z_i's presents no difficulties. The following notions are the deterministic analogues of "white noise" and of a "Markov process." The relation R on Z is said to be a *product relation* if $R = \prod_i P_{Z_i} R$. We will say that z_i *splits*

R if

$$R_{\{z_i=a_i\}} = P_{Z_1 \times \ldots \times Z_{i-1}} R_{\{z_i=a_i\}} \times P_{Z_{i+1} \times \ldots \times Z_{i+1}} R_{\{z_i=a_i\}}.$$

If z_i is splitting for all i then we will call R *Markovian*. Using these notions one can develop a systematic and novel approach to deterministic system theory and its realization problem which is a bit more general and in some applications much more appropriate than the existing input/output approach. However, in the present paper we will only pursue the realization problem for $T = \{1,2\}$. Then we have the following definitions:

Let R_e be a given relation on $Y_1 \times Y_2$. If R is a relation on $Y_1 \times X \times Y_2$ such that

(i) x splits (for simplicity we call R then *splitting*), and

(ii) $R_e = P_{Y_1 \times Y_2} R$,

then we will call R a (splitting) *realization* of R.

Many of the qualitative notions of classical deterministic and stochastic realization theory admit very natural generalizations to this framework: Let R_1 and R_2 be two realizations of the same R_e with respective splitting spaces x_1 and x_2. Consider now the pre-ordering $R_1 > R_2$ defined by

$$\{R_1 > R_2\} :\Leftrightarrow \{\exists \text{ a partial surjective set to point map } f: X_1 \to X_2 \text{ such that } \{(y_1,x_2,y_2) \in R_2\} \Rightarrow \{\exists x_1 \in f^{-1}(x_2) \text{ such that } (y_1,x_1,y_2) \in R_1\}\}.$$

Two realization will be called *equivalent* if there exists a bijection $f: X_1 \to X_2$ such that $\{(y_1,x_1,y_2) \in R_1\} \Leftrightarrow \{(y_1,f(x_1),y_2) \in R_2\}$. A realization R of R_e is said to be *irreducible* if any other realization $R' > R$ is necessarily equivalent to R'. It is said to be *attainable* if for all $x \in X$ there is $(y_1,y_2) \in R_e$ such that $(y_1,x,y_2) \in R$ and such that $\{(y_1,x',y_2) \in R\} \Rightarrow \{x'=x\}$. It is said to be *observable* if $a \mapsto P_{Y_2} R_{\{x=a\}}$ is injective as a map from X to 2^{Y_2}. It is said to be *reconstructible* if $a \mapsto P_{Y_1} R_{\{x=a\}}$ is injective. It is easy to show that irreducibility implies attainability, observability, and reconstructibility, but as we shall see shortly, the converse is not necessarily true.

Let R be a realization of R_e. Then it is said to be *output induced* if there exists $f: Y_1 \times Y_2 \to X$ such that

$$\{(y_1,x,y_2) \in R\} \Leftrightarrow \{x = f(y_1,y_2)\} \ ;$$

it is said to be *past (future) output induced* if

$$f: Y_1 \to X \ (f: Y_2 \to X) \ .$$

(Actually output induced realizations have multiplicity one. The *multiplicity* of a point $(y_1,y_2) \in R_2$ in the realization R is the cardinality of the set

$$\{a \in X \mid (y_1.y_2) \in R_{\{x=a\}}\} \ .$$

In output induced realizations every point of R_e is thus covered exactly once.)

Geometric Illustration: Let R_e be a subset of $\mathbb{R}^2$. A realization of R_e is simply a family (parametrized by elements of X) of rectangles which together cover R_e exactly. If every point is covered once then this realization is output induced. It is irreducible if no non-trivial recombinations or deletion of these rectangles results in a new realization. Thus the problem of finding an irreducible realization is the problem of filling up a given set by a (in this sense) minimal number of (non-overlapping) rectangles. We can also view the realization R as a relation on $\mathbb{R} \times X \times \mathbb{R}$ which has rectangular x-level sets and which projected down along X yields R_e. This realization is output induced if R is a "surface" with a "global chart" $\mathbb{R}^2$.

The following proposition links some of the concepts introduced above:

Proposition. *Let* R *be a realization of* R_e. *Then*

(i) $\{R$ *is irreducible*$\} \Rightarrow \{R$ *is attainable, observable, and reconstructible*$\}$;

(ii) $\{R$ *is irreducible*$\} \Leftrightarrow \{R$ *is attainble*$\}$ *and* $\{\{R_{\{x\in A\}}$ *is rectangular*$\} \Leftrightarrow \{A$ *consists of at most one point*$\}\}$.

It would be of much interest to give this last property in (ii) a satisfactory system theoretic interpretation. Actually the above proposition falls considerably short from the results of the classical literature on deterministic realization theory. This is due to the fact that the realizations considered there are all past output induced (actually in that context it is better to speak of "past input induced").

Proposition. *Let* R *be a past output induced realization of* R_e. *Then* {R *is irreducible*} $\Leftrightarrow$

(i) {R *is attainable*} *(which in this case means that for all* $x \in X$ *there exists* $y_1 \in Y_1$ *such that* $x = f(y_1)$, *i.e., reachability), and*

(ii) {R *is observable*} .

An analogous proposition holds for future induced realizations. It would be of interest to discuss the cases in which {irreducibility} $\Leftrightarrow$ {reachability, observability, and reconstructibility}, a situation which we shall have in the case of gaussian random variables. In general however ($\Leftarrow$) does not hold, not even for output induced realizations, as the following picture decisively illustrates:

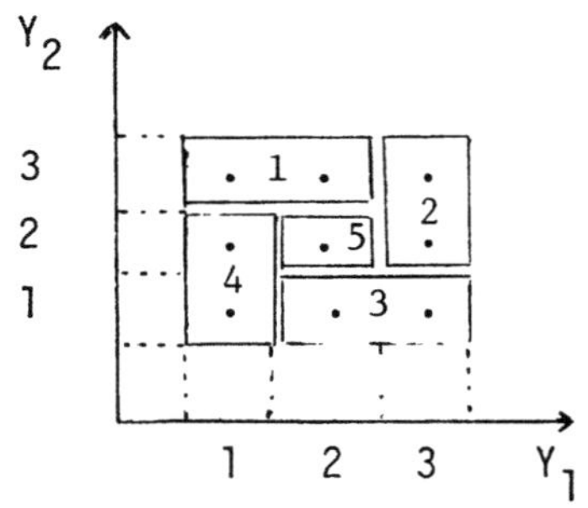

$Y_1 = Y_2 = \{1,2,3\}$

$X = \{1,2,3,4,5\}$

$R_e = Y_1 \times Y_2$

R: see picture:
(1,1,3) $\in$ R,
(2,1,3) $\in$ R, etc.

In trying to construct realizations there are three constructions which appear natural:

(i) by defining an equivalence relation E_1 on Y_1 by

$$\{y_1' E_1 y_1''\} :\Leftrightarrow \{\{(y_1', y_2) \in R_e\} \Leftrightarrow \{(y_1'', y_2) \in R_e\}\} ,$$

taking $X = Y_1/E_1$, and defining a (past output induced) realization from there in the obvious way as

$$R^+ = \bigcup_{(y_1, y_2) \in R_e} (y_1, y_1 (\text{mod } E_1), y_2) .$$

This realization is called the *forward canonical realization;*

(ii) using the same idea on the set Y_2, thus obtaining R^-, the (future output induced) *backward canonical realization*;

(iii) defining an equivalence relation E_{12} defined as the coarsest refinement of E_1 and E_2, i.e., the equivalence relation on $Y_1 \times Y_2$ defined by

$$\{(y_1', y_2') E_{12} (y_1'', y_2'')\} : \Leftrightarrow \{y_1' E_1 y_1'' \text{ and } y_2' E_2 y_2''\}$$

and proceeding in a similar fashion. The ensuing realization will be denoted by $R^{\pm}$.

The idea behind constructing R^+ is thus to view R_e as a map, f, from y_1 into 2^{Y_1} and to define the equivalence relation as the kernel of f. Hence every splitting element in R^+ can either be identified by a subset of Y_1 (the elements of the partition induced by f) or a subset of Y_2 (the elements of the range of f). The family of subsets of Y_1 form a partition and are thus non-overlapping while the family of subsets of Y_2 need not have such structure.

It is easy to see that R^+ and R^- are refinements of the realization in (iii), and thus in general this realization will not be irreducible. We have the following result which is a rather nice generalization of what can be obtained in the classical case:

Proposition .

(i) *The forward canonical realization* R^+ *is minimal in the class of past output induced realizations (in the sense that every other past output induced realization* R *satisfies* $R^+ < R$*) and thus irreducible.*

(ii) *All irreducible past output induced realizations are equivalent to* R^+ *and thus minimal and pairwise equivalent.*

(iii) *(see the previous proposition) A past output induced realization is irreducible iff it is reachable and observable.*

Needless to say that a similar proposition holds for R^-. Unfortunately general statements regarding the structure of the other (output induced) irreducible realizations appear hard to come by. We pose the following

Research Problem. Investigate whether every irreducible (output induced) realization R may be obtained from $R^{\pm}$ in the sense that $R < R^{\pm}$ and describe an effective procedure by which all irreducible realizations may be obtained from $R^{\pm}$.

2.2 Splitting Random Variables

The situation with random variables is much like the one with relations as explained in the previous section but where instead of having a yes-no situation on the elements in the relation one has a probability measure on the product space which expresses how likely it is for two elements to be related. Of course, all notions such as independence then need to be interpreted in a measure theoretic sense. However, in the context of realization theory problems, a new dimension is added in the problem. This is akin to the problem of output induced versus not output induced realizations as discussed in the previous sections, but is also related to the common dichotomy in probability theory which is concerned with the question whether the probability space $\{\Omega, \mathscr{A}, P\}$ is given (an impression which one gets from studying modern mathematical probability theory) or whether it is to be constructed (a point of view which appears much closer to what one needs in applications). In this section we will describe the random variable approach and the next section is devoted to the measure theoretic approach.

For a brief review of some relevant notions from probability theory, the reader is referred to the Appendix.

Let $\{\Omega, \mathscr{A}, P\}$ be a probability space, $(Y_1, \mathscr{Y}_1)$, $\{Y_2, \mathscr{Y}_2\}$, and $\{X, \mathscr{X}\}$ be measurable space, and $y_1, y_2, x: \Omega \rightarrow Y_1, Y_2, X$ random variables on Ω. We say that x *splits* y_1 and y_2 if y_1 and y_2 are conditionally independent given x, in which case we say that (y_1, x, y_2) realizes (y_1, y_2). A realization is said to be *irreducible* if

(i) $\{X, \mathscr{X}\}$ is Borel and x is surjective (see Appendix) (in which case we call the realization *attainable*), and

(ii) if $\{X', \mathscr{X}', x'\}$ is any other realization for which there exists a surjection $f: X \rightarrow X'$ such that the scheme

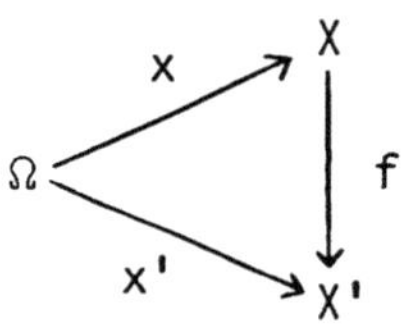

commutes, then f is injective (see Appendix).

Two realizations with respective splitting spaces X_1 and X_2 are said to be *equivalent* if there exists a bijection $f: X_1 \rightarrow X_2$ such that $x_2 = f(x_1)$. A realization is said to be *observable* if

$x \mapsto P(y_2|x)$ is injective and *reconstructible* if $x \mapsto P(y_1|x)$ is injective. It is easy to see that irreducibility implies attainability, observability, and reconstructibility, but the converse is in general not true, unless the random variables involved are all gaussian random vectors.

A realization is called (past, future) *output induced* if there exists $(f: Y_1 \to X,\ f: Y_2 \to X)$ $f: Y_1 \times Y_2 \to X$ such that $(f(y_1) = x,\ f(y_2) = x)$ $f(y_1, y_2) = x$. In an output induced realization the subset of $Y_1 \times X \times Y_2$ defined by

$$\{(y_1', x_1', y_2') \mid \exists\omega \text{ such that } (y_1', x', y_2') = (y_1(\omega), x(\omega), y_2(\omega))\}$$

is a surface parametrised by ω or by the "global chart" $Y_1 \times Y_2$.

Remark. The problem of finding a past output induced splitter is very akin to the Bayesian idea of sufficient statistic [22]. Formally: Let y_1, y_2 be random variables and assume that $f: Y_1 \to X$ is measurable. Then $x := f(y_1)$ is said to be a *sufficient statistic* for the estimation of y_2 through y_1 if x splits y_1 and y_2. Thus the problem of finding a sufficient statistic is the same as finding a past output based realization.

Proposition. *Let* (y_1, x, y_2) *be a past (future) output induced realization of* (y_1, y_2). *Then*

$$\{\textit{irreducibility}\} \Leftrightarrow \{\textit{attainability and observability (reconstructibility)}\}\ .$$

In general no such proposition is true for arbitrary realizations, not even for output induced ones. However for gaussian random variables we will show that {irreducibility} ⇔ {attainability, observability, and reconstructibility}. An example which shows where things can go wrong is the following:

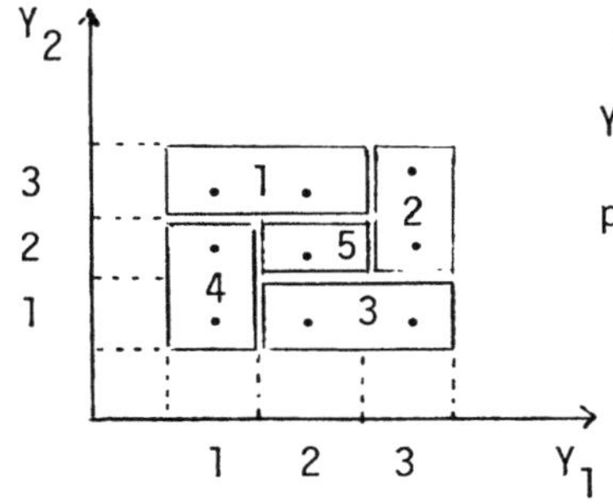

$\Omega = Y_1 \times Y_2$

$Y_1 = Y_2 = \{1,2,3\}$

$p(y_1, Y_2) = \frac{1}{9}$ for all y_1, y_2

$X = \{1,2,3,4,5\}$

The realization is the output induced realization defined by $f: Y_1 \times Y_2 \to X := \{1,3\} \mapsto 1,\ \{2,3\} \mapsto 1,\ \{3,3\} \mapsto 2,\ \{3,2\} \mapsto 2,\ \{3,1\} \mapsto 3,\ \{2,1\} \mapsto 3,\ \{1,1\} \mapsto 4,\ \{1,2\} \mapsto 4,\ \{2,2\} \mapsto 5$. It is easily shown that it is attainable, observable, and reconstructible. However the realization is not irreducible since $X' = \{1\}$, yields a reduced realization.

Similarly as in the deterministic case with relations one may construct the canonical past output induced realization R^+, the canonical future output induced realization R^-, and the join of both, $R^{\pm}$.

In constructing the forward canonical realization R^+, one considers the equivalence relation E_1 on Y_1, defined by

$$\{y_1' E_1 y_1''\} :\Leftrightarrow \{P(y_2\ y_1') = P(y_2\ y_1'')\}\ .$$

Defining now

$$X = Y_1/E_1, \quad \mathscr{X} = \mathscr{Y}_1/E_1 \quad \text{and} \quad x: \omega \to y_1(\omega)/E_1$$

leads to a past output induced realization of (y_1, y_2). Note that one may identify elements of X with subsets of Y_1 (those given by the partition of X induced by E_1) or with "random" probability measures on Y_2 (given by $P(y_2|y_1)$). However, in the second parametrization it is a bit more difficult to give the exact nature of the subset of probability measures on Y_2 which are thus obtained. It is possible to formulate a proposition which reads identical to the last proposition in Section 2.1. Its proof presents no difficulties, at least in the countable or smooth finite dimensional case. The technical details however still need to be worked out. We formulate this as a

Research Problem. Prove the stochastic analogon of the last proposition of Section 2.1 in the case that Y_1 and Y_2 are arbitrary measurable spaces. Investigate whether every irreducible (output induced) realization may be obtained from $R^{\pm}$ in a similar manner (a surjective set to point mapping on the splitting spaces) as is the case for (deterministic) relations.

2.3 Splitting measures

Much of what has been said in Section 2.2 may be repeated for the case in which a probability measure is given on $Y_1 \times Y_2$ or on $Y_1 \times X \times Y_2$ directly. We will not give all the relevant definitions in detail but restrict ourselves to that of a realization.

Let $\{Y_1, \mathscr{Y}_1\}$ and $\{Y_2, \mathscr{Y}_2\}$ be measurable spaces and let P_e be a probability measure on $Y_1 \times Y_2$. A *realization* of P_e

is defined by a measurable space $\{X,\mathscr{X}\}$ and a probability measure on $Y_1 \times X \times Y_2$ which induces P_e on $Y_1 \times Y_2$ and which is such that x splits y_1 and y_2.

The problem of realizing two given random variables (y_1,y_2) may hence be interpreted in the sense of the notions defined in Section 2.2 or in the sense of the above definition where we take the given measure on $Y_1 \times Y_2$ which is to be realized to be the one induced on $Y_1 \times Y_2$ by the probability measure on Ω. We now formalize these two possible types of realizations of given random variables.

Let $\{\Omega_1,\mathscr{A}_1,P_1\}$ and $\{\Omega_2,\mathscr{A}_2,P_2\}$ be two probability spaces, $\{Y,\mathscr{Y}\}$ a measurable space, and $z_1: \Omega_1 \to Y$, $z_2: \Omega_2 \to Y$ be two random variables on Ω_1 and Ω_2 respectively. We will say that z_1 and z_2 are *equivalent* if the measure induced by z_1 on Y is the same as the one induced by z_2 on Y.

Problem 1 *(the strong realization problem)*. Let (y_1,y_2) be given random variables defined on a probability space $\{\Omega,\mathscr{A},P\}$. The strong realization problem consists in finding a measurable space $\{X,\mathscr{X}\}$ and a random variable $x: \Omega \to X$ such that (y_1,x,y_2) is a realization of (y_1,y_2).

Problem 2 *(the weak realization problem)*. Let (y_1,y_2) be given random variables defined on a probability space $\{\Omega,\mathscr{A},P\}$. The weak realization problem consists in finding a probability space $\{\Omega',\mathscr{A}',P'\}$, a measurable space $\{X,\mathscr{X}\}$, and random variables $y_1',x,y_2': \Omega' \to Y_1,X,Y_2$ such that x splits y_1' and y_2' and such that (y_1',y_2') is equivalent to (y_1,y_2).

It is clear from these problem statements that Problem 2 is actually a problem which involves finding a splitting measure and it is best to think about it in these terms, without involving Ω at all, but starting from the measures induced on $Y_1 \times Y_2$. In considering Problem 1 it is unclear what Ω should be and usually one would take $\Omega = Y_1 \times Y_2$. The problem then becomes precisely the problem of finding output induced realizations as discussed in Section 2.2. A possible and meaningful generalization on which very little work has been done so far is to start with three random variables y_1, y_2, and z, defined on Y_1, Y_2, and Z respectively, taking $\Omega = Y_1 \times Y_2 \times Z$ and finding strong realizations of (y_1,y_2). One could also ask the question if there are z-induced realizations. In this context z would thus be the random variable which carries the information on which the splitter x has to be based.

As we shall already see in the next section there is very much of a difference in the specific solutions of the strong and the weak realization problems.

2.4 The gaussian case

In this section we will give a rather complete picture of the problems formulated in the previous sections in an important particular case, namely when all the random variables are jointly gaussian. We will thus assume that (y_1, y_2) is a real zero mean gaussian random vector with $y_1 \in \mathbb{R}^{n_1}$ and $y_2 \in \mathbb{R}^{n_2}$. Also, we will be looking for realizations (y_1, x, y_2) which are real zero mean gaussian random vectors with $x \in \mathbb{R}^n$. The measure of (y_1, x, y_2) is hence completely specified by

$$\Sigma = \left(\begin{array}{c|c|c} \Sigma_{11} & \Sigma_{1x} & \Sigma_{12} \\ \hline \Sigma_{x2} & \Sigma_{xx} & \Sigma_{x2} \\ \hline \Sigma_{21} & \Sigma_{2x} & \Sigma_{22} \end{array}\right)$$

where $\Sigma_{11} := \mathscr{E}\{y_1 y_1^T\}$, $\Sigma_{1x} := \mathscr{E}\{y_1 x^T\}$, etc.

The conditional independence condition is specified in the following:

Proposition. *The following conditions are equivalent:*

(i) x *splits* y_1 *and* y_2;

(ii) $y_1 - \mathscr{E}\{y_1 | x\}$ *and* $y_2 - \mathscr{E}\{y_2 | x\}$ *are independent;*

(iii) *(if* $\Sigma_{xx} > 0$*)* $\Sigma_{12} = \Sigma_{1x} \Sigma_{xx}^{-1} \Sigma_{x2}$.

Also the conditions for the irreducibility, attainability, etc., of a realization are easily established:

Proposition. *Assume that* (y_1, x, y_2) *is a realization of* (y_1, y_2). *Then*

(i) $\dim x =: n \geq \operatorname{Rank} \Sigma_{12} =: n_{12}$;

(ii) $n = n_{12}$ *iff the realization is irreducible;*

(iii) $\Sigma_{xx} > 0$ *iff the realization is attainable;*

(iv) $\operatorname{Rank} \Sigma_{2x} = n_{12}$ *iff the realization is observable;*

(v) $\operatorname{Rank} \Sigma_{1x} = n_{12}$ *iff the realization is reconstructible;*

(vi) $n = n_{12}$ *iff* $\Sigma_{xx} \quad 0$ *and* $\operatorname{Rank} \Sigma_{2x} = \operatorname{Rank} \Sigma_{1x} = n_{12}$.

As a consequence of property (ii) it is natural to call irreducible realizations *minimal*. Notice that for this gaussian case, even for realizations which are not output induced, we obtain the equivalence {irreducibility} ⇔ {attainability, observability, and reconstructibility}. Note also that here irreducibility means that whenever a surjective matrix S is such that $\{y_1, Sx, y_2\}$ is also a realization of (y_1, y_2), then S is necessarily square and invertible.

In most of the problems of realization theory the choice of the bases is immaterial and we may thus choose them to our convenience. The choice of the bases of the vector spaces in which y_1 and y_2 lie will be chosen so as to give us the canonical variable representation, as introduced by HOTELLING [23].

Lemma. *There exist nonsingular matrices* S_1 *and* S_2 *such that the covariance matrix of* $\bar{y}_1 := S_1 y_1$ *and* $\bar{y}_2 := S_2 y_2$, *defined by*

$$\bar{\Sigma} := \left[\begin{array}{c|c} \bar{\Sigma}_{11} & \bar{\Sigma}_{12} \\ \hline \bar{\Sigma}_{21} & \bar{\Sigma}_{22} \end{array}\right]$$

with $\bar{\Sigma}_{11} := \mathcal{E}\{\bar{y}_1 \bar{y}_1^T\}$, $\bar{\Sigma}_{12} = \bar{\Sigma}_{21}^T := \mathcal{E}\{\bar{y}_1 \bar{y}_2^T\}$, *and* $\bar{\Sigma}_{22} := \mathcal{E}\{\bar{y}_2 \bar{y}_2^T\}$ *takes the form*

$$\left[\begin{array}{cccc|cccc} I & 0 & 0 & 0 & I & 0 & 0 & 0 \\ 0 & I & 0 & 0 & 0 & \Lambda & 0 & 0 \\ 0 & 0 & I & 0 & 0 & 0 & 0 & 0 \\ 0 & 0 & 0 & 0 & 0 & 0 & 0 & 0 \\ \hline I & 0 & 0 & 0 & I & 0 & 0 & 0 \\ 0 & \Lambda & 0 & 0 & 0 & I & 0 & 0 \\ 0 & 0 & 0 & 0 & 0 & 0 & I & 0 \\ 0 & 0 & 0 & 0 & 0 & 0 & 0 & 0 \end{array}\right]$$

- indentical components
- correlated components
- independent components
- zero components

where $\Lambda = \text{diag}(\lambda_1, \lambda_2, \ldots, \lambda_{n_{21}})$ *with* $1 > \lambda_1 \geq \lambda_2 \geq \ldots \geq \lambda_{n_{21}} > 0$.

We will denote the various components of $\bar{y}_1$ and $\bar{y}_2$ in this basis by $\bar{y}_{11}$, $\bar{y}_{12}$, etc. and their dimension by n_{11}, n_{12}, etc. Moreover, since we will assume that this basis transformation has been carried out we will drop the bars on the y's.

The components of y_1 and y_2 in this representation are called *canonical variables*. They are very useful in statistical analyses. They are unique modulo the following transformation: an orthogonal transformation on y_{11} and y_{21}, one on y_{13}, one on y_{23}, one on y_{14}, and one on y_{24}. Moreover, if in the sequence of λ_i's there is equality:

$$\lambda_{i-1} > \lambda_i = \lambda_{i+1} = \dots = \lambda_{i+k} > \lambda_{i+k+1},$$

then one can also apply an orthogonal transformation on the component of y_{21} and y_{22} corresponding to these equal λ_i's.

The above lemma shows how the basis for y_1 and y_2 is chosen. We choose the basis of x as follows:

Proposition. *Assume that the* (y_1,x,y_2) *is irreducible. Then we may always choose the basis for* x *such that* $\mathcal{E}\{y_2|x\} = x$. *This implies together with conditional independence that* $\Sigma_{2x} = \Sigma_{xx}$ *and* $\Sigma_{1x} = \Sigma_{12}$.

The following two theorems are the main results of this section. The first theorem solves the "weak" realization problem for gaussian random vectors, while the second theorem solves the "strong" realization problem. The interpretation of the output induced irreducible realizations in terms of canonical variables is rather striking.

Theorem. *In the bases given,* (y_1,x,y_2) *will be a minimal weak realization of* (y_1,y_2) *iff the correlation matrix of* (y_1,x,y_2) *takes the form*

$$\left(\begin{array}{cccc|cc|cccc}
I & 0 & 0 & 0 & I & 0 & I & 0 & 0 & 0\\
0 & I & 0 & 0 & 0 & \Lambda & 0 & \Lambda & 0 & 0\\
0 & 0 & I & 0 & 0 & 0 & 0 & 0 & 0 & 0\\
0 & 0 & 0 & 0 & 0 & 0 & 0 & 0 & 0 & 0\\
\hline
I & 0 & 0 & 0 & I & 0 & I & 0 & 0 & 0\\
0 & \Lambda & 0 & 0 & 0 & \Sigma & 0 & \Sigma & 0 & 0\\
\hline
I & 0 & 0 & 0 & I & 0 & I & 0 & 0 & 0\\
0 & \Lambda & 0 & 0 & 0 & \Sigma & 0 & I & 0 & 0\\
0 & 0 & 0 & 0 & 0 & 0 & 0 & 0 & I & 0\\
0 & 0 & 0 & 0 & 0 & 0 & 0 & 0 & 0 & 0
\end{array}\right)$$

with Σ *any matrix satisfying*

$$\boxed{\Lambda^2 \leq \Sigma \leq I}$$

The above theorem seems more complicated than it is because we have taken a completely general case for (y_1,y_2). The point however is that in this choice of the bases it is exceedingly simple to see how the correlation matrix of (y_1,x,y_2) can look like. Note that the components which are common to y_1 and y_2 will appear in every realization as components of x. The uncorrelated components on the other hand do not influence x.

Theorem. *Assume that* (y_1,x,y_2) *is a minimal output induced realization of* (y_1,y_2). *Then there exists a choice of canonical variables for* y_1 *and* y_2 *such that in a suitable basis* x *is given by* $x = (y_{11} = y_{21},z_1,z_2)$ *with* z_1 *a vector consisting of some components of* y_{21} *and* z_2 *a vector consisting of the other components of* y_{22}. *Conversely, for every choice of the canonica- variables and every such choice of* z_1 *and* z_2, x *will be an output induced minimal realization of* (y_1,y_2).

Let us call two gaussian realizations (y_1,x_1,y_2) and (y_1,x_2,y_2) *equivalent* if there exists a non-singular matrix S such that $x_2 = Sx_1$. Otherwise they will be called *distinct*. From the above theorems the following corollary is immediate:

Corollary. *The number of distinct weak realizations is one in the case* $n_{21} = n_{22} = 0$ *and non-denumerably infinite otherwise. The number of distinct output induced minimal realizations is* $2^{n_{21}} = 2^{n_{22}}$ *if* $1 > \lambda_1 > \lambda_2 > \dots > \lambda_{n_{21}} > 0$, *and non-denumerably infinite otherwise.*

We close this section with some remarks:

1. Much of the structure of the realizations shown in the above theorem may be found in one form or another in the work of RUCKEBUSH [see e.g. 16,17, 18,19].

2. In order to generate x in a weak realization, one has to add a source of randomness which is external to (y_1,y_2). In fact, one needs exactly $\text{rank}(\Sigma-\Lambda^2) + \text{rank}(I-\Sigma) - n_{12}$ additional independent random variables to achieve a minimal weak realization.

3. It is of interest to develop the above theory for given gaussian vectors (y_1,y_2,z) and requiring x to be z-induced in the realization (y_1,x,y_2) of (y_1,y_2).

3. STOCHASTIC DYNAMICAL SYSTEMS

We will in this paper exclusively be concerned with stochastic systems without external inputs (with this we mean that the dynamics are influenced by a chance variable but not by other "external" inputs). We will therefore introduce the relevant concepts on this level of generality. Properly speaking we are concerned with *autonomous* stochastic systems. These are described by a stochastic process.

3.1 Basic Definitions

Definition. A *stochastic system (in output form)*, Σ_e, is defined by

(i) a probability space $\{\Omega,\mathcal{A},P\}$,

(ii) a *time index set* $T \subset \mathbb{R}$,

(iii) a measurable space $\{Y,\mathcal{Y}\}$ called the *output space*, and

(iv) a stochastic process $y: T \times \Omega \rightarrow Y$.

It is said to be *time-invariant* if T is an interval in $\mathbb{R}$ or $\mathbb{Z}$, and

(v) y is stationary.

Many of our comments are in the first place relevant to the time-invariant case.

Let $z: \Omega \times T \rightarrow Z$ be a process. We will denote by $z_t^- := \{z(\tau),\tau < t\}$ the *past* and by $z_t^+ := \{z(\tau),\tau > t\}$ the *future* of z. Let $z,r: \Omega \times T \rightarrow Z,R$ be two processes. We will say that r *splits* z if $r(t)$ splits z_t^- and $(z_t^+,z(t))$ for all t. Notice that this definition is not symmetric in time.

Definition. Let Σ be a stochastic system on $X \times Y$. Then it will be said to be in *state space form* with *state space* X and *output space* Y if x splits (x,y). The external behavior of Σ is simply the process y which we may of course consider to be a stochastic system in its own right. We will denote it by Σ_e and say that Σ *realizes* Σ_e, denoted by $\Sigma \Rightarrow \Sigma_e$.

Remarks.

1. It is obvious from the above definition that x will be a Markov process. Let R denote the time-reversal operator, i.e., $Rz(t) := z(-t)$. Note that $R\Sigma$ will in general not be a system in state space form. Thus, contrary to Markov processes, a state space system forward in time need not be a state space system backwards in time. However, in the continuous time case there is a weak condition under which we will have this time-reversibility,

i.e., when the σ-algebras induced by $\{y(\tau),\tau < t\}$ and $\{y(\tau),\tau \leq t\}$ are equal for all t. This will be the case whenever the sample paths of y are smooth in some appropriate sense. Actually, in that case, there exists a map $f: X \times T \to Y$ such that $y(t) = f(x(t),t)$.

2. It is likely that there are many applications (e.g. in recursive signal processing and in stochastic control) where the relevant property is that the process x splits y, and that the state property expressed in the fact that x splits (x,y) is not as crucial as we have learned to think.

The problem of (state space) realization is then simply the following: *Given* Σ_e, *find* Σ *such that* $\Sigma \Rightarrow \Sigma_e$. This problem is the strong realization problem. The alternative approach to stochastic systems where one starts with measures leads to the following

Definition. *A stochastic system (in output form) in terms of its measures* is defined by giving for all $t_1,t_2,\ldots,t_n \in T$ a probability measure on Y^n satisfying the usual compatibility conditions. Such a system defined on $X \times Y$ is said to be in *state space form* if for all $t_1 \leq t_2 \leq \ldots \leq t_n \leq t$ and for every bounded real measurable function f there holds

$$\mathscr{E}\{f(x(t),y(t)) \mid (x(t_i),y(t_i)),\ i = 1,2,\ldots,n\} =$$

$$\mathscr{E}\{f(x(t),y(t)) \mid x(t_n)\} .$$

Instead of specifying the measures directly as in the above definition one can do this in terms of appropriate kernels which express how a given state $x(t_0)$ will result in a state/output pair $(x(t_1,y(t_1))$ at $t_1 \geq t_0$. This construction requries only a slight extension from the usual construction of Markov kernels and we will not give them in detail here.

From the above definition it should be clear what one means by the externel behavior of a system defined in terms of its measures and hence by the weak version of the state space realization problem. If we also consider the fact that for every process defined in terms of its marginal probability laws one can construct a probability space and an equivalent stochastic process on it, then we see that this weak version of the realization problem may be expressed in the following, albeit somewhat indirect, way:

Definition. Two stochastic systems Σ_e' and Σ_e with the same time index set and output space are said to be *equivalent* if the defining processes are equivalent in the sense that they have the same marginal measures. We will call Σ' a *weak realization* of Σ_e

if Σ' is in state space form and if Σ'_e is equivalent to Σ_e.

All of the definitions (irreducibility, attainability, equivalence, observability, and reconstructibility) have obvious generalizations to the problem at hand. We will not give them explicitly. Instead we turn to a topic which we will only touch on very briefly but which fits very well in an exposition on the representation of stochastic systems.

3.2 White Noise Representation

Often, particularly in the engineering literature, one starts with a stochastic system which is actually defined in terms of a deterministic system drive by "white noise." This starting point may be introduced in our framework as follows:

Definition. Let T be a (possibly infinite) interval in $\mathbb{Z}$. A *discrete time white noise driven stochastic system* (denoted by Σ_r: r stands for "recursive") is defined by

(i) three processes, $w,x,y: T \to W,X,Y$ with w white noise and (x,y) a stochastic system in state space form, such that

(ii) x_t^- and $(w(t),w_t^+)$ are independent for all t, and

(iii) two maps $f,r: X \times W \times T \to X,Y$ called respectively the *next state map* and the *read-out map*, such that

$$x(t+1) = f(x(t),w(t),t)$$

$$y(t) = r(x(t),w(t),t) .$$

There is clearly an analogue of this definition for systems defined in terms of their measures. We will denote the system Σ in state space form induced in the obvious way by Σ_r by $\Sigma_r \Rightarrow \Sigma$. The *white noise representation problem* is the problem of finding for a given stochastic system in state space form Σ a white noise driven system Σ_r such that $\Sigma_r \Rightarrow \Sigma$. Thus in this problem one is asked to construct the white noise process w and maps f and g. There is also an obvious "weak" version of this problem.

The following definition gives a limited continuous time version of the above:

Definition. Let T be a (possibly infinite) interval in $\mathbb{R}$. A continuous time *gaussian white noise driven stochastic system* (denoted by Σ_d: d stands for "differential") is defined by

(i) three processes $w,x,y: T \to W,X,Y$ with w an m-dimensional Wiener process on T, X a subset of $\mathbb{R}^n$, and (x,y) a stochastic system in state space form, such that

(ii) x_t^- and $(w - w(t))_t^+$ are independent for all t, and

(iii) three maps $f,h,r: X \times T \to \mathbb{R}^n, \mathbb{R}^{n\times m}, Y$ respectively called the *local drift*, the *local diffusicn* and the *read-out map*, such that

$$dx(t) = f(x(t),t)dt + h(x(t),t)dw(t),$$

$$y(t) = r(x(t),t) .$$

There are obvious generalizations of these notions to independent increment processes and, more to the point, to more general stochastic differential equations admitting, for example, also jump processes.

Note that if in the discrete time (or the continuous time) case T has a lower bound, say $t = 0$, then $x(0),w,f$ and r (resp. f,h and r) determine completely x and y. A similar property occurs when T has no lower bound but when the differential equation for x has certain asymptotic stability properties. In such case a white noise representation of a realization of a given dynamical system in output form will yield maps by which $y(t)$ is expressed, in a non-anticipating manner, as a function of the independent random variables $x(0),w(0),w(1).,...,w(t)$.

3.3 Research Problems

In this section we will indicate the sort of questions one asks in realization theory and, as far as this is known, how one expects the answers to look. In Section 4 it will be shown how some of these questions may be answered for stationary finite-dimensional gaussian processes. The claims made in the various questions will all require certain regularity conditions which we will not be concerned with here.

1. Let Σ_e be a given stochastic system. The basic problem is to construct a state space representation of it. Actually, if no other properties are asked then this problem is somewhat trivial. The following question however is much more useful and interesting: Find all (weak and strong) irreducible realizations of Σ_e. Give conditions under which Σ_e admits a finite state realization or a finite dimensional realization. Give algorithms for going from a numerical specification of Σ_e to a numerical specification of an irreducible realization.

2. Describe the construction of the canonical past output and future output induced realizations. Show in what sense they are unique. Prove (or disprove) that all irreducible realizations may be derived from the join of the past and the future induced realizations, and describe how all irreducible realizations could this way be constructed.

3. Let Σ_e be given and assume that z splits y. It is easily seen how to define irreducibility of z as a splitting process. It is not unreasonable to expect that an irreducible splitting process z would yield an (irreducible) realization (z,y) of y. It would appear from [15] that this will not always be true. It is of interest to settle this problem and give the additional conditions which z needs to satisfy.

4. When does a discrete time system admit a white noise representation? The expectation is that for weak representations this will always be possible but that strong white noise representability is rather special. An appropriately general concept of white noise representability must undoubtedly allow the space W where the white noise takes its values to be x-dependent, pretty much like the situation with vector fields and bundles as a description of flows on manifolds. It is of much interest to clarify the probabilistic versions of such recursive stochastic models.

5. When does a continuous time system admit a white noise representation? The expectation here is that for weak representations this ought to require only some smoothness of the random process x as a function of t while for strong realizations it will also involve some sort of constant rank condition on the local covariance of x. What sort of uniqueness results can be gotten here? Some partial results for scalar equations are available in [5,24]. Extend these representation problems to differential equations involving jump processes.

6. Clarify the relations between the stochastic realization problem and the problem of white noise representation with the problem of representing a given stochastic process as a non-anticipating function of a sequence of independent random variables. To be more specific, assume that a stochastic system has a state space realization which admits a white noise representation. This representation will often lead, as explained before, to a representation $y(t) = G_t(w(t),w(t-1),\dots,w(0),x(0))$ when T has a lower bound or $y(t) = G_t(w(t),w(t-1),\dots)$

when T has no lower bound. Clearly this last representation is not always possible. The question is to prove when and how it is possible: e.g., is it possible for ergodic processes? In what sense are such representations unique?

7. Show the relation and the interaction between forward time and backward time realizations and white noise representations. Investigate possible applications to filtering, prediction, smoothing, and stochastic control.

8. Generalize the concepts given here to stochastic systems which are influenced not only by a chance variable w, but also by deterministic external inputs.

4. REALIZATION OF GAUSSIAN PROCESSES

In this section we will treat gaussian processes. Most of the results which we present have been known since the work of FAURRE [7,8,9]. We will briefly touch on the output induced realizations which have been discovered rather recently by LINDQUIST and PICCI [12] and RUCKEBUSH [16]. Actually, this whole area is still in a lot of motion and it is difficult to do justice to all the ideas which have recently been put forward. The importance of these results in recursive signal processing and filtering are discussed in [9,21]. For simplicity and in view of the space limitation, we will only consider the continuous time case.

Let $y: \Omega \times \mathbb{R} \to \mathbb{R}^p$ be a zero mean stationary gaussian process, defined on a probabiltiy space $\{\Omega, \mathcal{A}, P\}$. We assume y to be mean square continuous (since, as we shall see later, we are really interested in the case that y has a rational spectral density, this implies no real loss of generality). As we have already argued in Section 3, this assumption allows us to compensate on state space realizations of the form $y(t) = f(x(t))$ with x on a Markov process. In particular, forward time realizations are thus automatically backward time realizations, a situation which is no longer valid in the discrete time case. The main problems which one considers in this area are the following:

Problem 1 *(the weak realization problem)*. Under what conditions does there exist a zero mean stationary $\mathbb{R}^n$-valued Gauss-Markov process x and a matrix $C: \mathbb{R}^n \to \mathbb{R}^p$ such that y is equivalent to Cx (which in this case simply means that y and Cx should have the same autocorrelation function)? What is the minimum dimension of x for which this is possible? Develop algorithms for deriving C and the parameters of the Markov process x.

Problem 2 *(the strong realization problem).* If Problem 1 is solvable, will there also exist such Gauss-Markov processes x which are output induced (i.e., for which there exists a map $y \to x$)? Develop algorithms for deriving the parameters of the Gauss-Markov process in this case.

Problem 3 *(the white noise representation problem).* Under what conditions does there exist an m-dimensional Wiener process w, an n-dimensional Gauss-Markov process x, both defined on $\mathbb{R}$, and appropriately sized matrices A,B,C such that

(i) x_t^- is independent of $(w - w(t))_t^+$,

(ii) $dx(t) = Ax(t)dt + Bdw(t)$, $y(t) = Cx(t)$.

Develop algorithms for computing the matrices A,B,C.

The marginal probability measures of y (and thus of all processes which are equivalent to it) are completely specified by its *autocorrelation* function $R: \mathbb{R} \to \mathbb{R}^{p\times p}$, defined by

$$R(t) := \mathcal{E}\{y(t)y^T(0)\} .$$

The restriction of R to $[0,\infty)$ will be denoted by R^+. Of course, R^+ specifies R completely since $R(t) = R^T(-t)$. Let $\Phi: \mathbb{R} \to \mathbb{R}^{p\times p}$ denote the *spectral density function* of y. Very roughly speaking Φ is defined as the Fourier transform of R. The following lemma is well-known:

Lemma. *The following conditions are equivalent:*

(i) R^+ *is Bohl (i.e., every entry of* R^+ *is a finite sum of products of a polynomial, an exponential, and a trigonometric function);*

(ii) Φ *is rational;*

(iii) *there exist matrices* $\{F,G,H\}$ *such that* $R^+(t) = He^{Ft}G$, *with* $F \in \mathbb{R}^{n\times n}$, $G \in \mathbb{R}^{n\times p}$, *and* $H \in \mathbb{R}^{p\times n}$.

In addition:

(iv) *there is a minimal* n, n_{min}, *for which the factorization in* (iii) *is possible, called the McMillan degree of* R^+ *or* Φ, *and* $n = n_{min}$ *iff* (F,G) *is controllable and* (F,H) *is observable. The triple* (F,G,H) *is then called minimal;*

(v) *all* (F,G,H)*'s with* $n = n_{min}$ *are obtainable from one by the transformation group*

$$(F,G,H) \xrightarrow[\det S\neq 0]{S} (SFS^{-1},SG,HS^{-1}).$$

Assume that x is an n-dimensional zero mean stationary Gauss-Markov process with $\mathcal{E}\{x(t)x^T(t)\} =: \Sigma > 0$. Then x is mean square continuous and hence $\mathcal{E}\{x(t) \mid x(0)\}$ for $t \geq 0$ is of the form $e^{At}x(0)$ for some A. Moreover it is easily seen that Σ and A completely specify the marginal measures of this Markov process. Hence the marginal measures of the process Cx are completely specified by (A,Σ,C) and it makes sense to talk about this triple as defining a weak realization of y. We will call a weak realization (x,y) with x n-dimensional and n as small as possible a *minimal* realization.

Proposition. *Consider the stochastic system defined by* (A,Σ,C). *Then*

(i) *it is a weak realization of* y *iff* $Ce^{At}\Sigma C^T = R(t)$ $(t \geq 0)$;

(ii) *it is attainable iff* $\Sigma > 0$;

(iii) *it is observable iff* (A,C) *is observable;*

(iv) *it is reconstructible iff* $(A^T, C\Sigma)$ *is observable;*

(v) *it is irreducible iff* (A, C^T, C) *is a minimal realization of* R^+;

(vi) *it is irreducible iff it is minimal and* n_{min} = *the McMillan degree of* R^+.

Part (i) of the proposition may be verified by direct calculation. The claim about reconstructibility may be seen from considering the stochastic system $R(x,y)$ with R the time-reversal operator. It is easily calculated that this stochastic system has parameter matrices $(\Sigma A^T\Sigma^{-1},\Sigma,C)$. This yields, after applying (iii), the condition in (iv).

Two (gaussian) stochastic systems (x_1,y) and (x_2,y) will be called *linearly equivalent* if there exists a nonsingular matrix X such that $x_2(t) = Sx_1(t)$. The following theorem classifies all minimal weak realizations of y up to linear equivalence:

Theorem

(i) *There exists a finite dimensional weak realization of* y *iff* y *has rational spectral density (for various equivalent conditions, see the previous lemma);*

(ii) *all minimal realizations* (A, Σ, C) *can, up to linear equivalence, be obtained from a minimal factorization triple* (F,G,H) *of* R^+ *by taking* $A = F$, $C = H$, *and solving the equations*

$$\boxed{F\Sigma + \Sigma F^T \leq 0, \ \Sigma H^T = G}$$

for $\Sigma = \Sigma^T$.

The problem thus becomes one of solving a set of inequalities. Actually, it may be shown [9,25,26] that there exist solutions Σ_-, Σ_+ such that every solution Σ satisfies $0 < \Sigma_- \leq \Sigma \leq \Sigma_- < \infty$. Moreover, the solution set is convex and compact. A great deal of additional information on the structure of the solution set of these equations may be found in the above references.

Note that choosing $A = F$ and $C = H$ in the above theorem corresponds to fixing the basis in state space. Indeed, since $\mathcal{E}\{y(t) \mid x(0)\} = Ce^{At}x(0)$ for $t \geq 0$, this choice of the basis of the state space is very much like the situation in Section 3.3 where we also fixed the basis for x this way. Once the basis has been picked, it is only the covariance of x, Σ, which remains to be chosen.

There seems to be some applications, even in filtering, where <u>any</u> (also an indefinite) solution of the equality $\Sigma H^T = G$ can be used. This point was raised in Faurre's thesis [7] but seems to have been ignored since. That may have been a pity, since it is the inequality part which makes these equations hard to solve.

The strong realizations are covered in the following theorem:

<u>Theorem</u>

(i) *There exists a finite dimensional output induced realization of* y *iff* y *has rational spectral density (for various equivalent conditions, see the previous lemma);*

(ii) *the parameter matrices* (A,Q,C) *of all minimal output induced realizations of* y *may, up to linear equivalence, be obtained from a minimal factoriation triple* (F,G,H) *of* R^+ *by taking* $A = F$, $C = H$, *and solving the equations*

$$\boxed{F\Sigma + \Sigma F^T \leq 0; \ \text{rank}(F\Sigma+\Sigma F^T) = \text{minimum}; \ \Sigma H^T = G}$$

for $\Sigma = \Sigma^T$.

The matrices Σ^- and Σ^+ mentioned above actually correspond to output induced minimal realizations, and thus satisfy the equations of the above theorem. In fact, they correspond to the (up to linear equivalence) *unique* past output and future output induced minimal realizations. Thus in the corresponding realizations (x_-,y) and (x_+,y), the state $x_-(t)$ may be viewed as the parametrization of the conditional probability measure $P(y_t^+|y_t^-)$. Of course, $x_+(t)$ admits a similar interpretation. Actually, both the solution set of the Σ's of the above theorem and the corresponding output induced realizations (x,y) have a great deal of very appealing structure. It would take us too long to explain all that here. The reader is referred to [12,26,27] for details. We just like to mention one item: namely if (x_-,y) and (x_+,y) are the canonical past and future output induced realizations and (x,y) is any other output induced minimal realization, then there will exist a projection matrix P such that $x = Px_- + (I-P)x_+$. In fact, "generically," there are a finite number of such matrices possible and in a suitable basis this states that every output induced irreducible realization may be obtained from x_-,x_+ by picking certain components from x_- and the remaining components from x_+ (see [12,27] for a more precise statement to this effect). These results show that the situation described in Section 2.3 is rather representative for the general case.

Remarks

1. It is clear from the above theorem that, contrary to what one sees happening for deterministic finite dimensional time-invariant linear systems, there is no unique (up to linear equivalence) minimal realization of stochastic finite dimensional stationary gaussian processes. However, there is one thing which one can say, namely that if there exists a finite dimensional realization, then there will also exist a finite dimensional realization (z,y) having the property that all minimal realizations (x,y) may be deduced from it by a surjective matrix $S: z \to x$. Among the realizations having this property there are again irreducible (i.e., minimal) elements and it may be shown that such an irreducible realization is unique up to linear equivalence. Such realizations could be called *universal*. One such universal realization may be deduced from (x_-,x_+) by defining

$$z(t) := (x_-(t),x_+(t)/\mathrm{Ker}\ Q$$

with

$$Q := \mathscr{E}\left\{\begin{pmatrix} x_-(0) \\ x_+(0) \end{pmatrix}\begin{pmatrix} x_-(0) \\ x_+(0) \end{pmatrix}^T\right\}.$$

We expect that this type of universal realizations may have some as yet undiscovered applications for example in two-sided data processing, e.g. smoothing.

2. An interesting problem which has recently been investigated is how time-reversibility of y may be reflected in its realization (x,y). The basic result obtained in [28] and, independently, in [29] states that if y is equivalent to Ry and if one chooses the weak realization (x,y) right, then in a suitable basis $x = (x_1,x_2)$ with (x_1,x_2,y) equivalent to $R(x_1,-x_2,y)$. The sign reversal of x_2 is the striking element in this result. This type of time-reversibility has been referred to as "dynamic reversibility" in the stochastic processes literature.

3. There is a very interesting duality between the problems of realization of gaussian processes and that of passive electrical network synthesis. The basic synthesis with resistors, capacitors, inductors, transformers, and gyrators is dual to the weak realization problem with minimality referring to the minimality of the number of inductors plus the number of capacitors. The output induced realization problem turns out to be dual to the synthesis with a minimal number of resistors, and the time reversibility problem discussed in the previous paragraph turns out to be dual to the gyratorless synthesis problem. It would be nice to have a more fundamental understanding of this situation!

We close this section with an almost trivial theorem on white noise representability.

Theorem. *Let* (x,y) *be a finite dimensional zero mean gaussian dynamical system. Then there exists a white noise representation of the form:*

$$dx = Axdt + Bdw; \quad y = Cx.$$

In fact, when $\mathcal{E}\{x(0)x^T(0)\} =: Q$, *then* (A,Q,C) *are precisely the parameter matrices of* (x,y) *as a realization of* y *and* B *is a solution of the equation:*

$$\boxed{AQ + QA^T = -BB^T} \; .$$

Moreover, if (x,y) *is a minimal realization of* y *then* $\{(A,B)$ *controllable*$\} \Leftrightarrow \{y$ *is ergodic*$\}$.

Note that the output induced realizations are precisely those which allow a white noise representation with a minimal dimensional driving Wiener process.

5. CONCLUSION

In this paper we have attempted to give an exposition of the problems of stochastic realization theory. We have been rather detailed on the abstract setting of the problem but somewhat scant on the realization of gaussian processes for which much more is available than we have covered here. Actually, we believe that the conceptual "set theoretic level" definitions for stochastic systems have been neglected and that putting these problems on a sound footing would provide an important pedagogical, theoretical, and practical contribution in stochastic system and control theory.

APPENDIX

Let $\{\Omega,\mathcal{A},P\}$ be a probability space with Ω a set, $\mathcal{A}$ a σ-algebra of subsets of Ω, and $P:\mathcal{A}\to[0,1]$ a probability measure on Ω. With a *random variable* we mean a measurable mapping $f:\Omega\to F$ with $\{F,\mathcal{F}\}$ a measurable space, i.e., F is a set and $\mathcal{F}$ is a σ-algebra on F. The measure $\mu(F') := P(f^{-1}(F'))$ defined for all $F'\in\mathcal{F}$ is called the *induced measure*. Note that our random variables need not be real. If F is a topological space and $\mathcal{F}$ is the smallest σ-algebra containing all open sets, then $\{F,\mathcal{F}\}$ is called a *Borel space*. We will delete $\mathcal{F}$ in the notation of a random variable whenever it is unimportant or clear what $\mathcal{F}$ exactly is.

A random variable f on a Borel space $\{F,\mathcal{F}\}$ is said to be *surjective* if the complement of $f(\Omega)$ contains no non-trivial open sets, i.e., if the closure of $f(\Omega)$ is F. Let f_1,f_2 be random variables on $\{F_1,\mathcal{F}_1\}$ and $\{F_2,\mathcal{F}_2\}$ and $h: F_1\to F_2$ a measurable map. Let μ_1 and μ_2 be the induced measures. In this context h is said to be *injective* if

$$\mu_1(\{a \mid \exists b\neq a \text{ such that } h(a)=h(b)\}) = 0.$$

Let f be a random variable on $\{F,\mathcal{F}\}$. The sub σ-algebra of $\mathcal{A}$ defined by $f^{-1}(\mathcal{F})$ will be called the *sub σ-algebra induced by* f. We call two sub σ-algebras $\mathcal{A}_1,\mathcal{A}_2$ of $\mathcal{A}$ *independent* if $P(A_1\cap A_2)=P(A_1)\cdot P(A_2)$ for all $A_1\in\mathcal{A}_1$ and $A_2\in\mathcal{A}_2$. Two random varibles are *independent* if their induced sub σ-algebras are independent.

Let $\mathcal{A}_1,\mathcal{A}_2,\mathcal{A}_3$ be sub σ-algebras of $\mathcal{A}$. Then we will say that $\mathcal{A}_1$ and $\mathcal{A}_3$ are *conditionally independent* given $\mathcal{A}_2$ if for all P-integrable real random variables f_1, f_3 which are respectively $\mathscr{A}_1$ and $\mathscr{A}_3$-measurable, there holds

$$\mathscr{E}(f_1 f_3)|\mathscr{A}_2) = \mathscr{E}(f_1|\mathscr{A}_2) \cdot \mathscr{E}(f_3|\mathscr{A}_2).$$

The random variables f_1 and f_3 are said to be *conditionally independent* given f_2 if the induced sub σ-algebras are conditionally independent. We also say that $\mathscr{A}_2$ (resp. f_2) is *splitting*.

It is well-known how conditional expectation is defined and what one means with regular versions of conditional probabilities [30, p. 139]. We will with explicit mention assume that these exist whenever needed, and denote by $P(f_2|f_1)$ the conditional measure of f_2 given f_1, where f_1 and f_2 are two random variables. We will use a similar notation for conditional expectation.

Let T be a set and $f(t) : T \times \Omega \to F$ be a family of random variables. Then the family of measures induced for all n on F^n by the random variables $(f(t_1), f(t_2), \ldots, f(t_n))$ are called the *marginal probability measures* of f. If $T \subset \mathbb{R}$ then f is called a *random process*. The smallest σ-algebra containing all the σ-algebras induced by $f(\tau)$ for $\tau < t$ is called the σ-algebra induced by the past at time t. Whenever we are conditioning or taking conditional expectation with respect to the past this should be understood in this sense. We will denote the strict past (future) at t by $f_t^-(f_t^+)$. A random process is said to be a *white noise process* if, for all t, f_t^- and $(f(t), f_t^+)$ are independent. It is said to be a *Markov process* if, for all t, f_t^- and f_t^+ are conditionally independent given $f(t)$. An $\mathbb{R}^n$-valued process is said to be *gaussian* if all its marginal measures are gaussian and a *Gauss-Markov process* if it is both Markov and gaussian.

Finally, let $\{\Omega_1, \mathscr{A}_1, P_1\}$ and $\{\Omega_2, \mathscr{A}_2, P_2\}$ be two probability spaces and $f_1, f_2 \colon \Omega_1, \Omega_2 \to F$ be two random variables defined on the same outcome space $\{F, \mathscr{F}\}$. Then they are said to be *equivalent* if they induce the same measure of F. Similarly two random processes are said to be *equivalent* if they induce the same marginal probability measures.

REFERENCES

[1] Picci, G.: 1977. 1978, *On the internal structure of finite-state stochastic processes*, in Recent Developments in Variable Structure Systems, Economics and Biology (eds. R. R. Mohler and A. Ruberti), Proc. of US-Italy Seminar,

Taormina, Sicily, pp. 288-304. Springer-Verlag, Lecture Notes in Economics and Mathematical Systems, Vol. 162.

[2] Brockett, R. W.: 1977, *Stationary covariance generation with finite state Markov processes*, Proceedings 1977 IEEE Conference on Decision and Control.

[3] Brockett, R. W. and Blankenship, G. L.: 1977, *A representation theorem for linear differential equations with Markovian coefficients*, 1977 Allerton Conference on Circuits and System Theory.

[4] Van Putten, C., and Van Schuppen, J. H.: 1979, *On stochastic dynamical systems*, Preprint BW 101/79, Mathematical Centre, Amsterdam, Proc. of the Fourth International Symposium on the Mathematical Theory of Networks and Systems, Delft, July 1979, pp. 350-356.

[5] Doob, J. L.: 1944, *The elementary Gaussian processes*, Ann. Math. Stat., 15, pp. 229-282.

[6] Kalman, R. E.: 1965, *Linear stochastic filtering--reappraisal and outlook*, Proc. Polytechnic Inst. of Brooklyn, Symposium on System Theory, pp. 197-205.

[7] Faurre, P.: 1967, *Representation of Stochastic Processes*, Thesis, Stanford University.

[8] Faurre, P.: 1973, *Réalisations Markoviennes de Processus Stationnaires*, Report IRIA No. 13.

[9] Faurre, P., Clerget, M., and Germain, F.: 1979, *Opérateurs Rationnels Positifs*, Dunod.

[10] Anderson, B. D. O.: 1973, *The inverse problem of stationary covariance generation*, J. of Stat. Physics, 1, pp. 133-147.

[11] Picci, G.: 1976, *Stochastic realization of gaussian processes*, Proc. IEEE, 64, pp. 112-122.

[12] Linquist, A., and Picci, G.: 1979, *On the stochastic realization problem*, SIAM J. Control and Optimization, vol. 17, pp. 365-389.

[13] Lindquist, A., and Picci, G.: 1978, *A state-space theory for stationary stochastic processes*, Proc. 21st Midwest Symposium on Circuits and Systems, Ames, Iowa.

[14] Lindquist, A., and Picci, G.: 1979, *Realization theory for multivariate stationary gaussian processes I: State space*

construction, to appear in the Proc. of the Fourth International Symposium on the Mathematical Theory of Networks and Systems, Delft, July 1979, pp. 140-148.

[15] Lindquist, A., Picci, G., and Ruckebusch, G.: 1979, *On minimal splitting subspaces and Markovian representations,* Math. Systems Theory, 18, pp. 271-279.

[16] Rickebusch, G.: 1975, *Représentations Markoviennes de Processus Gaussiens Stationnaires,* Thesis, Un. de Paris VI.

[17] Ruckebusch, G.: 1977, 1978, *Représentations Markoviennes de Processus Gaussiens stationnaires et applications statistiques,* Journées de Statistique de Processus Stochastiques (eds. D. Dacunha-Castelle and B. van Cutsem), Proc. of Grenoble Symposium, 1977, pp. 115-139. Springer-Verlag, Lecture Notes in Mathematics, Vol. 636.

[18] Ruckebusch, G.: 1977, 1978, *On the theory of Markovian represenation,* in Measure Theory Applications to Stochastic Analysis (eds. G. Kallianpur and D. Kölzow), Proc. of Oberwolfach Tagung, pp. 77-88. Springer-Verlag, Lecture Notes in Mathematics, Vol. 695.

[19] Ruckebusch, G.: *A geometric approach to the stochastic realization problem,* submitted to SIAM J. Control and Optimization.

[20] Akaike, H.: 1975, *Markovian representation of stochastic processes by canonical variables,* SIAM J. Control, 13, pp. 162-173.

[21] Willems, J. C.: 1978, *Recursive filtering,* Statistica Neerlandica, 32, 1, pp. 1-39.

[22] Skibinsky, M.: 1967, *Adequate subfields and sufficiency,* Ann. Math. Stat., 38, pp. 155-161.

[23] Hotelling, H.: 1936, *Relations between two sets of variates,* Biometrika, 28, pp. 321-377.

[24] Gihman, I. I., and Skorohod, A. V.: 1972, *Stochastic Differential Equations,* Springer-Verlag.

[25] Willems, J. C.: 1972, *Dissipative dynamical systems,* Archive for Rational Mechanics and Analysis, 45, pp. 321-392.

[26] Willems, J. C.: 1971, *Least squares stationary optimal control and the algebraic Riccati equation,* IEEE Trans. on Aut. Control, AC-16, pp. 621-634.

[27] Pavon, M.: *Stochastic realization and invariant directions of the matrix Riccati equation*, submitted for publication.

[28] Willems, J. C.: 1977, 1978, *Time reversibility in deterministic and stochastic dynamical systems*, in Recent Developments in Variable Structure Systems, Economics and Biology (eds. R. R. Mohler and A. Ruberti), Proc. of US-Italy Seminar, Taormina, Sicily, pp. 318-326. Springer-Verlag, Lecture Notes in Economics and Mathematical Systems, Vol. 162.

[29] Anderson, B. D. O., and Kailath, T.: 1978, *Forwards, backwards and dynamically reversible Markovian models of second order processes*, Proc. IEEE Int. Symp. on Circuits and Systems.

[30] Gikhman, I. I., and Skorokhod, A. V.: 1969, *Introduction to the Theory of Random Processes*.

INDEX

Batch number: 10607222

Printed by Printforce, the Netherlands